Autodesk Civil 3D 2025 Unleashed

Elevate your civil engineering designs and advance
your career with Autodesk Civil 3D

Stephen Walz

Tony Sabat

‹packt›

Autodesk Civil 3D 2025 Unleashed

Group Product Manager: Rohit Rajkumar

Publishing Product Manager: Urvi Shah

Book Project Manager: Sonam Pandey

Senior Editor: Rashi Dubey

Technical Editor: Reenish Kulshrestha

Copy Editor: Safis Editing

Indexer: Manju Arasan

Production Designer: Alishon Mendonca

DevRel Marketing Coordinators: Nivedita Pandey and Anamika Singh

First published: July 2024

Production reference: 1120624

Published by Packt Publishing Ltd.

Grosvenor House

11 St Paul's Square

Birmingham

B3 1RB, UK

ISBN 978-1-83546-774-9

www.packtpub.com

To my colleagues, mentors, friends, family, and the readers of my previous book, your unwavering support and invaluable feedback have been instrumental in shaping this follow-up endeavor, aimed at propelling our careers to new heights. A heartfelt thank you to my wife, Danna, and daughters, Alexis and Addison, for their boundless love and unwavering support throughout my career, and during the development of this new book.

– Stephen Walz

Thank you to my wife, who has supported me throughout this book's development as well as throughout my career.

– Tony Sabat

Contributors

About the authors

Stephen Walz has been working with a multitude of design, collaboration, and visualization products and platforms in the **Architecture, Engineering, and Construction** (AEC) industry since early 2003. His primary focus is on civil/environmental engineering fields in which he has held varying levels of design support and CAD/BIM/CIM management roles. Currently, he is HDR's Corporate Digital Design Lead. He works with their business group and technology leadership, vendors, and Information Technology Group to evaluate and implement new technology solutions and strategies that support digital design and delivery services, build awareness and drive consistency with how digital design and delivery tools and platforms are being utilized, and build skillsets through mentorship, guidance, and training of our technologists.

Tony Sabat is a consultant, adviser, and writer focusing on improving the built environment. He primarily consults with teams to develop innovative and disruptive technologies and processes that vary from early-stage start-ups to Fortune 500 companies. Together, they develop digital twin strategies and implement emerging technologies such as reality data modeling, virtual construction, and even distributed ledger technology.

He began his career with the early iterations of building information modeling working on integrating such concepts and technologies into the civil infrastructure space. Today, Tony focuses on partnering with companies looking to adopt and optimize emerging technologies including artificial intelligence, reality technologies, building information modeling, digital twin strategies, as well as virtual design and construction.

About the reviewers

Justin Brooks, PE, PMP, has been an avid user of Civil3D and other Autodesk products over his 20-plus-year career in civil engineering. He holds both undergraduate and graduate degrees in civil engineering and an associate degree in computer-aided design. He has held multiple roles within the civil engineering industry, from designer to project engineer and project manager, with his current role being Design Technology Manager with Civil & Environmental Consultants, Inc. Along with his time in industry, he has also spent a portion of his career in post-secondary education as a professor and curriculum developer in construction management and technology programs.

Ian Chapman has been a passionate user of Civil 3D for many years and has utilized all aspects of the software across several civil engineering sectors within building services and transportation.

His experience of delivering digital content for projects spans 18 years at design consultants across London and the South East of England, where Ian studied civil engineering. He has dedicated himself to AEC industry software as an Autodesk Certified Professional in Civil 3D and AutoCAD.

As Principal Digital Designer for JBA Consulting, Ian provides a leadership role for BIM and digital delivery across the group's UK southern region, championing the development and implementation of new design tools as part of the company's future digital strategy.

Table of Contents

3

Advanced Design and Analysis Capabilities within Civil 3D 2025   37

4

Rail Design Capabilities within Autodesk Civil 3D 2025   65

Part 2: Improving Our Civil BIM Designs with Civil 3D Extensions and Customized Design Workflows

5

Harnessing Reality Capture to Enhance Civil Projects within Autodesk Civil 3D 2025   97

Part 3: Managing Information Models and Automating Workflows

12

Automating Routine Workflows with Dynamo and Scripting   237

Part 4: Extending Infrastructure Projects beyond Civil 3D

13

Preparing and Extending the Purpose of Our BIM Designs for Collaboration and Visualization   261

Preface

Autodesk Civil 3D 2025 Unleashed is a comprehensive guide that equips civil engineers and designers with advanced skills to unlock new levels of efficiency in their projects and careers. Divided into four parts, this book addresses different aspects of Civil 3D capabilities and extensions. Starting with elevating Civil 3D designs using **building information modeling (BIM)** principles, you'll build a strong foundation in BIM and its integration into civil engineering projects.

By focusing on design customization with Civil 3D extensions, this book will empower you to harness reality capture technologies, optimize grading designs, and explore content catalog customization. You'll delve into information management with Civil 3D, covering property sets, Project Explorer, and workflow automation using tools such as **Dynamo for Civil 3D (D4C3D)** and scripting. You'll also get the direction to prepare BIM designs within Civil 3D for a multitude of downstream uses. Finally, the book will teach you how to extend infrastructure projects beyond Civil 3D and prepare BIM designs for integration into collaborative and visualization products such as Revit, Navisworks and InfraWorks.

By the end, you'll be able to prepare and utilize BIM designs within Civil 3D and several other products for easier project creation and management.

Who this book is for

This book is for civil engineers, designers, BIM managers, modelers, and technicians seeking to advance their designs using Civil 3D's complex workflows and tools. Those interested in integrating workflows with other major design and collaboration tools to enhance overall project coordination and collaboration will also benefit from this book's approach and insights, which will equip you with the skills needed to prepare BIM designs for a multitude of downstream uses that contractors, clients, owners, and operators can utilize in a seamless and iterative manner.

What this book covers

Chapter 1, Taking Civil 3D to the Next Level, explores what it means to actually take our Civil 3D BIM designs to the next level. We'll explore strategies that will set our design teams up for successful design and collaboration and put in our homework to understand full design requirements before a project even starts.

Chapter 2, Building Blocks for Civil 3D Designs, will recap several objectives and learning paths discussed in our previous book, *Autodesk Civil 3D 2024 From Start to Finish.* We will summarize a lot here, but just focus on setting the stage for our next leg up.

Chapter 3, Advanced Design and Analysis Capabilities within Civil 3D 2025, begins to expand on our design capabilities and explores more recently added tools and functionality Autodesk has made available in Civil 3D 2025.

Chapter 4, Rail Design Capabilities within Autodesk Civil 3D 2025, explores how we'll continue building onto our Roadway Modeling Toolbelt and how we can apply similar workflows to design railroads/railways within Civil 3D 2025.

Chapter 5, Harnessing Reality Capture to Enhance Civil Projects within Autodesk Civil 3D 2025, looks at the capture methods that can enhance our civil projects. As an industry, civil engineering is based upon the capture of existing conditions data. This was typically done with traditional methods of capturing points of interest on the project site, whether curb and gutter or the center locations of manholes for utility capture, and so on. The industry primarily responsible for this capture is surveying and it has dramatically expanded and changed over the past few years with new technologies for a more expansive capture of reality. Now, the surveying industry can be compiled into the broader category referred to as reality capture, which has primarily been taken over by techniques such as terrestrial laser scanners or aerial **Light Detection and Ranging** (**LiDAR**), and even photogrammetry techniques. This chapter will dive briefly into the capture methods beyond typical point capture in surveying and into how to handle such immense amounts of data and make them manageable for our design purposes.

Chapter 6, Streamlining Design with Grading Optimization, jumps into the world of advanced cloud-computing grading tools that, when used properly, can increase design efficiencies and save teams' valuable time earlier on in the project as alternative designs are being considered.

Chapter 7, Exploring Content Catalog Editor, explores the Content Catalog Editor. Content Catalog Editor allows design teams to develop custom solutions where out-of-the-box Civil 3D utility parts can be limiting for design teams.

Chapter 8, Empowering Utility Modeling with Infrastructure Parts Editor, explores the Infrastructure Parts Editor. Infrastructure Parts Editor allows design teams to develop custom solutions where out-of-the-box Civil 3D utility parts can be limiting to design teams.

Chapter 9, Custom Roadway Design with Subassembly Composer, looks at the world of customized roadway design. Subassembly Composer allows design teams to develop custom solutions where out-of-the-box Civil 3D subassemblies can be limiting for design teams.

Chapter 10, Information Modeling with Property Sets, explores the ways we can bolster our models and objects to inform project stakeholders of pertinent information related to our civil BIM designs. Information related to our models and objects can greatly improve collaboration, construction processes, and asset management and maintenance after design is complete.

Chapter 11, Introduction to Project Explorer, dives into the world of Project Explorer. Project Explorer has many benefits that can improve and speed up model adjustments, design changes, design reviews, and reporting capabilities.

Chapter 12, Automating Routine Workflows with Dynamo and Scripting, dives into the world of Dynamo for Civil 3D. Utilizing Dynamo for Civil 3D has many benefits for design teams, from streamlining design workflows to drive consistency in how we develop consistent designs, to improving collaboration across all project stakeholders, to simply spending less time on routine and mundane tasks.

Chapter 13, Preparing and Extending the Purpose of our BIM Designs for Collaboration and Visualization, explores how we can integrate BIM designs in a multi-disciplined environment. We'll also learn how we can extend our civil BIM designs and integrate directly into collaboration and visualization tools. By doing this, we'll find new ways to work more efficiently and become even better civil BIM model managers.

To get the most out of this book

You will need to have a basic understanding of civil engineering and surveying workflows, as well as a foundational understanding of Autodesk's Civil 3D and AutoCAD products, to make the most of this book.

Software/hardware covered in the book	Operating system requirements
Autodesk Civil 3D 2025	Windows, macOS, or Linux
Dynamo for Civil 3D	Windows, macOS, or Linux
Content Catalog Editor 2025	Windows, macOS, or Linux
Infrastructure Parts Editor 2025	Windows, macOS, or Linux
Subassembly Composer 2025	Windows, macOS, or Linux
Autodesk ReCap 2025	Windows, macOS, or Linux
Autodesk Revit 2025	Windows, macOS, or Linux
Autodesk Navisworks 2025	Windows, macOS, or Linux
Autodesk InfraWorks 2025	Windows, macOS, or Linux

Note: All the examples in this book are compatible with Autodesk Civil 3D 2024.

Download the exercise files

You can download the example code files for this book at `https://packt.link/gbz/9781835467749`. If there's an update to the exercise files, it will be updated to this link.

Conventions used

There are a number of text conventions used throughout this book.

`Code in text`: Indicates code words in text, database table names, folder names, filenames, file extensions, pathnames, dummy URLs, user input, and Twitter handles. Here is an example: "With our `Utility Model Start.dwg` file open and view set up, let's put on our model manager hat for a bit and run through some various design analysis tools we have available to us within Autodesk Civil 3D."

Bold: Indicates a new term, an important word, or words that you see onscreen. For instance, words in menus or dialog boxes appear in **bold**. Here is an example: "Just as we did when we ran the design check earlier, if you hover your cursor over each of the warning symbols, you'll get some additional details that indicate the location and current actual coverage of your pressure pipe, along with the value that we defined as the minimum and/or maximum depth of cover in the **Run Depth Check** dialog box."

> **Tips or important notes**
> Appear like this.

Get in touch

Feedback from our readers is always welcome.

General feedback: If you have questions about any aspect of this book, email us at customercare@ packtpub.com and mention the book title in the subject of your message.

Errata: Although we have taken every care to ensure the accuracy of our content, mistakes do happen. If you have found a mistake in this book, we would be grateful if you would report this to us. Please visit www.packtpub.com/support/errata and fill in the form.

Piracy: If you come across any illegal copies of our works in any form on the internet, we would be grateful if you would provide us with the location address or website name. Please contact us at copyright@packt.com with a link to the material.

If you are interested in becoming an author: If there is a topic that you have expertise in and you are interested in either writing or contributing to a book, please visit authors.packtpub.com.

Share Your Thoughts

Once you've read *Autodesk Civil 3D 2025 Unleashed*, we'd love to hear your thoughts! Scan the QR code below to go straight to the Amazon review page for this book and share your feedback.

https://packt.link/r/1835467741

Your review is important to us and the tech community and will help us make sure we're delivering excellent quality content.

Download a free PDF copy of this book

Thanks for purchasing this book!

Do you like to read on the go but are unable to carry your print books everywhere?

Is your eBook purchase not compatible with the device of your choice?

Don't worry, now with every Packt book you get a DRM-free PDF version of that book at no cost.

Read anywhere, any place, on any device. Search, copy, and paste code from your favorite technical books directly into your application.

The perks don't stop there, you can get exclusive access to discounts, newsletters, and great free content in your inbox daily

Follow these simple steps to get the benefits:

1. Scan the QR code or visit the link below

https://packt.link/free-ebook/9781835467749

2. Submit your proof of purchase
3. That's it! We'll send your free PDF and other benefits to your email directly

Part 1:
Next-Level Civil 3D Capabilities

In this part of the book, we'll advance our Civil 3D design, modeling, coordination, and management skills based on the foundational knowledge we already have (or acquired from our previous book, *Autodesk Civil 3D from Start to Finish*) by tapping into many of the enhanced solutions and integrations we have available to us. Everything covered herein will step up our BIM game in the Civil 3D world and put us on a path to becoming an expert model manager and leading design teams, as well as becoming as efficient and productive as possible.

This part contains the following chapters:

- *Chapter 1, Taking Civil 3D to the Next Level*
- *Chapter 2, Building Blocks for Civil 3D Designs*
- *Chapter 3, Advanced Design and Analysis Capabilities within Civil 3D 2025*
- *Chapter 4, Rail Design Capabilities within Autodesk Civil 3D 2025*

1
Taking Civil 3D to the Next Level

For professionals in the civil engineering world, Autodesk Civil 3D has become the industry standard for designing, collaborating, and delivering civil infrastructure projects efficiently. Civil 3D is a powerhouse program for all of a civil engineer's needs, whether on an individual basis or at a team level. Civil 3D encompasses all aspects of a design project from manipulating and importing existing conditions from various formats, through utility and grading design, and all the way to the finish when it comes to plan production. But while Civil 3D can handle all these evolving aspects of a project, there is much more that Civil 3D can enable beyond itself for civil design projects that require more collaboration, customization, or even communication beyond typical plan production. This book will take you and your skills with Civil 3D and other platforms, such as Infraworks, Navisworks, and more, to the next level and truly bring your civil designs to the forefront of the industry.

In this book, we will not only look at Civil 3D and its advanced capabilities as it relates to utilizing existing conditions, grading, utility design, and others, but will also examine how this software can be integrated into the portfolio of tools that would be utilized by a BIM model manager Now as a refresher, **BIM** is short for **Building Information Modeling**, and the role of BIM model manager has recently grown in popularity as software and technology have become so prominent in the **Architecture, Engineering, and Construction (AEC)** industry these days. A BIM model manager's role begins with a foundational understanding of specific software they are to utilize and can blend into the management of that software as it applies to templates and standards amongst a group. Also, a BIM model manager can be a specialist in industry-specific software, such as Civil 3D and others, which we will cover later in this book. This book is designed to leverage a civil BIM model manager's understanding and educate on the numerous other tools that can take the solid foundation of Civil 3D and expound upon it for further project collaboration amongst other project disciplines, visualizations for winning more project work from clients, and automating project information for a stronger overall working environment as a company.

Within this chapter, we will be focusing on the overall enhanced functionality that is possible within Civil 3D, then emphasize where and why we can take our civil design projects to the next level for maximal value creation amongst our teams and stakeholders. Many books about Civil 3D do a great job enabling a deep understanding of the program itself, whereas this book is meant to build upon a preexisting Civil 3D understanding and unleash the details of all of its components. This book is guaranteed to amplify you and your team's practices with more agile workflows for custom part modeling and more intelligent corridor modeling, to leveraging your civil information into databases and visual presentations.

The following are the topics we will dive into, and further on in the chapter we will learn the importance of each element:

- Next-level Civil 3D capabilities
- Civil information modeling
- Customization of grading designs
- Customization of utility designs
- Information management and automation
- Extending infrastructure projects beyond Civil 3D

Technical requirements

As we know, the programs we use in the industry are incredibly powerful and as such require significant hardware and knowledge to leverage fully. Autodesk Civil 3D requires professional-grade computing horsepower, but this book will dive into other programs beyond Civil 3D that will prefer even more horsepower.

Though we have listed the minimum required hardware for this book, we have also indicated preferred guidelines on hardware to bear in mind as you look to incorporate these tools into your existing career. The minimum recommended hardware will give you a place to start and handle these tools, but we would push you to incorporate higher-capacity hardware when available to ensure fast working operations and to avoid any potential data or progress loss. Here, we'll review the minimum requirements that Autodesk recommends, with a few of our suggestions added to increase efficiency and speed throughout the BIM design process:

- **Operating system** – 64-bit Microsoft Windows 10
- **Processor** – 4+ GHz
- **Memory** – 16 GB RAM+ (32 GB or 64 GB preferred)
- **Graphics card** – 4+ GB
- **Display resolution** – 1980 x 1080 with True Color

- **Disk space** – 20 GB
- **Pointing device** – Microsoft compliant mouse

Next-level capabilities within Civil 3D

Our first book, *Autodesk Civil 3D 2024 from Start to Finish*, acted as a foundational piece in the civil engineer's technology journey. Civil engineers are entrenched in the design aspect of a project as well as many other nuances that come along with design, but one facet of design is quickly evolving and becoming a new standard for efficient, collaborative projects: Civil 3D and many more civil engineering technologies. In this book, we dive deeper into the concepts detailed in our previous book. We branch out from Civil 3D and look into other tools for taking these civil designs and making them more collaborative with other disciplines, along with leveraging them into stakeholder meetings for better project understanding or visualization to win more work. We also look into taking the information portion of BIM to the next level with better utilization and project awareness in the other phases of a project besides design.

We will kick off with a recap of the basics of Civil 3D and refamiliarize ourselves with the overall processes within the platform. We will not dive back into all of the details, but it will serve as a refresher for those looking to take their existing Civil 3D knowledge to the next level. After we recap, we will dive deeper into the advanced design and analysis capabilities of Civil 3D 2025 that will serve as a comprehensive guide to taking our Civil 3D game up another level.

We'll begin with a residential subdivision design by applying advanced roadway and utility design and analysis workflows to improve our design and collaboration. We'll also introduce a new dataset to get a basic understanding of how Civil 3D 2025 can be utilized in rail design as well.

Next, we'll dive in the world of cloud computational design, along with drone and laser scanning integration to supplement our existing conditions model and develop skills that will allow us to customize our utility network designs through manual 3D-part modeling and parametric-part content integration.

From there, we'll dive into the world of the information modeling, automation, and CIM management tools available in Civil 3D 2025. We'll learn how we can prepare our models for future technology solution integrations advantageous to construction, asset management, and owner/operator purposes. We'll get a better understanding of how we can streamline information tagging within our models to fit the requirements of project stakeholder purposes.

We'll then complete our BIM model manager learning path with an overview of how we can integrate designs from major design authoring tools to improve our design collaboration. With our coordinated design established, we can then take our comprehensive Civil 3D design to a whole other level by reviewing workflows used to integrate into collaboration and design review tools and learn how we can extract the information from our models and further utilize them for design collaboration, quantifications, and even cost estimation purposes. Finally, we'll close out with a review of how we can leverage visualization tools to better convey our design intent to all project stakeholders. All of these capabilities are essential facets of civil information modeling, the idea of leveraging the data created in dynamic civil engineering projects for further downstream purposes and more effective projects.

Civil Information Modeling

In this section, we'll embark on a journey into the realm of **Civil Information Modeling** (**CIM**), a transformative approach that has reshaped how we design and execute civil engineering projects. We'll share experiences, explore processes, and navigate the successes and challenges encountered while implementing full CIM designs. Our focus extends beyond individual projects to the ambitious goal of fully integrating CIM across entire organizations.

But before we delve into the intricacies of CIM implementation, let's take a moment to demystify some common acronyms that have become synonymous with the AEC industry. You're likely familiar with BIM, a term that has long been associated with intelligently designing buildings and structures. However, on the civil engineering front, we've been crafting and modeling everything outside of buildings in a 3D environment for as long as, if not longer than, BIM has been in circulation.

Throughout this journey, we've mastered the art of building surfaces, creating corridor models, designing pipe networks, generating dynamic profiles and cross sections, performing clash detections, calculating earthwork quantities, and producing cost estimate reports. However, despite our advancements, our final designs often comprise a blend of 2D and 3D elements, which has led to a divergence in the perception of our work as true BIM among the majority.

To address this divergence and clarify the role of civil engineering in this evolving landscape, the industry has embraced the acronym CIM. While CIM shares processes with BIM, its scope extends beyond the creation of a 3D model. It involves harnessing additional dimensions that enable us to extract and analyze the intelligent components embedded in our designs.

As clients and owners increasingly recognize the value of CIM, it's becoming commonplace for them to mandate the inclusion of 3D models alongside traditional plan sets in design deliverables. Some are even making the leap to going fully digital, considering the **Model as the Legal Document** (**MALD**) approach. Furthermore, we're witnessing a growing trend of contractors engaging and collaborating in the early stages of project design, leveraging design models for construction intelligence updates. This practice, known as **Virtual Design and Construction** (**VDC**), fosters seamless collaboration, leads to improved as-built outcomes, and results in substantial cost savings on the back end.

Now, as we delve into the world of CIM, we'll shift our focus to understanding some of the complexities involved when beginning your CIM journey or even deploying CIM on your 100th project.

CIM journey

Creating a genuine CIM design is a transformative journey. It's important to anticipate challenges along the way. Initial frustrations may arise, leading some to revert to old 2D habits. However, embracing the model's details is crucial, especially if your company intends to deliver electronic models or as-built models in the future. The devil truly lies in the BIM and CIM details.

Costs and efforts in a fully implemented CIM design are front-loaded, with a tapering effect as design development progresses. Clients and companies must grasp this shift in resource allocation. A 3D model environment encourages thoughtful design, resulting in quicker adjustments during later phases. The benefits of generating a BIM or CIM design become evident at this stage.

During project initiation, defining the final product's representation and planning the path to achieve it is vital. 3D models offer extensive information but can expose potential errors and omissions. BIM model managers and project managers must collaborate to manage these risks. BIM and CIM execution plans, which outline workflows and uses, are increasingly common and require comprehensive training for the entire project team.

Now, let's explore the critical aspect of CIM execution plans, also known as **Digital Deliverable Plans** (**DDP**). These plans provide a structured framework to guide CIM implementation effectively. They define workflows, uses, and processes that are instrumental in achieving the desired final product. Proper training and understanding are essential for the entire project team to navigate the complexities of BIM and CIM design successfully.

CIM uses

Let's delve into the realm of CIM uses. A multitude of distinct CIM uses can be brought to bear on a project, each possessing its unique purpose and utility, suitable for deployment across singular or multiple phases. Traditionally, any given project traverses four key phases: the planning phase, design phase, construction phase, and operations phase.

The array of potential CIM uses that can be employed within each of these phases is expansive, with some even exhibiting crossover into subsequent project phases. Drawing from our experience, the strategic approach involves adhering to CIM uses essential for realizing the project's envisioned final outcome.

To illustrate this, consider the scenario of designing a roadway for a client. Typically, the expected final product encompasses delivering a hardcopy plan set alongside a 3D model of the design. In light of this, the use of CIM is indispensable for achieving this objective, encompassing design authoring (responsible for 3D design model generation), drawing generation (including production), 3D coordination (commonly known as clash detection), and design reviews, itself encompassing **quality assurance and quality control** (**QA/QC**). While other tasks such as modeling existing conditions, automation, and traffic analysis may be considered, their inclusion should align with the specific project requirements.

In essence, plugging the use of CIM into your execution plan essentially commits you to fulfilling all the associated requirements of that particular use of CIM. However, if comprehensive analysis is deemed necessary or aligns with your company's standards, there is merit in the incorporation, as they can significantly enhance the thoroughness of the design.

Ultimately, there are four core CIM uses that merit universal application across projects, regardless of their market segment. These foundational CIM uses encompass design authoring, drawing generation, 3D coordination, and design reviews.

Design authoring

Design authoring involves using software to create a 3D model based on the information necessary for an accurate representation of the design. One important aspect to consider in this CIM use is that quality control is a critical component in every design phase and every CIM use. Quality control in the context of design authoring typically concentrates on the precision and completeness of CIM elements at a detailed level during interdisciplinary design. While 3D coordination and design review offer a broader perspective on interdisciplinary design, they specifically focus on identifying errors and ensuring completeness. In practice, these aspects are interconnected, and quality assurance is an integral part of CIM, as it is in all phases of the project lifecycle.

When it comes to modeling existing conditions, one of the most crucial considerations is determining the **level of development (LOD)**, sometimes referred to level of detail, required in the model. This aspect is typically addressed during scope and fee development and should involve discussions between the project manager and a lead modeler or BIM model manager. It's also essential to note that designing based on an imprecise existing model can lead to constructability issues and pose risks to your company. On the contrary, excessive detail can drive up survey costs and prolong processing times.

To mitigate these challenges, it's important to identify the appropriate level of detail and set boundaries to prevent overwork. Determining the level of detail depends on the project type and client requirements, which can naturally vary significantly from one project to another. Let us discuss a few examples of factors to consider when deciding the level of detail to incorporate into your existing models.

For instance, if you're working on a linear roadway design, you should assess whether there will be new pavement connecting to existing pavement. This evaluation may entail a review of the contract terms, client specifications, and design guidelines to establish accuracy requirements. It's crucial to recognize that the design's precision or accuracy cannot surpass that of the underlying survey data.

Alternatively, in a project involving site civil grading, different areas of the project site may necessitate varying levels of accuracy. Landscaping, for example, typically doesn't demand the same level of precision and detail as pavement design. Additionally, considerations related to drainage and surface elevation points should be factored in.

Likewise, for a utility relocation project, a thorough assessment of the number and types of utilities within the project area is essential. If there are high-risk utilities near proposed structures, using borings as a data source might be more appropriate than ground-penetrating radar. Furthermore, if a clash detection analysis is part of the design process, it's imperative that all utility information covers the area designated for clash detection.

Drawing generation

Next up is drawing generation, which essentially entails the process of taking our CIM components and producing drawings and comprehensive drawing sets, encompassing schematic design, development, and construction documents. The overarching objective here is to establish model elements once and utilize that information without compromising accuracy or integrity in the generation of drawings. This approach enhances the overall quality of the drawings and concurrently minimizes the effort invested while reducing errors in the design process.

In most projects, the primary deliverables typically revolve around plans, making drawing generation a well-established concept. However, the transformation brought about by CIM implementation lies in the ability of designers to swiftly present information in a 2D plan view with minimal to no loss in precision or accuracy. This means that plans, profile sections, and detailed sheets can all be efficiently generated from your model.

3D coordination

Moving on to 3D coordination, commonly referred to as clash detection, it entails the utilization of 3D design software to pinpoint spatial interferences among objects such as utilities, drainage pipes, pilings, and more, within one or multiple 3D models. The primary objective of clash detection is to preemptively eliminate major system conflicts before the commencement of construction. This process is an ongoing cycle throughout the design phase and is ideally initiated before reaching the initial project milestone, in accordance with the guidelines outlined in our BIM or CIM execution plan.

However, it's worth noting that the timing of commencing 3D coordination and its frequency carry budgetary implications. Initiating 3D coordination during the conceptual drawing phase may yield false positives due to the approximate dimensions of elements and a lack of thorough investigation into design challenges in constrained areas. Additionally, excessive clash detections can lead to unnecessary consumption of time and project resources.

A crucial point to emphasize here is that solely relying on 3D coordination software to confirm the absence of conflicts within your project is not advisable. While clash detection tools are valuable for identifying physical conflicts, they are not a foolproof basis for verifying your design. The ultimate responsibility for conflict avoidance rests with the design team, and a conflict-free report generated by a program should not be the sole defense against design issues.

For optimal 3D coordination results, it's essential that the elements being verified are organized on appropriate layers or levels within the drawing itself. As part of our standard practice, we maintain a record of the base files used for comparison. We export a read-only 3D coordination model complete with a date stamp and store it in the same directory as our report. Upon compiling a comprehensive report, we identify the individuals responsible for resolving each conflict, whether through redesign or subsequent engagement with other disciplines or the project owner. It's important to emphasize that coordination remains incomplete until the responsible leads or designers acknowledge the conflicts and take ownership of the conflict resolution process.

Design review

The essence of the design review lies in sharing our 3D engineering drawings and their content with various stakeholders. The design review typically lacks formalization but serves two primary objectives:

- The first objective pertains to quality verification, with a particular emphasis on detecting issues that may be challenging to identify otherwise

- The second aim revolves around swiftly and accurately conveying design information to fellow designers and stakeholders

For all projects, we bear the responsibility of delivering services that adhere to a standard of care, demanding plans devoid of errors and omissions. Errors, relatively speaking, are more conspicuous as they manifest on our drawings or computation sheets. Omissions, on the other hand, pose a greater challenge in identification since they do not surface in the plans. In the realm of geometric design, CIM stands uniquely qualified to aid in addressing omissions by simplifying the process of granting a comprehensive project view to a wider audience. This perspective also facilitates the recognition of constructability concerns.

The communication of design information introduces two distinct challenges:

- Firstly, not all audiences possess the universal software or technical proficiency essential for navigating 3D files. This accessibility challenge is an area where the design review can contribute by facilitating in-person meetings, teleconferencing, and content sharing through videos or images.

- Secondly, different groups of designers and stakeholders express varying interests in specific design elements, posing a content challenge. For instance, attendees at public information meetings may prioritize topics such as access control or artistic elements on retaining walls, while professional designers may focus more on geometric adherence and compliance with standards. The task of tailoring content to these diverse preferences is managed through effective execution.

The application of the design review with CIM is designed to assist in this endeavor.

The design review frequently takes place during staff-level design meetings, with a common emphasis on verifying the design's constructability and completeness, as discussed in preceding sections. One of the most valuable uses of CIM lies in comparing two alternatives, enabling design staff to comprehend the repercussions of choices at a detailed level within minimal time. It is advisable to employ a consistent vantage point or perspective to facilitate the ease of comparing these alternatives.

As we can see, CIM has many use cases and applications where, when properly applied early enough, can provide tremendous enhancements for projects. This section outlines a portion of the potential of CIM, and we will dive further into this concept later in the book. Next, we will dive into advancing the grading aspects of our civil designs and how to tailor them to our unique project needs.

Customization of Grading Designs

Grading is a heavy topic as it involves generating existing conditions and verifying the accuracy for which the entire design will be based upon, as well as leveraging those existing conditions to minimize the amount of earthwork needed and balancing that with a successful design. In our previous book, we delve into all things grading and topography, from existing conditions to proposed gradings, and the many workflows for generating the successful design that our project needs.

Autodesk Civil 3D 2025 Unleashed will begin with the existing conditions of a project site and will also take a look at some of the advanced functionality of Civil 3D. Civil 3D can utilize existing surface data from numerous sources, whether basic survey points from traditional surveying methods, elevation files from previous data captures, or even LiDAR data captures. **LiDAR** is short for **Light Detection and Ranging** and is a form of a growing industry known as **reality capture** – the capturing of real-world conditions. Reality capture is also known as **remote sensing**, which means sensing environmental variations remotely, without having to be in physical contact to measure. So, whether we are capturing an existing road that will be redone for expansion or measuring a gorge that a new bridge design will cross, they all require the use of reality capture to quantify the existing conditions to be used as the basis for the design.

Now, traditional surveying with a survey pole and the capturing of points one at a time would still be considered reality capture, albeit a method that captures less data than more modern methods, such as LiDAR. We can envision traditional surveying being able to capture hundreds of data points per day, where LiDAR can capture millions of points per second. This will give us some scale as to what can be captured depending upon the method, but also does not necessarily mean the data capture is more accurate or easier to work with. We will dive into the nuances of this topic later in *Chapter 5*, but this is all to say that this industry of reality capture is rapidly evolving and there are many ways to go about generating the format you need to begin your proposed grading designs.

Once existing conditions have been captured, the next step is moving into the proposed design and optimizing your gradings. Civil 3D has some basic workflows for generating proposed terrain and matching existing conditions, but in this book, we will dive into the depths of Civil 3D's functionality. In our previous book, we covered many of the traditional grading tools and workflows that have been included in the software for quite some time. In this book, we'll explore the **Grading Optimization** tool that has been included as an add-in over the past few releases of Civil 3D. Grading Optimization is a cloud computing tool that optimizes and significantly speeds up the grading process based on the design criteria and parameters we input into our design model. We can then integrate grading model outputs directly into our designs.

Now that we've covered advanced grading optimization, we'll move on to the next layer of our design, customizing and learning more of the functionality we have with utility designs.

Customization of Utility Designs

After grading customization, the book dives into the ability of Civil 3D to apply similar concepts to utility design. Civil 3D allows users to go beyond the basic parts and materials that come out of the box with the software. The program allows for completely customizable parts for your unique design needs, as well as atypical piping sizes and materials, and much more. Most projects require the basic materials and dimensions, but where Civil 3D can keep a project accurate and progressing is with its ability to generate and accurately design custom structures or materials and dimensions. Later in *Chapters 7* and *8*, we will dive into the details of a few complimentary programs made specifically for this customization.

Content Catalog Editor and Infrastructure Parts Editor are two programs that complement Civil 3D and both have very specific purposes, with both providing us with the ability to generate custom utility parts for both gravity and pressure networks. As we dive into each, we'll begin to understand which tool should be utilized based on the given design scenario and requirements. After our custom parts have been created in each program, we'll be able to publish and share our custom catalogs across design teams and the wider organization as needed, integrating them into our designs with relative ease.

Once we have understood how to create custom and unique parts and materials within Civil 3D, it is as important to quantify and capture the data created. This leads us into the next part of the book, looking specifically at converting this data into information and the management of it all as projects grow in complexity.

Information Management and Automation

As we dive deeper into the customization of Civil 3D elements, we transition to the hidden piece that many overlook in infrastructure projects – the information and data component. All the work dimensioning and definitions of materials, amongst other data points we created with our native elements, can be leveraged further into the design portion of the project and even further into the lifecycle of the designed asset itself. This is where we begin to fully understand the power of the "I" in BIM and take the information piece of our work to the next level.

In this section of the book, we focus on surfacing this data and optimizing the collection and analysis of it for smarter project assessments and better application downstream of the design phase of a project. We will learn about information modeling through property sets within Civil 3D for a more holistic understanding of the project and its components. From there, we begin practicing with a relatively new functionality in Civil 3D, called Project Explorer. Project Explorer is a powerful addition to the Civil 3D portfolio and has become a standard toolbelt for the industry-expert BIM model manager.

Once we understand the amount of data created in Civil 3D and learn how to compile it and interpret it for stronger decision-making, we then look into some advanced functionality in the automation of routines and creating repetitive workflows with scripting, as well as integrating Autodesk's developer program known as Dynamo.

With this section focused entirely on the information facet of BIM, by the end we will have a foundational understanding of how to leverage BIM for civil projects and begin to work smarter, not harder, within our projects as a whole.

Extending Infrastructure Projects beyond Civil 3D

This part of the book is focused on extending and drilling into the depths of what Civil 3D can do for infrastructure projects. So far, a majority of the chapters in previous parts of this book have delved deeper into Civil 3D itself, but this part will look at Civil 3D and branch away from it, showing how to carry the accurate designs it produces to other software programs for further leverage of the work done.

The first chapter of this part carries Civil 3D into the world of design coordination with Navisworks, a program that can act as a nimble aggregator of many different file formats for the merging and comparison of data to understand how designs from other disciplines interact with each other. This can be incredibly useful when coordinating infrastructure projects that have an interior and an exterior component and when utility handoffs must be made between the two. Navisworks is the program used to meld all this data, present it with a holistic view, and ensure an accurate design throughout the project.

From there, we carry designs toward the public eye and focus on the visualization of designs with Infraworks. Many projects require stakeholder approval, whether public or private clientele, and Infraworks is a heavy rendering engine that can not only aggregate large amounts and types of data but can also visualize and create awe-inspiring demonstrations for different clients to better understand designs away from the traditional 2D plan production required in construction. The chapter dives into the best ways to visualize the challenging work created within Civil 3D both before the program is even opened, and afterward, once the final design is completed.

Summary

This chapter serves as an educational and awareness chapter for briefly explaining the new topics that will be covered in this book. They are all relevant to our previous book and will be used in your daily role as a civil engineer, but in varying ways. This chapter was intended to give you a broad understanding of each and then the following chapters will begin branching off and dive deeper into each topic, such as reality capture or custom utility designs.

With a brief overview of the contents of the next level of Civil 3D, let's refamiliarize ourselves with an introduction to Civil 3D and establish the foundation to begin diving into the full extent of Civil 3D and its counterpart programs. The next chapter serves as a brief refresher for level setting on what was previously covered, and then as we progress, we will follow a similar pattern and dive into the corresponding elements. Feel free to reach back into the previous book or chapters to fill in any gaps and ensure we continue building upon the foundation laid out in the first book.

2

Building Blocks for Civil 3D Designs

In this chapter, we will begin our journey from designer to BIM model manager with a quick overview of several key items to set a foundation for ourselves as we advance our skill sets and knowledge of how we can take our Civil 3D designs to the next level. This chapter is intended to build upon the knowledge you have of Civil 3D already with a focus on how you can access different capabilities and tools and better manage the advanced elements that will be covered throughout the remainder of this book.

That said, the key topics that we will cover in this chapter are the following:

- User interface
- Sharing Civil 3D data
- Design toolbelts for everyday use

Without further ado, let's jump into a quick overview to make sure you have a solid understanding and foundation established before we begin exploring new tools and workflows that will elevate our skill sets and careers.

Technical requirements

We will be using the same hardware and software requirements as discussed in the *Technical requirements* section of *Chapter 1*.

Let's go ahead and jump into Civil 3D. We'll begin by familiarizing ourselves with the user interface to get an idea of where to locate all the tools and functionality that will be covered in this book. We will use the dataset found in the Civil 3D 2025 Unleashed\Chapter 2\Model subfolder, starting with the file titled Grading Model_Start.dwg. Once opened, you should see the grading design model displayed, as shown in *Figure 2.1*:

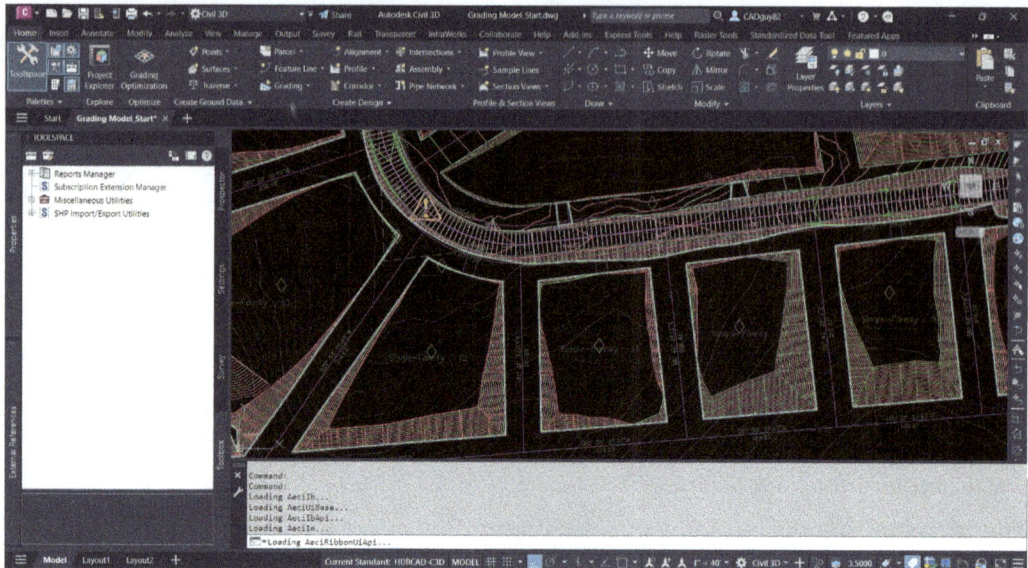

Figure 2.1 – Grading model

User interface

In this section, we'll understand how to navigate around Civil 3D and learn where the main tools are to continue refining and/or recreating portions of our design developed in *Autodesk Civil 3D 2024 from Start to Finish*. Understanding when and where to use these advanced tools will elevate our BIM knowledge significantly and put us on a trajectory to becoming a Civil 3D expert and BIM model manager. After launching Civil 3D and opening the `Grading Model.dwg` file, you'll notice that along the very top of the program, you are presented with three levels, or tiers, of tools and functionality, as shown in *Figure 2.2*:

Figure 2.2 – Top three levels of tools and functionality within the user interface

Each of these levels provides varying access to the tools and functionality available within Autodesk's Civil 3D platform. Much of this will become a bit clearer by the end of the section, as we dive into more detail about what is available at each level. For our benefit, we'll focus only on the second and third tiers to access tools that will be utilized along our journey.

The first ribbon in the second tier is the **Home** ribbon (refer to *Figure 2.3*). Here, we have quick access to many of the major design tools and functionality that we will be leveraging throughout the BIM design life cycle. As we move through different tabs next to the **Home** tab, we will be able to access other tools, such as **Project Explorer**, **Grading Optimization**, and **Infrastructure Parts Editor**.

Figure 2.3 – Home

The next ribbon in the second tier that we'll pay some attention to is the **Modify** ribbon (refer to *Figure 2.4*):

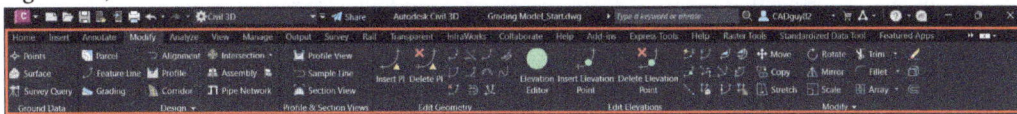

Figure 2.4 – Modify

Here is where we can quickly adjust design elements within the drawing. We'll utilize this ribbon and the tools available within it to advance our Civil 3D designs and explore advanced transitional capabilities within our corridor models.

Next up is the **Analyze** ribbon (refer to *Figure 2.5*):

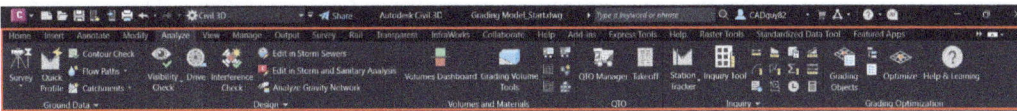

Figure 2.5 – Analyze

Here is where we can quickly access various reporting and computational tools and functionality available to us. Within this ribbon, we'll focus on some of the utility coordination capabilities available within Civil 3D to validate design elements that we've included in our models along with accessing our advanced grading optimization functionality.

The next ribbon that we'll look at is the **Rail** ribbon (refer to *Figure 2.6*).

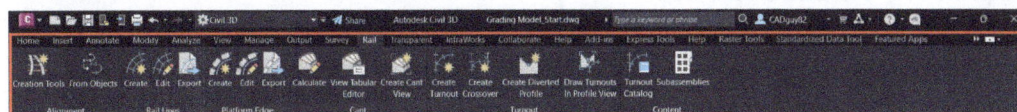

Figure 2.6 – Rail

Here is where we can quickly access BIM design capabilities related to railroads/railways. As we explore these tools and capabilities, we will use a different dataset that includes existing rail components that we can tie our design into.

The last ribbon we'll review is the **InfraWorks** ribbon (refer to *Figure 2.7*).

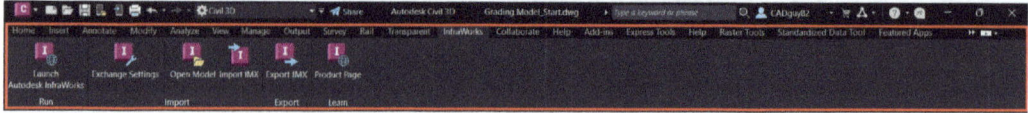

Figure 2.7 – InfraWorks

Here is where we can quickly connect to Autodesk InfraWorks by importing/exporting design elements generated within either the Civil 3D or InfraWorks product with relative ease. InfraWorks is another product offered by Autodesk that allows us to take our designs to another level, where we can quickly generate visualizations to help communicate design intent to additional project stakeholders. InfraWorks is also a great tool to be used for conceptual design purposes.

As we make our way through the chapters, locating the tools needed to advance our designs and skill sets will become second nature. Once we have acclimated to the user interface of Civil 3D, we will move on to focus on better collaboration within the program, which is another fundamental aspect for effective civil modelers and BIM model managers.

Sharing Civil 3D data

When we design projects, it's crucial to handle our design files and objects in Civil 3D in a smart way. While it might sound good to put all our Civil 3D stuff into one file for a project, it's not really practical for most projects, especially ones with lots of objects that need to interact dynamically. This gets even trickier with big projects, so engineers tend to divide the work among teams to make design and production tasks run faster and more efficiently.

At first, dealing with designs and objects in Civil 3D might seem tough, but once we look closer, it's not as complicated as it seems. This section is all about understanding how we set up our dataset to use Civil 3D effectively. We'll focus on something called **data shortcuts**. These shortcuts are like secret weapons that help us handle bigger projects by involving more team members without making things messy. The idea is to make our work more straightforward and keep our design files and models manageable.

Organizing surface models

Going back to our Grading Model_Start.dwg file, let's go up to our second tier, or level, of tools and activate the **Home** ribbon. Once activated, select the **Toolspace** (refer to *Figure 2.8*) icon in the **Palettes** section and then select the **Prospector** tab in **Toolspace**.

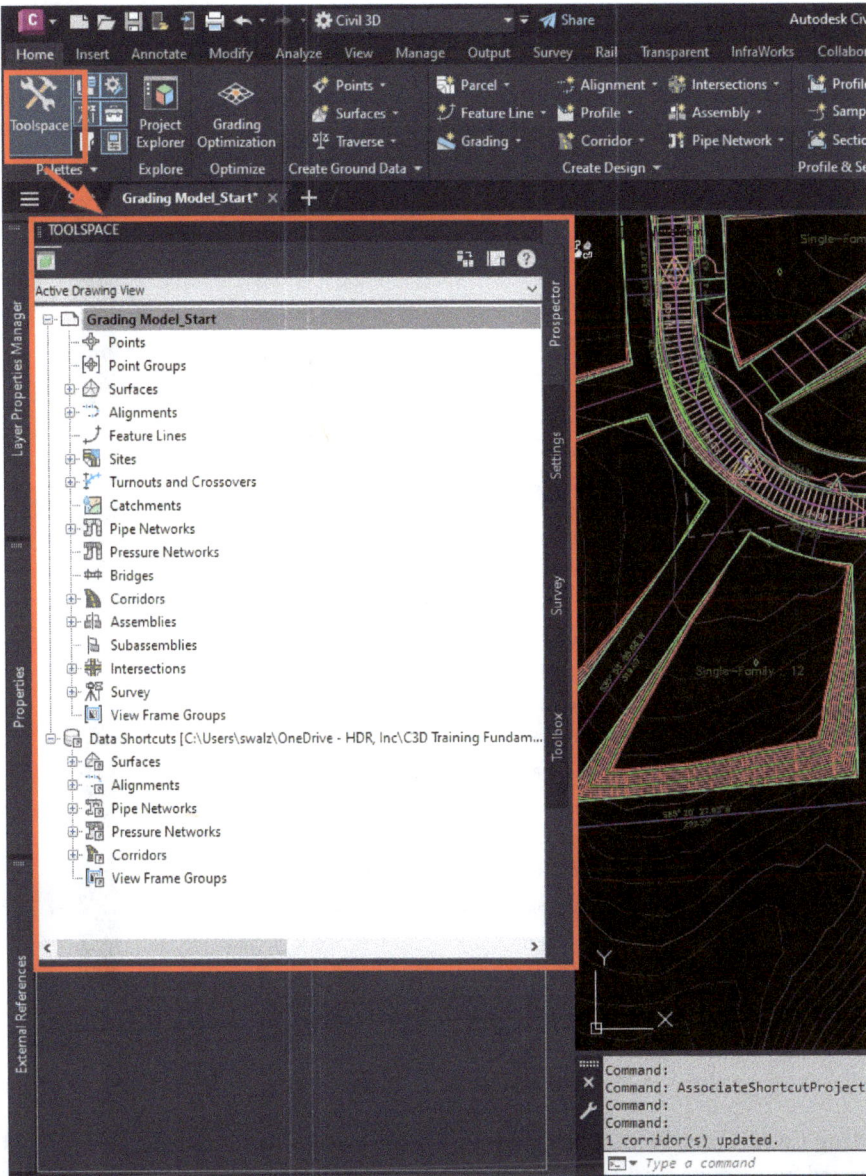

Figure 2.8 – Activate Toolspace

Sharing Civil 3D data while maintaining intelligent data links throughout the project and design files is a critical component of maintaining efficiency across project design teams from a collaboration standpoint. Proper structure and deployment of how Civil 3D modeled objects are organized is another component to maintaining higher levels of efficiency across a project design team due to consistency and natural intuition.

If we were to expand **Surfaces** in our **Toolspace | Prospector**, we'd notice that we have three subfolders within it, each containing surface models that have been used to create our dataset and overall design (refer to *Figure 2.9*).

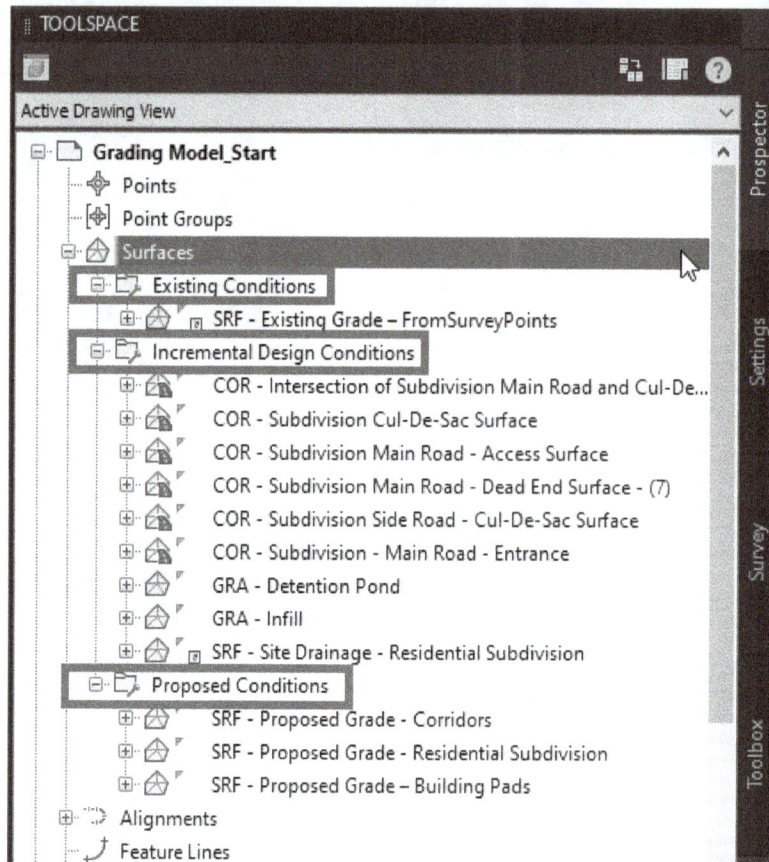

Figure 2.9 – Surface organizational structure

As indicated, we've gone ahead and organized our Surface models into the following three categories:

- **Existing Conditions**: Within this subfolder, we have our existing surface model that was generated from points received from the surveyor. If you have additional sources that need to be separated into additional surface models representing the existing built environment, you can place these here as well.

- **Incremental Design Conditions**: Within this subfolder, we have our individual design surface models that have been generated using corridor models (indicated with COR -), grading objects (indicated with GRA -), and incremental surfaces that have been used on a temporary basis for analysis purposes (indicated with SRF -).

- **Proposed Conditions**: Within this subfolder, we have our combined surfaces that will ultimately be used to display our final grading design. Most of these surfaces have been generated by merging a combination of those surfaces stored in our **Incremental Design Conditions** subfolder.

Moving down to **Alignments**, let's go ahead and expand each of the types of alignment categories that we're able to. We should have a tree breakdown similar to that shown in *Figure 2.10*:

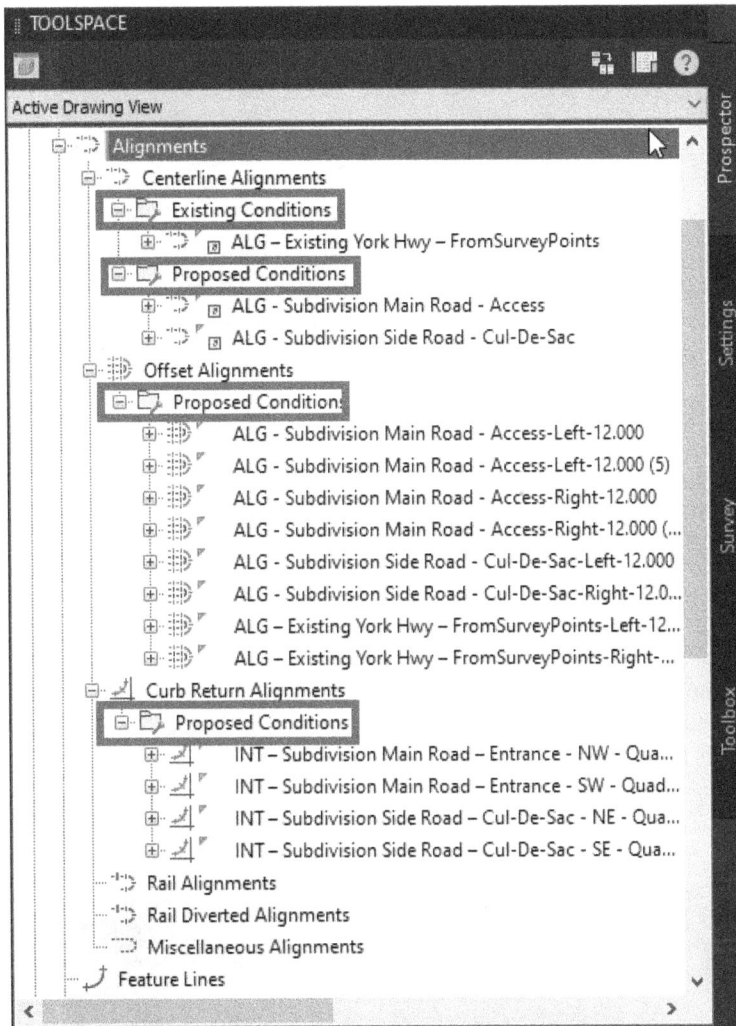

Figure 2.10 – Alignment organizational structure

As we expand the **Centerline Alignments**, **Offset Alignments**, and **Curb Return Alignments** Civil 3D object categories, we can see that there is a **Proposed Conditions** subfolder in each of these and an **Existing Conditions** subfolder only in the **Centerline Alignments** Civil 3D object category.

In most cases, we're not going to need to categorize and create existing conditions alignments for our offset and curb return alignments as these are typically associated with design, or proposed, conditions related to corridor modeling workflows. Although we won't see an Existing Conditions Offset or Curb Return Alignment in most of our projects, there could be specific scenarios where this would be required, or potentially the need to generate an **Incremental Design Conditions** subfolder of sorts. That said, it's always good practice to keep workflows as consistent as possible to minimize learning curves and increase familiarity for additional and possibly new design production staff.

Managing subfolders

With regard to subfolders and how they are created, there are two schools of thought on how to manage our Civil 3D-modeled objects and data shortcuts. Subfolder creation for **Existing Conditions**, **Incremental Design Conditions**, and **Proposed Conditions** can occur at either the current drawing level or the data shortcuts level. In the previous book, *Autodesk Civil 3D 2024 from Start to Finish*, we indicated that, out of the two options, our personal preference is to create folders and manage them at the data shortcuts project level. This is due to the learning curve involved when just starting out with using and deploying Autodesk Civil 3D as it's more of a one-and-done workflow.

At a fundamental level, this makes perfect sense. Although Civil 3D objects are not automatically categorized as they're created, it really shines a light on proper data shortcut management and driving consistency across your entire project design.

Workflow at the data shortcuts level

If we were to decide to manage folders at the data shortcuts level, and not within each individual drawing, the workflow would be as follows:

1. Once your **Data Shortcuts** project has been associated with your drawing, right-click on the Civil 3D modeled objects in your **Data Shortcuts** project and select the **Create Folder** option, as shown in *Figure 2.11*:

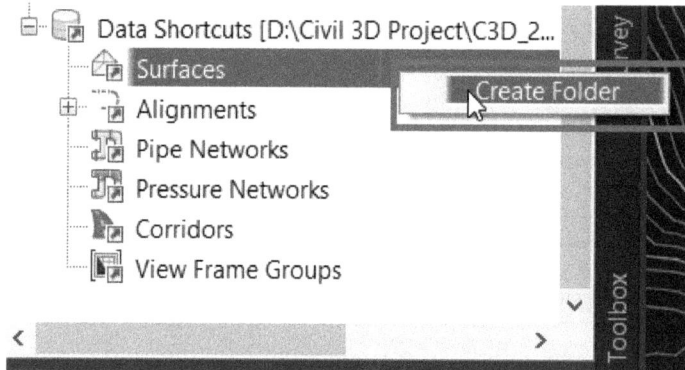

Figure 2.11 – Creating a folder in the Data Shortcuts project

2. After you create the folder and create the data reference again, you'll notice that any surfaces in the current drawing can be added but will be uncategorized. As shown in *Figure 2.12*, we'll then need to select **SRF - Existing Grade –FromSurveyPoints** in the **Data Shortcuts** project and drag and drop it into the **Existing** folder.

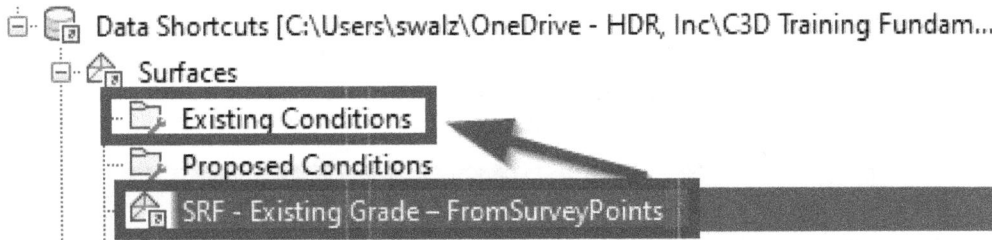

Figure 2.12 – Reorganizing items in the Surfaces folder in the Data Shortcuts project

As we begin to expand the teams involved in these projects and introduce more files and data management requirements, data shortcuts project management has the potential to become a full-time job if not handled properly. So, it is ideal to handle folder structures and the proper organization of Civil 3D modeled objects in one location (the **Data Shortcuts** project), rather than having all individual files created for the project that contain Civil 3D modeled objects.

If we look through the Advanced level lens, it is worth noting that we can certainly find a way to manage this at the drawing level very efficiently as well. In the book *Autodesk Civil 3D 2024 from Start to Finish*, we talked at great length about the importance of creating and updating your company's Civil 3D template, especially as we continue to evolve as professionals and explore more advanced workflows and methodologies related to managing Civil 3D projects and designs. That said, if we are confident that our teams will continue to utilize our company's Civil 3D template when new files are created, we can certainly set up these subfolders within each of our Civil 3D modeled objects listed in **Toolspace | Prospector**, which will be included in all new drawing files created that use said drawing template.

All that said, and with this being an advanced-level book, we recommend that you take on the challenge of deploying this methodology of creating subfolders at the drawing level and maintaining these updates in your company's drawing template. As an advanced user and BIM model manager of your company's designs, you are likely responsible for providing direction and training to design and production staff, especially those involved in the projects you're managing the design for. With that, you have the ability to direct and influence teams on how they should be deploying workflows from file creation to modeling designs within the Civil 3D environment.

Workflow at the drawing level

To set your company's template up, let's jump back into Civil 3D and open up the `Company Template File.dwt` file available in our `Civil 3D 2025 Unleashed\Chapter 2` subfolder. Once opened, let's make sure **Toolspace** is launched again and that the **Prospector** tab is active. You'll probably notice that we started developing this template back in *Autodesk Civil 3D 2024 from Start to Finish*. We're going to continue building this out to increase design efficiencies and make sure we're building a robust template for our company. You'll notice also that our company template is currently lacking organizational structure (refer to *Figure 2.13*).

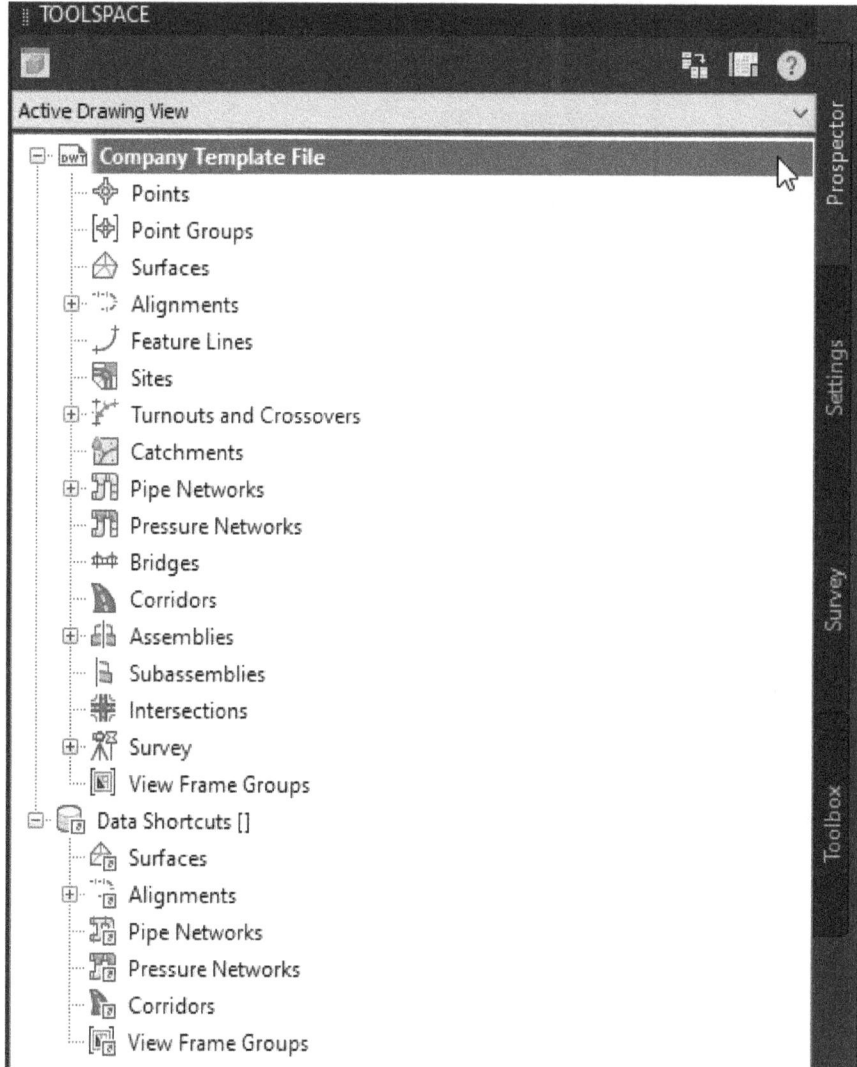

Figure 2.13 – Clean Company Template File in Prospector view

Now that we've decided to manage folders at the drawing level, and not within each data shortcut, the workflow is as follows:

1. Right-click each of the Civil 3D modeled objects listed in **Prospector** and select the **Create Folder** option (refer to *Figure 2.14*). It is worth noting that we may need to expand a few of the objects in the list first to get down to the individual modeled objects (e.g., `Alignments`, `Pipe Networks`, etc.).

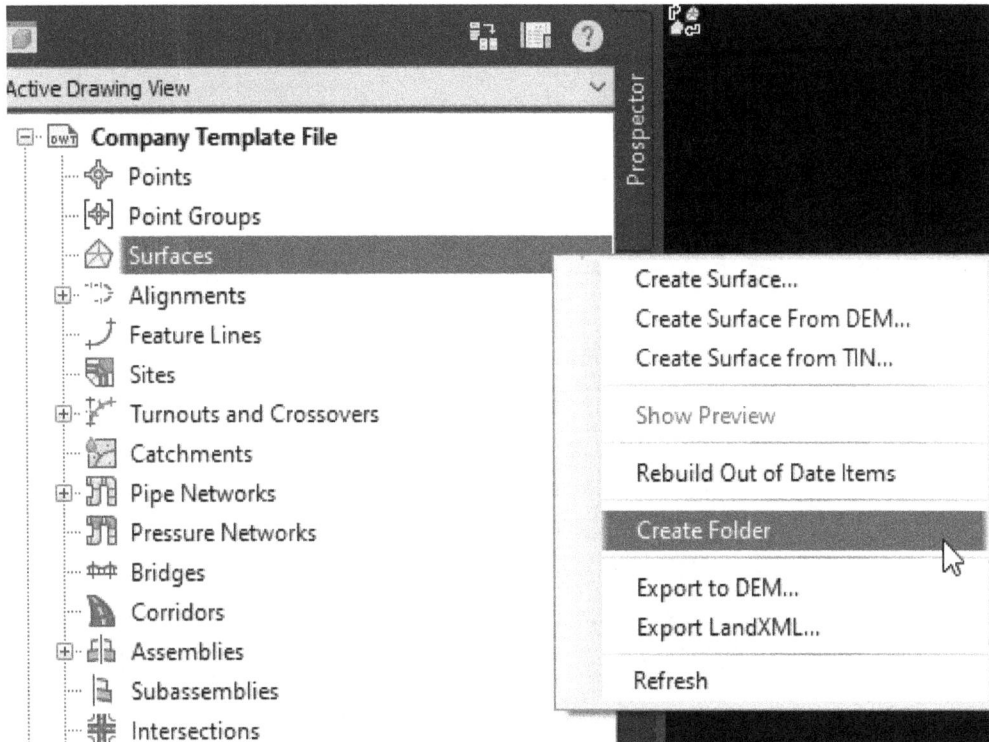

Figure 2.14 – Creating folders for each Civil 3D object

2. Once selected, we'll type in the new folder name. In this case, we'll start by typing `Existing Conditions`, and then select the **OK** button (refer to *Figure 2.15*).

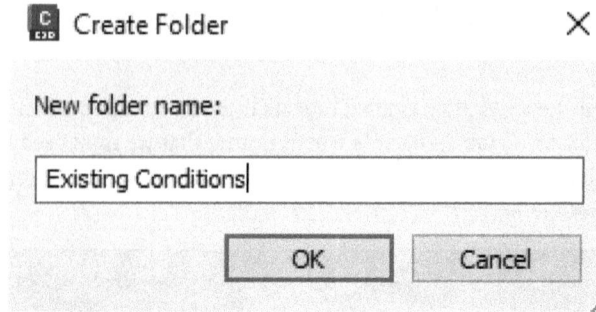

Figure 2.15 – Naming folders for each Civil 3D object

3. Repeat *steps 1 and 2* to create folders for both **Incremental Design Conditions** and **Proposed Conditions** (refer to *Figure 2.16*).

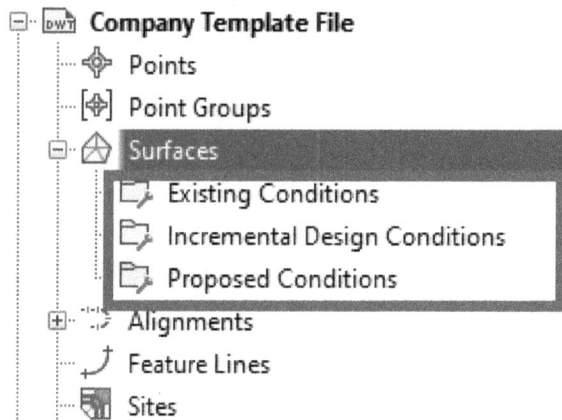

Figure 2.16 – Folders created to keep surfaces organized

Important to note is that we cannot currently create folders for all objects listed. If you were to right-click on **Points** or **Point Groups**, you would not be presented with an option to create a folder like you were with **Surfaces**. That said, let's apply the preceding workflow to each of the remaining Civil 3D objects that we're able to, with the final product appearing similar to the following, each with **Existing Conditions**, **Incremental Design Conditions**, and **Proposed Conditions** folders (refer to *Figures 2.15* and *2.16*).

Figure 2.17 displays our templates for organization:

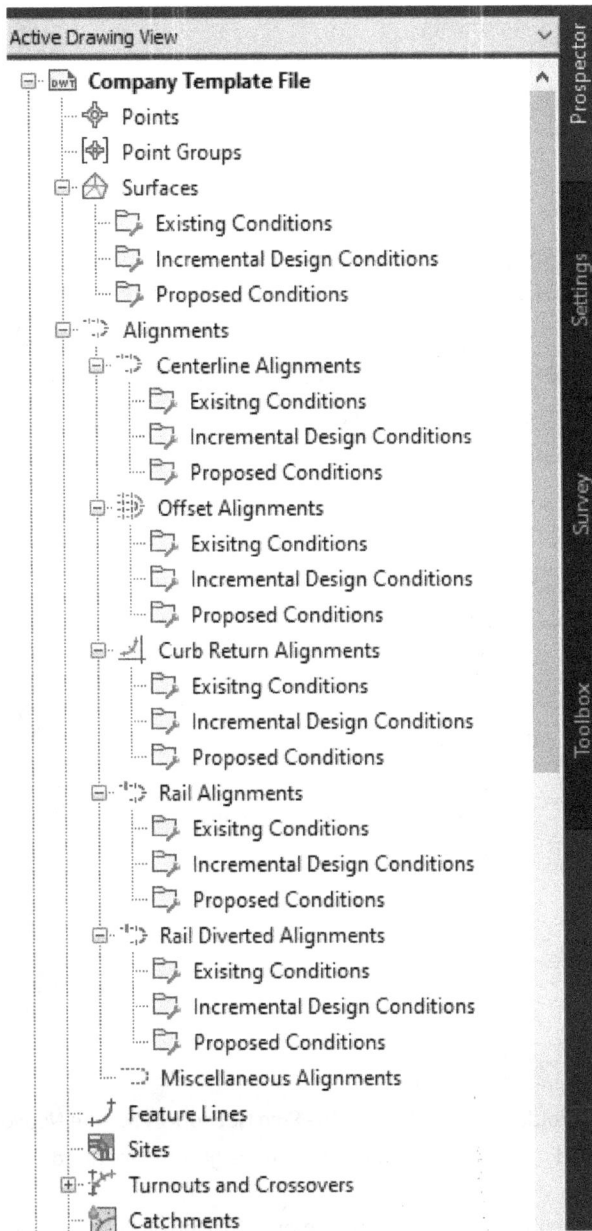

Figure 2.17 – Folders created to keep surfaces and alignments organized

Figure 2.18 shows our templates applied to the current working file:

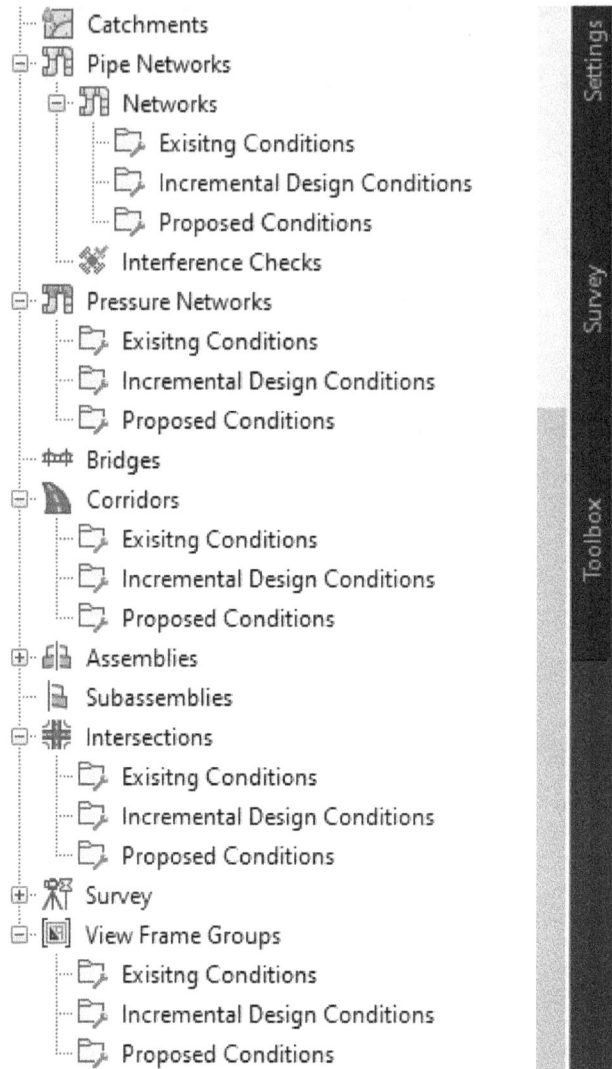

Figure 2.18 – Folders created to keep the Pipe Networks, Pressure Networks,
Corridors, Intersections, and View Frame groups organized

After all the folders have been created, we can resave our `Company Template File.dwt` file, overwriting the previous version in our `Civil 3D 2025 Unleashed\Chapter 2` subfolder.

Properly managing **Data Shortcuts** in our design projects is only half the battle. Properly managing the content contained within our files is a whole other side of the equation that needs to be considered and properly managed as well.

Managing Civil 3D files

To manage Civil 3D files effectively, it's crucial to organize them by model and the types of content contained within. In our previous book, *Autodesk Civil 3D 2024 from Start to Finish*, we reviewed and provided recommendations for splitting our files into categories based on the content contained within each. These three types are model files, reference files, and sheet files. By splitting our files into these categories, we're setting our design teams up to perform at higher levels of efficiency by providing more flexibility for multiple team members to update design files and content. Let's explore each file type.

Model files

A model file in Civil 3D, like that displayed in *Figure 2.19*, serves the purpose of housing both existing conditions and proposed design modeled objects in separate files. These files utilize data references and external files (attached as overlays) to provide context and help further progress our designs.

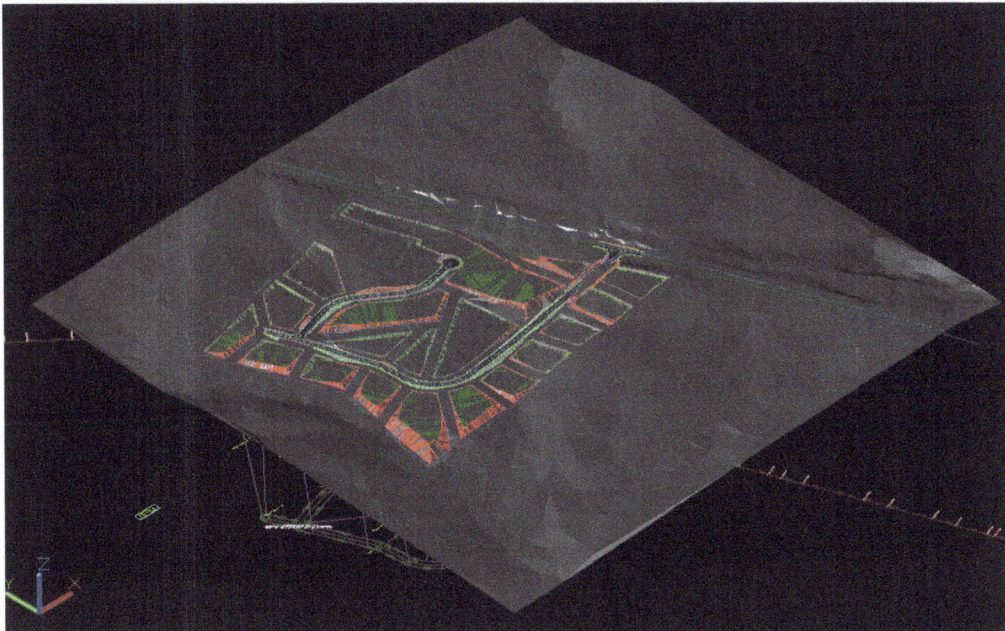

Figure 2.19 – Model file example display

Examples of model files are the following:

- Survey model, which preserves the original survey file while creating a new model using our company's Civil 3D template to apply standards, styles, and settings while employing data shortcuts for organization and continuation through to our design model files. There could be a need to create an alignment model that focuses on designing alignments for various road and site design layouts, excluding gravity and pressure pipe network alignments, and employing a logical naming convention for proposed alignments.

- Another model file that could be necessary to support your design is a grading model, which is dedicated to designing proposed surfaces, with names reflecting their purpose, such as **Site Grading Subdivision**.

- Another model file that could be needed is a utility model, which would be used to design both gravity and pressurized pipe networks, which could result in parsing out or the development of multiple Utility Model files for different network types and purposes.

Reference files

A reference file in Civil 3D, like that displayed in *Figure 2.20*, serves the purpose of housing 2D geometry, static elements, annotations, and data-referenced design objects from our model files.

Figure 2.20 – Reference file example display

Examples of a reference file are the following:

- One example is the site plan reference file, which contains points for coordinate tables, data-referenced alignments, and optional topographic data

- Another example of a reference file is a grading reference file, in which data references both existing and proposed surfaces from survey and grading model files

- Another example of a reference file is a utility plan reference file, which is used for data referencing pipe networks, alignments, and surfaces as needed

- We may also need to create a profile and/or section reference file, which is used to reference alignments, design profiles, pipe networks, and surfaces for profile and section views alike

Sheet files

Finally, the project culminates in sheet files, representing the final product that will typically represent the deliverable, similar to that displayed in *Figure 2.21*.

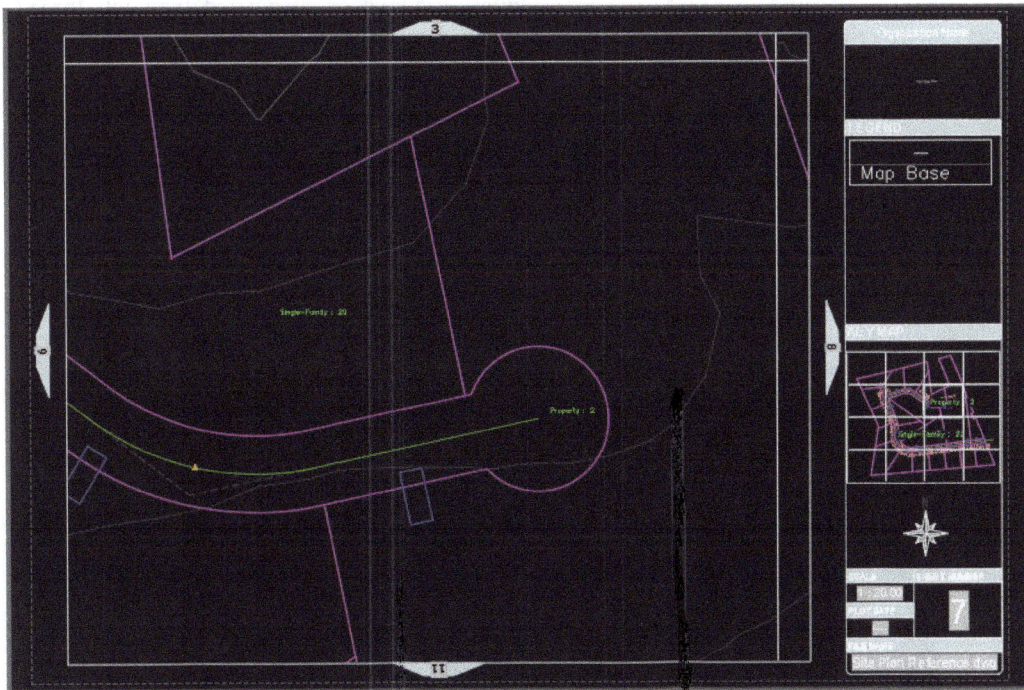

Figure 2.21 – Sheet file example display

The assembly of these sheets involves external referencing of our reference, while adding final sheet-specific information such as borders, title blocks, general notes, key notes, north arrows, and any additional sheet-specific information.

Adopting a project structure of this nature offers numerous advantages, from optimizing the overall design file to fostering collaborative opportunities across project design teams and your organization. Regardless of your project's size and the teams involved, implementing this methodology is certain to enhance the efficiency of your design process. This structured approach ensures that project teams can concentrate on and tackle different aspects of design and final production tasks, allowing multiple team members to work simultaneously on various files and design models.

Furthermore, as we introduce more advanced tools throughout this book, we'll really see the value of separating content by file types and generating additional files to continue expanding the design and collaboration team that can access design content and data being modeled. Next up, we will dive into our design toolbelts to look at how we will use them on a daily basis, along with how we can delve deeper into their capabilities to unleash our civil designs.

Design toolbelts for everyday use

In *Autodesk Civil 3D 2024 from Start to Finish*, we covered most of the tools we have at our disposal within Civil 3D. With those tools discussed, we essentially grouped, or categorized, them into three toolbelts based on the major types of design applications they support, and also discussed a practical implementation of these toolbelts.

Types of toolbelts

Here are the three toolbelts based on the major types of design applications:

- The Land Development toolbelt primarily focuses on foundational site design components, involving the creation and management of parcels and sites and the application of grading tools. These tools are essential for establishing the foundation and groundwork of a residential subdivision design (refer to *Figure 2.22*).

Figure 2.22 – Example of a design using the Land Development toolbelt

- The Roadway Modeling toolbelt extends the design capabilities into transportation and civil infrastructure, with tools available for creating and managing subassemblies, assemblies, and corridors and designing intersections and cul-de-sacs. This toolbelt emphasizes the importance of dynamic links between foundational and roadway model objects, facilitating a seamless design progression (refer to *Figure 2.23*).

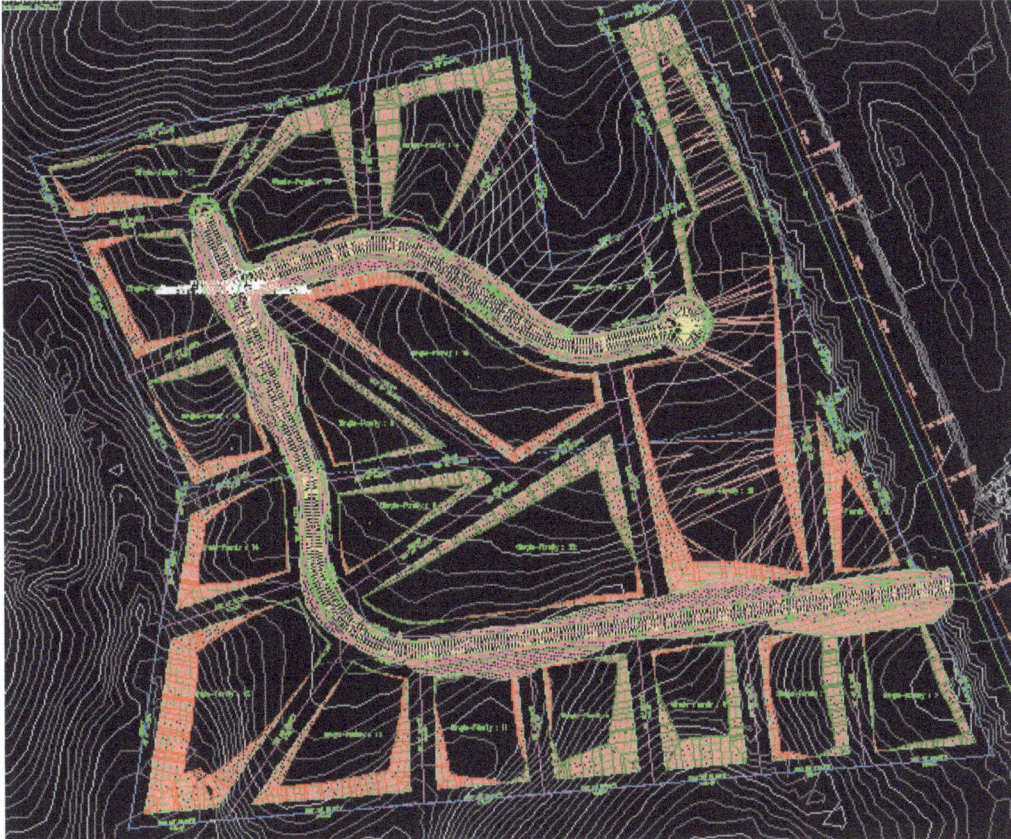

Figure 2.23 – Example of a design using our Roadway Development toolbelt

- The final toolbelt is the Utility Modeling toolbelt, which integrates utility design applications, including refining surface models for proper site drainage and creating storm drainage and sanitary sewer pipe networks, as well as pressurized networks for domestic water services. The tools in this belt ensure we have a comprehensive approach to utility design (refer to *Figure 2.24*).

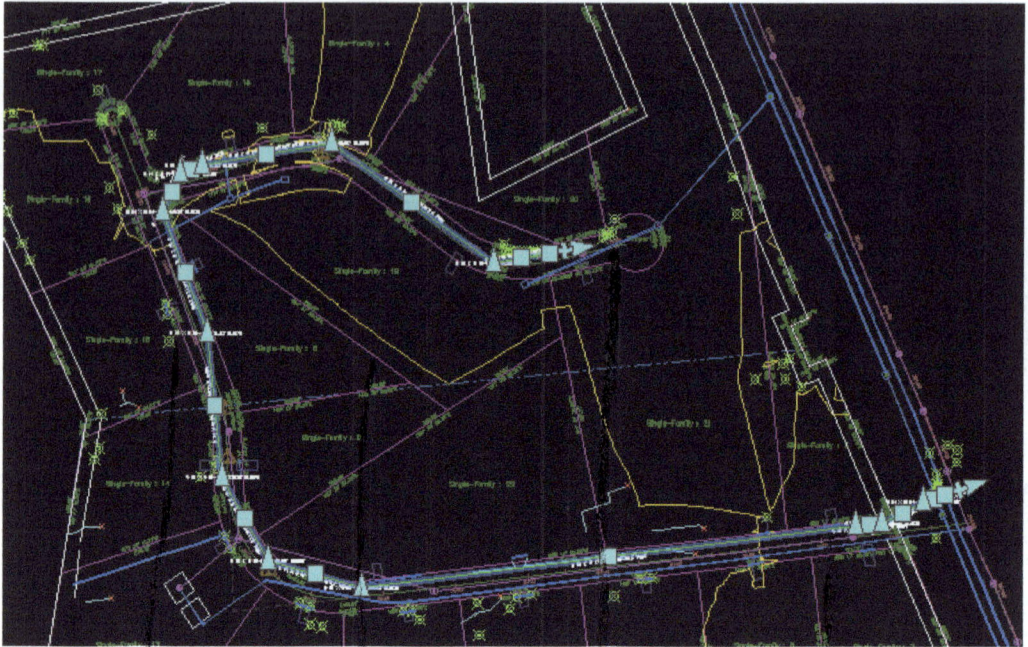

Figure 2.24 – Example of a design using our Roadway Development toolbelt

Together, these toolbelts cover a wide spectrum of design applications crucial for a holistic approach to residential subdivision design in Civil 3D. Thinking outside the box a little bit, we can also quickly pick up that these tools and toolbelts can offer us endless design solutions for a multitude of design scenarios outside of just residential subdivision designs. We can utilize our Land Development, Roadway Modeling, and Utility Modeling toolbelts fairly extensively for stream and wetland design and restoration projects, highway improvement projects, landfill design projects, or even airport expansion projects.

Examples using the three toolbelts

To expand upon this thought and concept a little more, if we were to use Civil 3D to design a stream and wetland design/restoration project, we would have the potential to utilize each of the toolbelts for the following applications within our design scenario:

- **Land Development toolbelt**: We can apply **Parcel** tools available to us within Civil 3D to manage and create existing and proposed parcels, zoning requirements, easements, and so on. From a grading standpoint, we can utilize these tools to manage water flows and storage, improve stability and erosion conditions, and so on. In more intricate designs, we could even utilize the grading tools to design a waterfall or regenerative stormwater conveyance system.

- **Roadway Modeling toolbelt**: We can apply various subassemblies, assemblies, and corridors to lay out and design the routing of existing and proposed flows, streams, rivers, and so on. We can also utilize intersection tools where we may have a stream or river convergence to maintain proper water flows.

- **Utility Modeling toolbelt**: We can even apply our stormwater design tools within our Utility toolbelt to improve stormwater conditions feeding into, or even exiting, our stream or wetland, helping us design a solution that will effectively manage stormwater runoff and prevent eroding conditions.

In this book, and future publications, we will begin to explore, expose, and highlight the application of these tools – along with advanced tools, covered in this book specifically – on more project types. We will be able to take our skill sets and careers to the next level as we continue mastering fundamental and new toolbelts throughout this book.

Summary

As we worked through this chapter, we mostly reviewed key foundational items (also covered in our previous publication, *Autodesk Civil 3D 2024 from Start to Finish*) to ensure that we are fully prepared to take that next step in elevating our careers, skill sets, and overall understanding of this intricate design authoring tool. We also highlighted and discussed a few new approaches to expanding our design teams from both a personnel and an organizational standpoint. Now we have the mindset and thorough understanding of the foundational tools and functionality that Civil 3D offers us.

In the next chapter, we'll not only build on our comprehension and understanding of our toolbelts but truly elevate our skill sets to accommodate different situations we encounter throughout the design process, making us better prepared to update designs on the fly and improve our final product.

3

Advanced Design and Analysis Capabilities within Civil 3D 2025

Over the past couple of chapters, we've both recapped and begun to understand some of the more advanced capabilities and workflows that we can deploy when utilizing **Autodesk Civil 3D** for our BIM designs. Moving forward, we'll explore some of the newer tools and workflows that are available to us in this software so that we can elevate and advance our careers in the civil BIM arena, and expedite our learning curves to bridge this career development path. In future chapters, we will dive deeper into the specific areas Civil 3D offers and highlight those areas as we progress.

We'll continue to build on top of all of the foundational skills and knowledge we've developed thus far, allowing us to take our civil BIM designs to the next level. In this chapter specifically, we'll focus on advanced modeling and coordination techniques that will help us along the way.

The key topics that we will cover in this chapter include the following:

- Advanced corridor modeling techniques and workflows
- Advanced utility analysis techniques and workflows for **gravity networks**
- Advanced utility analysis techniques and workflows for **pressure networks**

Technical requirements

We will be using the same hardware and software requirements as discussed in the *Technical requirements* section of *Chapter 1*.

With that, let's go ahead and open up our `Grading Model Start.dwg` file located within the `Civil 3D 2025 Unleashed\Chapter 3\Model` location. Once opened, you'll notice that we'll be starting this chapter pretty much where we left off in the previous chapter, with the display of our model looking similar to that shown in *Figure 3.1*.

Figure 3.1 – Grading Model Start.dwg

Advanced corridor modeling techniques and workflows

We covered quite a bit of basic and some advanced corridor modeling techniques and workflows in our previous book, *Autodesk Civil 3D 2024 From Start to Finish*. As we develop our careers as civil designers, we'll realize at some point that these techniques and workflows aren't a one-size-fits-all type of design scenario.

There will be times throughout our careers when **lane transitions** will be required to be designed. In most cases, this will be quickly realized if our design focus is mostly on projects supporting transportation. Frequently, there is a need to understand lane transitions when designing residential subdivisions along with commercial-based designs. Occasionally, we may even apply "lane transition" workflows, or widening designs, for environmentally based projects such as stream or river restoration or improvement designs. We'll continue to utilize our previous dataset, taking the middle-of-the-pack approach, to learn how to account for lane transitions in our corridor and intersection models.

Adding a lane

Now, let's begin exploring a very simple process to add a lane, to account for a left- and right-turning exit lane, by adding a transition within our **INT – Subdivision Main Road – Entrance** intersection model, with the eventual update of our **SRF – Proposed Grade – Residential Subdivision** surface model located in **Surfaces | Proposed** within our **TOOLSPACE | Prospector** tab.

Using the following steps, we'll want to start with updating our intersection model:

1. Jumping to our **TOOLSPACE | Prospector** tab, we'll expand our **Intersections** Civil 3D object category (refer to *Figure 3.2*).

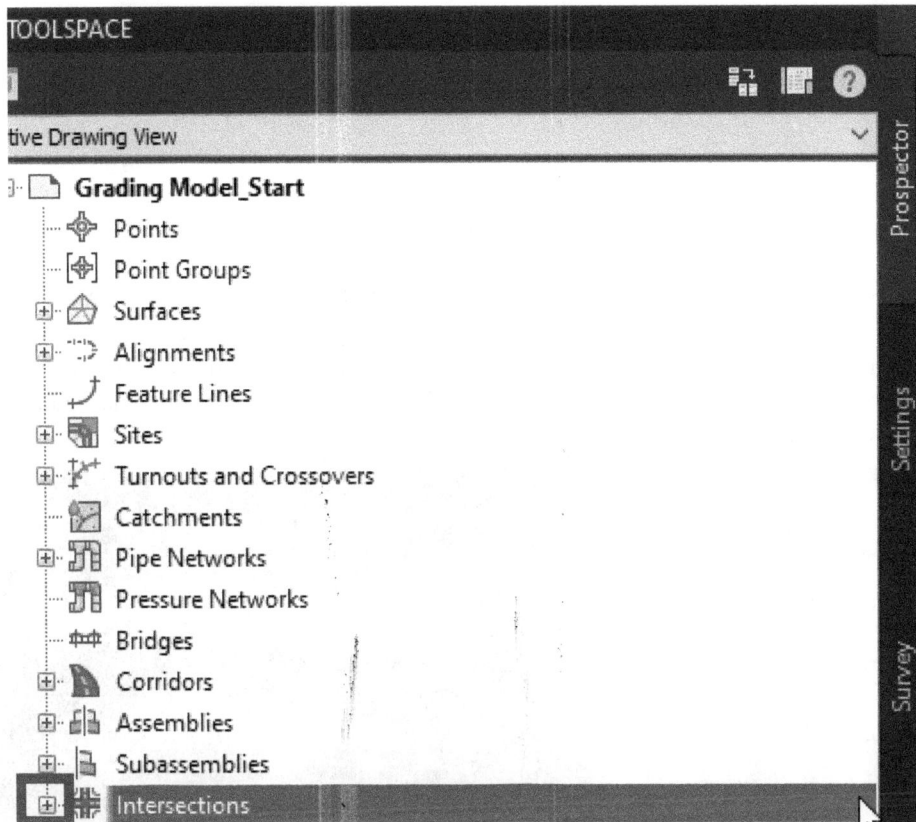

Figure 3.2 – Expand Intersections in TOOLSPACE | Prospector

2. Once expanded, go ahead and right-click with your mouse on the **INT – Subdivision Main Road – Entrance** and select **Edit Offset Parameters…** (as shown in *Figure 3.3*).

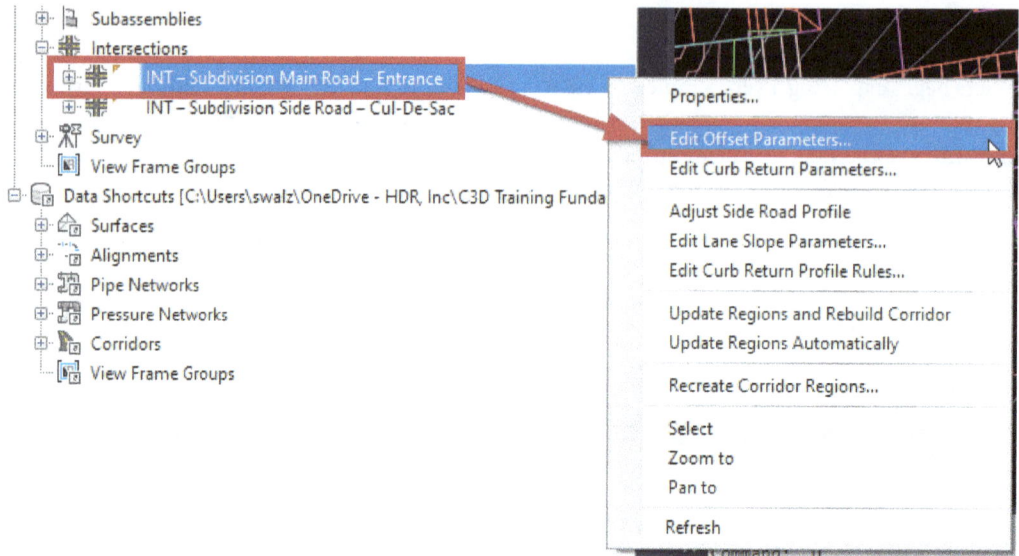

Figure 3.3 – Edit the offset parameters of intersections

3.   Next, we'll have the **INTERSECTION OFFSET PARAMETERS** dialog box appear. As we select either the **Primary Road** or the **Secondary Road** geometry, along with the subcomponents listed within, we'll notice that, in the model space of our design, an alignment or subcomponent is highlighted (refer to *Figure 3.4*). This automatic highlighting gives us a great quick indication as to which parameter and value we're updating or modifying.

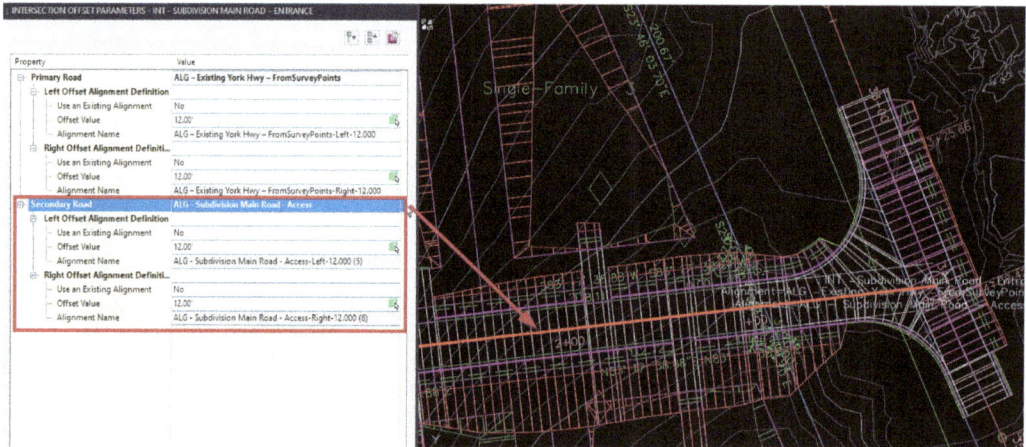

Figure 3.4 – Highlighting of design component / subcomponent

4. Now that we've quickly identified which alignment is our primary versus secondary road geometry, let's go ahead and change the offset value of the **Left Offset Alignment Definition** from `12.00'` to `24.00'` to double the value, essentially accounting for an extra lane of traffic for turning right versus left in the exiting process of the residential subdivision (refer to *Figure 3.5*). You'll notice that as we update the value in the **Offset Value** box, both the highlighted geometry, along with the curve, have also updated to represent its anticipated location.

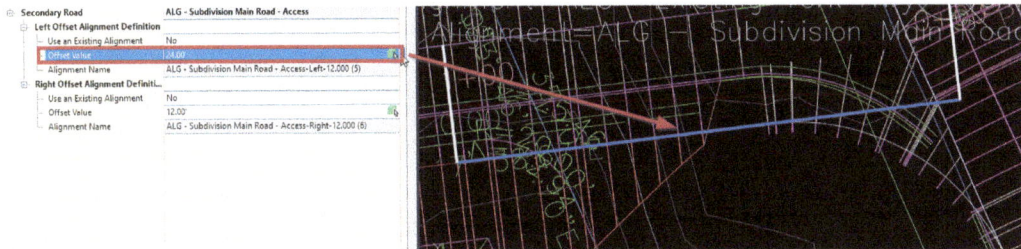

Figure 3.5 – Highlighting of design component / subcomponent

5. After updating the offset value to `24.00'` to account for the additional exit lane, let's go ahead and close out of our **INTERSECTION OFFSET PARAMETERS** dialog box.

6. Once closed, we'll notice that in our **TOOLSPACE | Prospector** tab, the **INT – Subdivision Main Road – Entrance** intersection model has an out-of-date icon next to it. Additionally, you may notice that the main **Intersections**, **Corridors**, and **Surfaces** Civil 3D objects have out-of-date icons next to them as well (refer to *Figure 3.6*). It is worth noting that this can be either good or bad depending on either the **level of detail (LOD)** required from a design standpoint or possibly a model health standpoint based on the sizes of the files being managed.

If either a higher LOD is required, or model health may be a concern on your project, we recommend that **Rebuild – Automatic** be toggled off; otherwise, toggle them on if these are not of concern. An additional note is that this decision should be a civil BIM model manager-level decision, which is now in your hands. If you have already set these Civil 3D objects to **Rebuild – Automatic**, you may skip *Steps 8* through *10*.

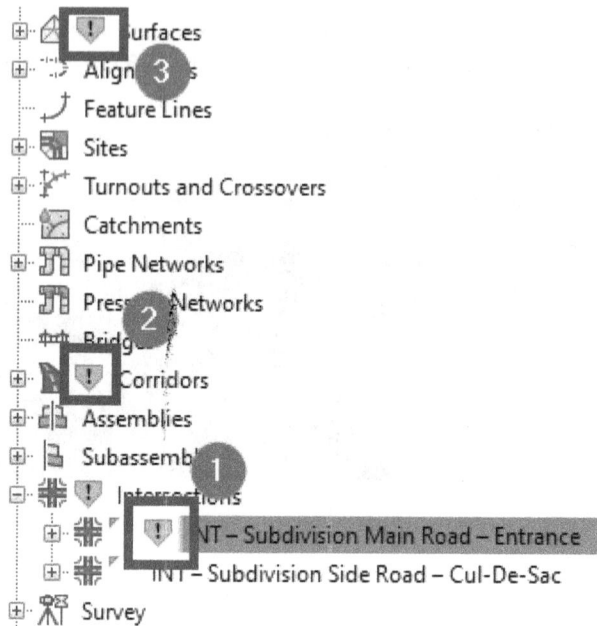

Figure 3.6 – The highlighting of outdated components

7. If not already toggled off due to LOD or model health purposes, or just simply overlooked, we'll want to start by right-clicking on the **INT – Subdivision Main Road – Entrance** intersection model and select the **Update Regions and Rebuild Corridor** option (refer to *Figure 3.7*).

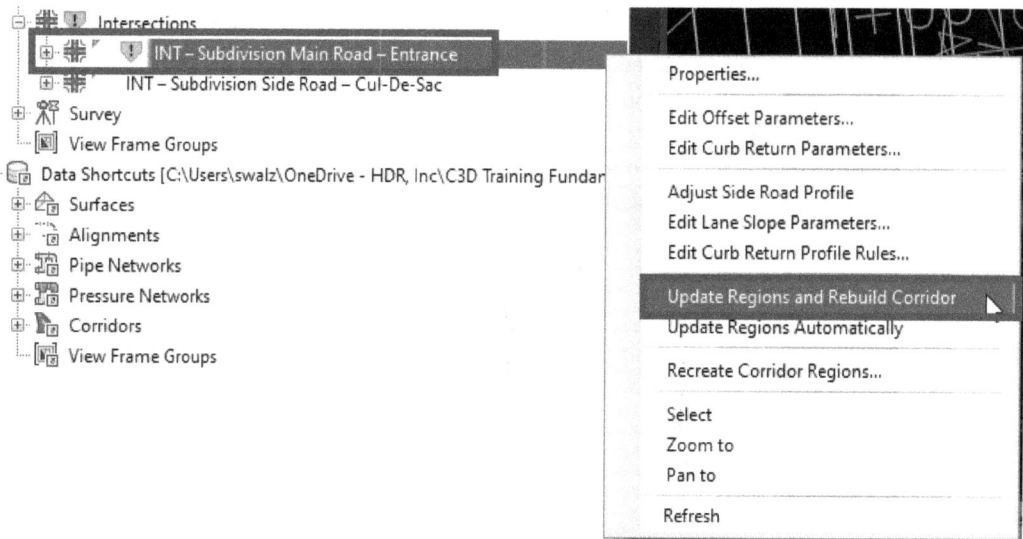

Figure 3.7 – Update Regions and Rebuild Corridor

8. We'll then be prompted with another dialog box asking us if we want to continue or not, giving us the **Continue with update** or **Cancel this command** options, at which point we'll want to select **Continue with update** (refer to *Figure 3.8*).

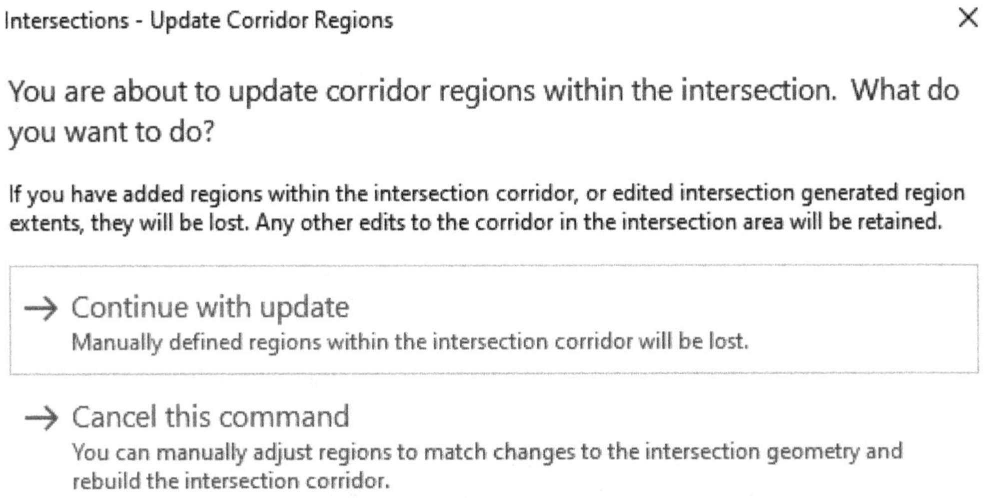

Intersections - Update Corridor Regions   ✕

You are about to update corridor regions within the intersection. What do you want to do?

If you have added regions within the intersection corridor, or edited intersection generated region extents, they will be lost. Any other edits to the corridor in the intersection area will be retained.

→ Continue with update
   Manually defined regions within the intersection corridor will be lost.

→ Cancel this command
   You can manually adjust regions to match changes to the intersection geometry and
   rebuild the intersection corridor.

Figure 3.8 – Intersections | Continue with update

9. Once our intersection has been updated, we'll next want to move up to our Corridor Model Civil 3D object classification in **TOOLSPACE | Prospector**, right-click or **Corridors**, and select the **Rebuild All** option, as shown in *Figure 3.9*).

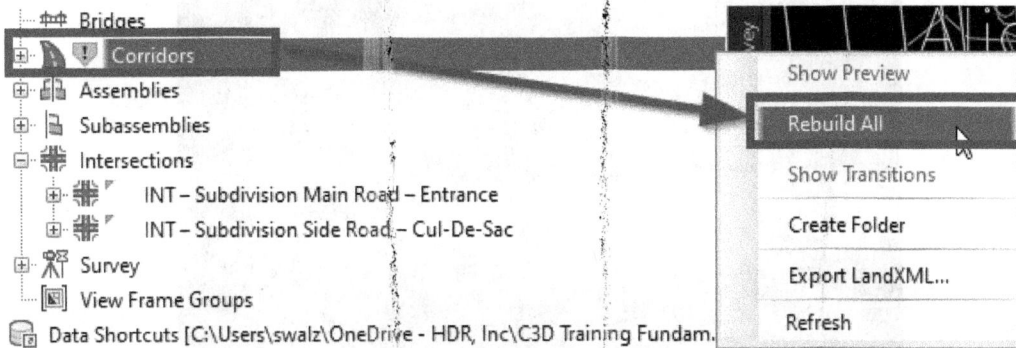

Figure 3.9 – Rebuild corridors (all corridors)

10. Next, we'll want to move up to our Surface Model Civil 3D object classification in **TOOLSPACE | Prospector**, right-click on **Surfaces**, and select the **Rebuild Out of Date Items** option, as shown in *Figure 3.10*).

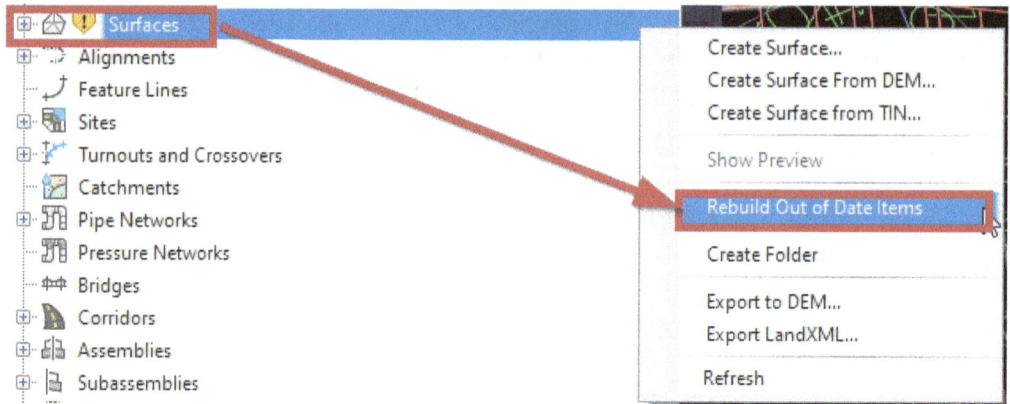

Figure 3.10 – Rebuild out of Date Items (All Surfaces)

With these steps followed, we should end up with an intersection design similar to that displayed in *Figure 3.11*.

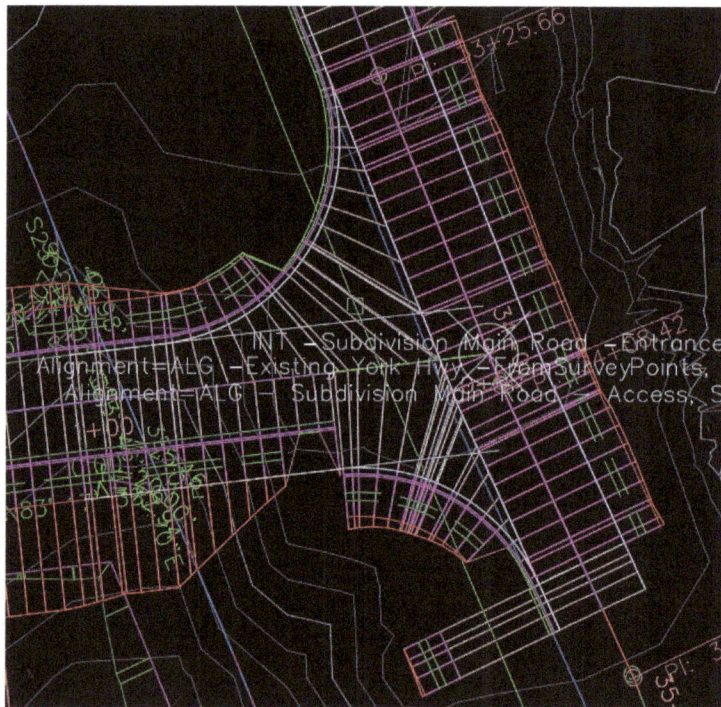

Figure 3.11 – Updated intersection parameters

Now that we have successfully completed our intersection, we can continue to build upon this element by adding a transition to it for more accurate modeling of real-world conditions within Civil 3D.

Adding a transition

Next, we'll want to add a transition to our **INT – Subdivision Main Road – Entrance** intersection model so that it ties smoothly into the updated version of our **INT – Subdivision Main Road – Entrance** intersection model.

We'll use the following steps to accomplish this:

1. Select our **INT – Subdivision Main Road – Entrance** intersection model in model space to bring up the Corridor Contextual Ribbon and select the **Edit Corridor Transition** tool located in the **Modify Corridor** palette (refer to *Figure 3.12*).

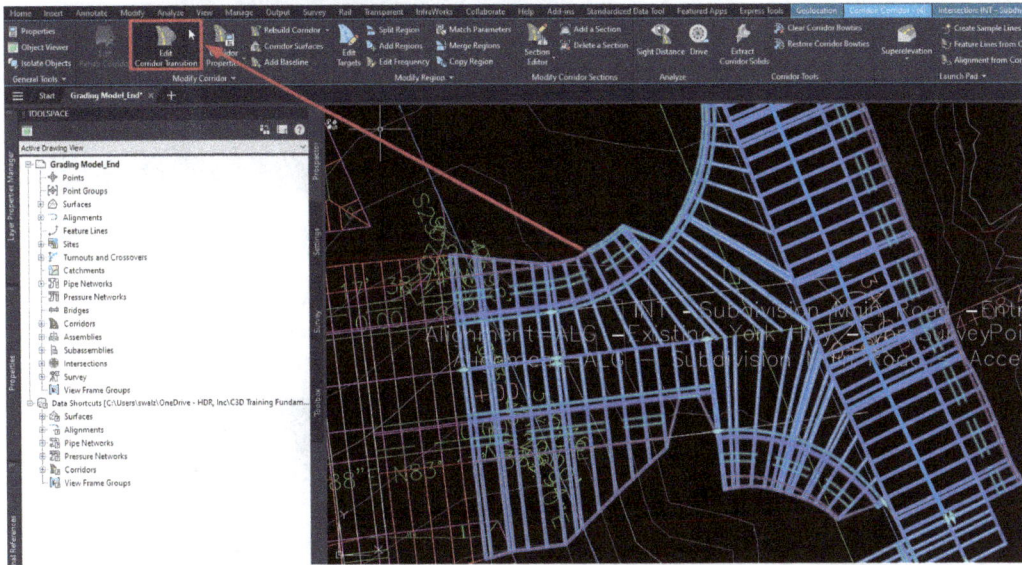

Figure 3.12 – Edit Corridor Transition

2. Once the **Edit Corridor Transition** tool has been selected, you'll be prompted at the command line with the **Select a Baseline** prompt, at which point we'll want to hover over the centerline alignment of our intersection along the main entrance into the subdivision design and click with our mouse when the alignment is highlighted in red, as shown in *Figure 3.13*.

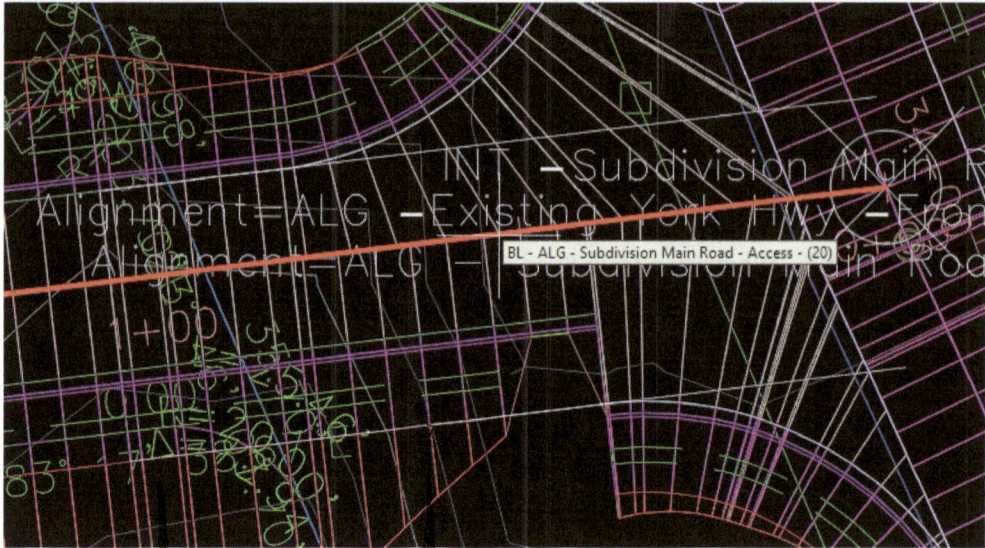

Figure 3.13 – Select a baseline of the intersection model

3. Next, we'll be prompted at the command line with the **Select a subassembly** prompt by *subassembly name* or *[Class name]*, at which point we'll want to hover over the exit lane that approaches the updated curb return assembly, as shown highlighted in bright blue in *Figure 3.14*.

Figure 3.14 – Select the subassembly within the intersection model

4. After selecting the subassembly we wish to apply the transition to, we'll be prompted to specify the parameter to transition (refer to *Figure 3.15*). In our case, we'll select the **Width** parameter.

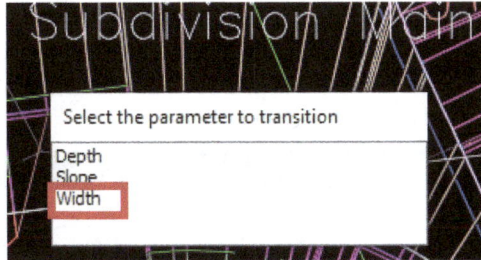

Figure 3.15 – Select Width for our transition prompt

5. After selecting **Width**, we'll be prompted at the command line to select a start station of the transition for the **Width** parameter. At this stage, we'll snap to the beginning of the subassembly that we just selected in *Step 3*, as shown in *Figure 3.16*.

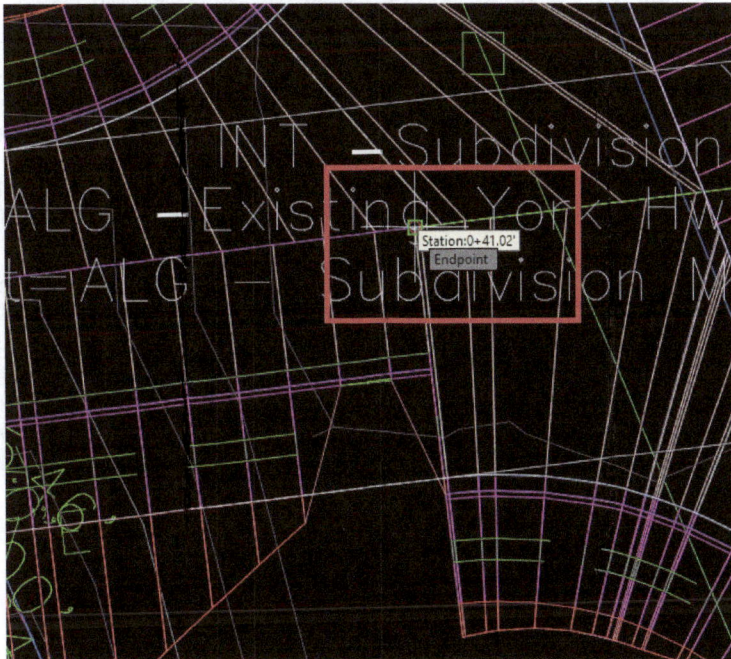

Figure 3.16 – Select the start station for the transition

6. Next, we'll be prompted to select the parameter value at `Station:0+41.02'`. Here, we'll want to specify `24.00'` as the value that we'll start the width transition.

7. Next, we'll be prompted to select the end station of the transition for **Width**, at which point we'll enter `81.02` as the station value to account for a `40'` long transition.

8. Next, we'll be prompted to enter the parameter value at `Station:0+81.02'`. Here, we'll want to specify `12.00'` as the value to transition back into a single lane.

9. Finally, we'll be prompted to enter the transition type, at which point we'll select **Reverse Parabolic** and then hit the *Enter* key again to give us a smooth curved transition from a single lane to two lanes as we exit our subdivision. We should have an updated Intersection view in model space that looks similar to that shown in *Figure 3.17*.

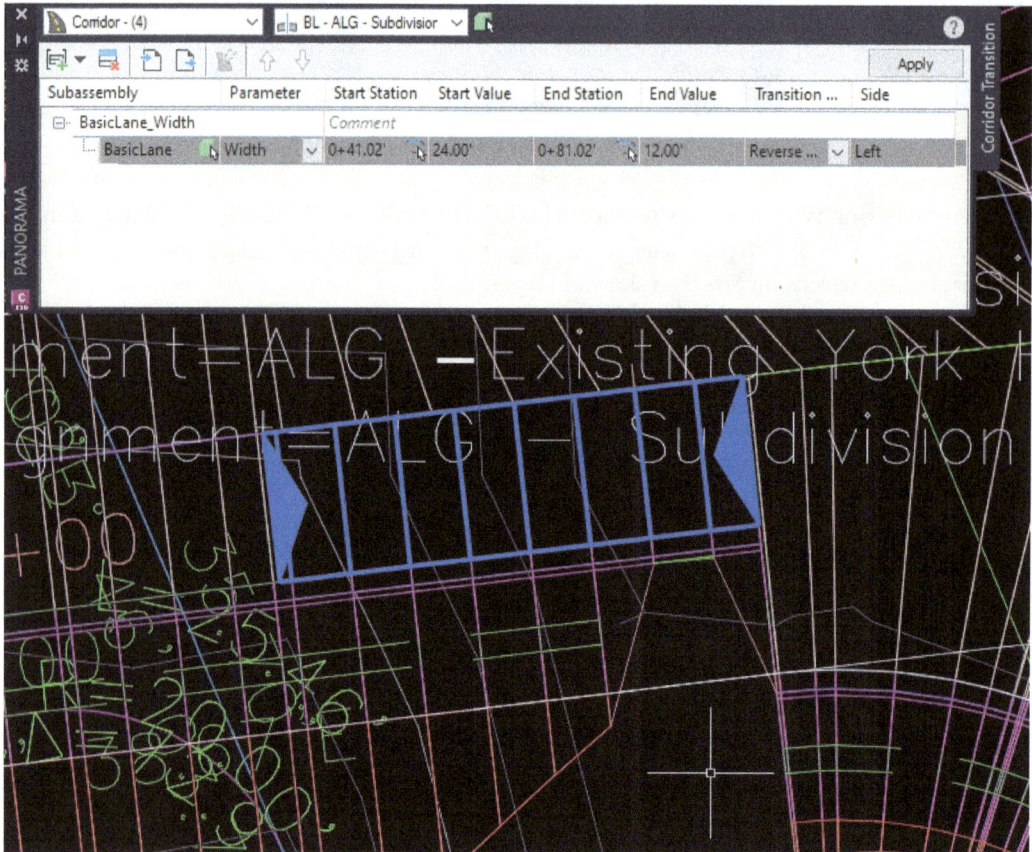

Figure 3.17 – Transition applied in model space to our intersection model

10. Now, go ahead and either click the **Apply** button in the top right of the panorama or close the panorama and select **Apply the transition data and close the dialog box** option when prompted (refer to *Figure 3.18*).

Warning   ✕

⚠ The transition data has not been applied to the
    corridor. What do you want to do?

    → Apply the transition data and close the dialog
       box

    → Close the dialog box without applying the
       transition data

☐ Always perform my current choice   Cancel

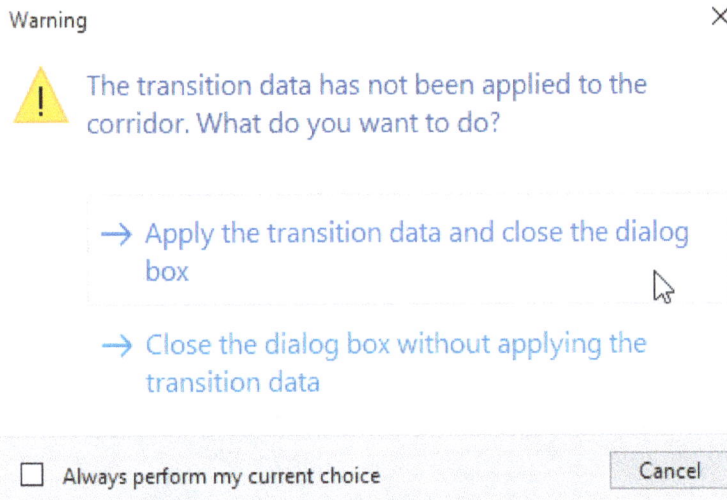

Figure 3.18 – Apply the transition data and close the dialog box

Our transition should now be applied to our intersection model, with the final result appearing similar to that shown in *Figure 3.19*.

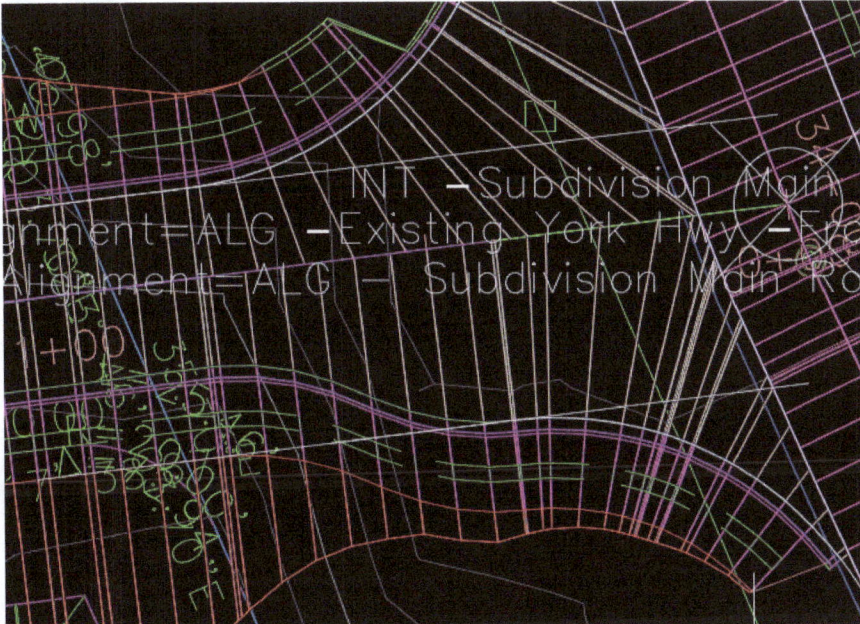

Figure 3.19 – The final transition as it should appear

This workflow demonstrates one of the most basic applications of adding a transition. In a more detailed situation, such as applying a taper of a curb where we have driveways located throughout our subdivision, we'd mostly follow similar steps, but specify a different subassembly and parameter to transition. In this instance, we'd select the `BasicCurbAndGutter` subassembly, then select the `CurbHeight` parameter, define our start and end station points, curb height transition from `0.75'` to `0.00'`, and finally apply the **Linear** Transition Type (refer to *Figure 3.20* for Settings and updated display).

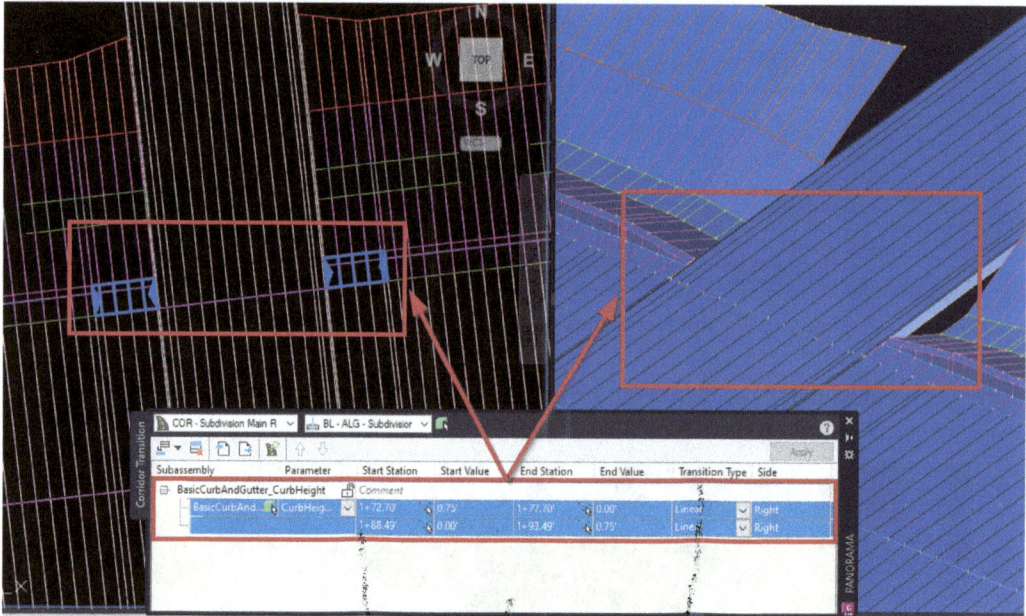

Figure 3.20 – Apply the transition to curbs at driveway locations

In the early days of Autodesk Civil 3D, a lot of these processes were very tedious and time consuming, and could take a good four to eight hours (sometimes longer) to complete if done correctly the first time. Nowadays, using the workflows detailed previously, we can knock this task out in a matter of minutes.

We've now officially upped our corridor-modeling game by applying advanced workflows all while maintaining dynamic links throughout all Intersection, Corridor, and Surface models within our files.

Advanced utility analysis techniques and workflows for gravity networks

Now that we have utilized some advanced workflows for corridors, let's follow this pattern and apply it to our project's utilities, specifically gravity networks. As we move on to our advanced utility analysis capabilities, which are available to us within Autodesk Civil 3D, let's go ahead and open up our `Utility Model Start.dwg` file located within our `Civil 3D 2025 Unleashed\ Chapter 4\Model` location.

In this file, we'd recommend isolating only the gravity and pressure networks in the view for ease of navigation and analysis. To do so, you'll want to right-click on each network in our **TOOLSPACE | Prospector** and pick the select option. Once all network objects have been selected, we'll then right-click in model space, select **Isolate Objects**, and then select the **Isolate Selected Objects** option, as shown in *Figure 3.21*.

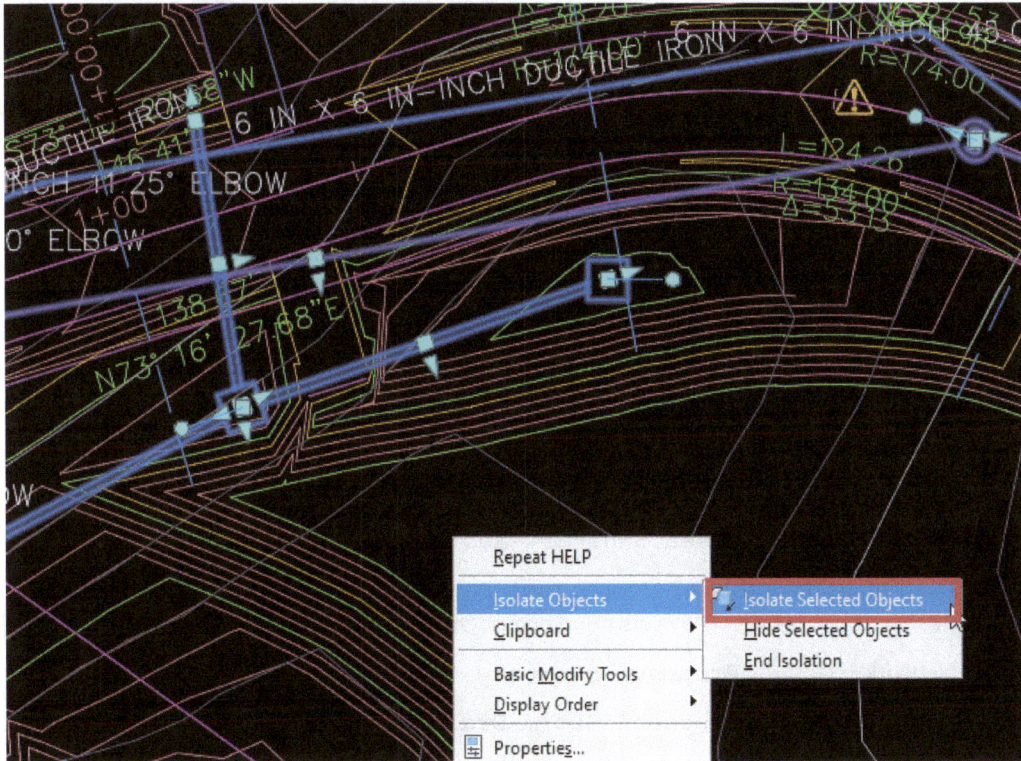

Figure 3.21 – Isolate Selected Objects

With our `Utility Model Start.dwg` file open and view set up, let's put on our BIM model manager hat for a bit and run through some various design analysis tools we have available to us within Autodesk Civil 3D. As our utility networks become more complex, with multiple networks crossing paths or running in parallel, we aren't always designing to the exact minimum separations (both horizontally and vertically).

We'll want to utilize the gravity **Interference Check** tool to ensure that our gravity networks at a minimum have the correct design separation requirements applied. To utilize this tool, we'll jump up to our **Analyze** ribbon, navigate to the **Design** panel, and select the **Interference Check** tool as shown in *Figure 3.22*.

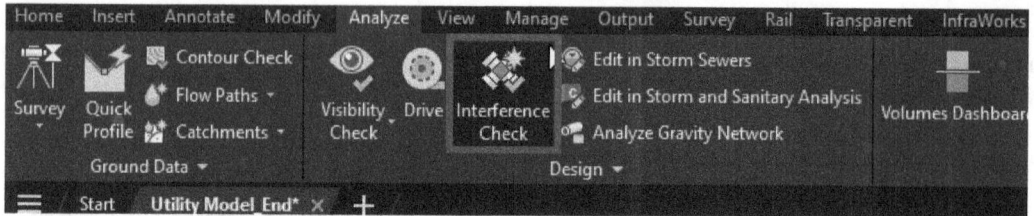

Figure 3.22 – Interference Check within the Analyze tab

Once selected, we'll use the following steps to run our interference check between our Gravity Storm and Sewer Networks:

1. We'll be prompted at the command line to select a part from the first network, at which point we'll select one of the pipes or structures in **GPN – Proposed Sanitary Sewer Gravity Network**.

2. Next, we'll be prompted to select another part from the same network or a different network, at which point we'll select one of the pipes or structures in **GPN – Proposed Storm Drainage Gravity Network**.

3. After both networks have been identified, we'll be presented with the **Create Interference Check** dialog box. Here, we'll fill the fields available as follows (also displayed in *Figure 3.23*):

    * **Name:** GHC - Proposed Sanitary and Storm

    * **Description:** Horizontal Comparison between GPN - Proposed Sanitary Sewer and GPN - Proposed Storm Drainage

    * **Network 1:** GPN - Proposed Sanitary Sewer

    * **Network 2:** GPN - Proposed Storm Drainage

    * **Layer:** C-STRM

    * **Interference style:** Basic

    * **Render material:** ByBlock

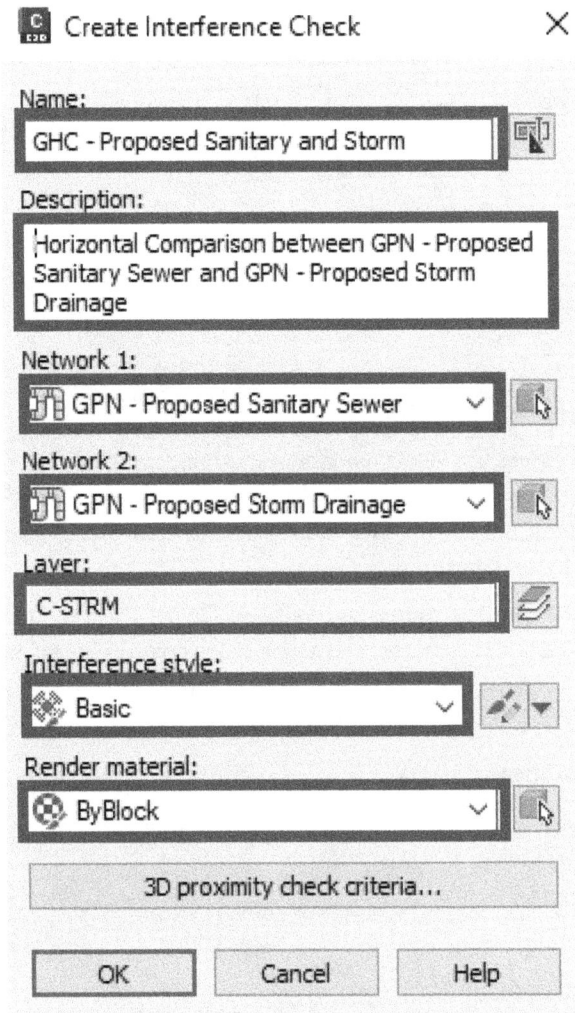

Figure 3.23 – Create Interference Check

4. Prior to pressing the **OK** button at the bottom left of the **Create Interference Check** dialog box, let's select the **3D proximity check criteria…** first to define our horizontal separation values.

5. In the **Criteria** dialog box, let's go ahead and check the **Apply 3D Proximity Check** option and set the **Use distance:** value to 10.00', as shown in *Figure 3.24*.

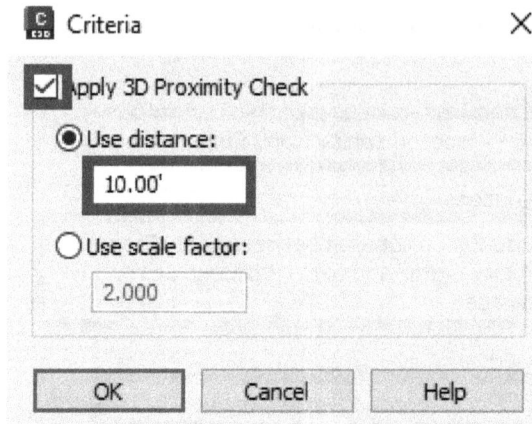

Figure 3.24 – The Criteria Proximity Check settings

6. Once defined, go ahead and click the **OK** button in the **Criteria** dialog box as well as the **Create Interference Check** dialog box.

    After all dialog boxes have been approved, Autodesk Civil 3D will run the interference check on the identified gravity pipe networks and a new dialog box will appear, indicating how many interferences have been found. In our dataset, there should have been **5** interferences found (refer to *Figure 3.25*).

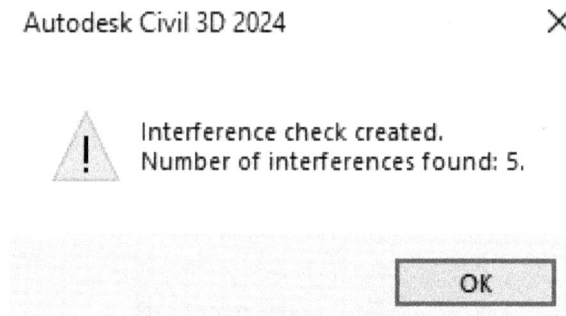

Figure 3.25 – Interferences found

As we navigate around our model now, we should see five orange nodes placed in locations where interferences have been located (refer to *Figure 3.26*).

Figure 3.26 – Interferences found

It's important to note that these interference checks will identify interferences using a 3D buffer around the Gravity Pipe Networks objects (pipes and structures). We'll need to perform two checks, one for horizontal and one for vertical interferences. Typical design requirements recommend that we maintain a 10' horizontal separation and a 1.5' vertical separation between these particular utility network objects.

We'll need to perform a little cleanup of the first interference check we ran in our file by removing the interference nodes placed where it accounted for a 10' vertical separation, leaving us with only 2 interferences where our design does not meet the 10' horizontal separation design requirement.

7. Next, we'll run the vertical-focused interference check by running through similar steps outlined previously and filling out the **Create Interference Check** dialog box as follows (also displayed in *Figure 3.27*):

- **Name:** GVC - Proposed Sanitary and Storm

- **Description:** Vertical Comparison between GPN - Proposed Sanitary Sewer and GPN - Proposed Storm Drainage

- **Network 1:** GPN - Proposed Sanitary Sewer

- **Network 2:** GPN - Proposed Storm Drainage

- **Layer:** C-STRM

- **Interference style:** Basic

- **Render material:** ByBlock

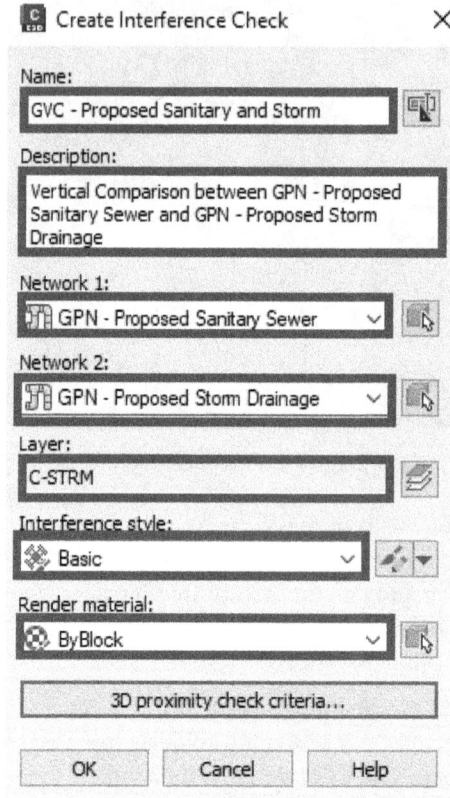

Figure 3.27 – Create Interference Check

8. Just as in our previous *Step 5*, let's go ahead and click on the **Apply 3D Proximity Check** button again and set the distance to 1.5', as shown in *Figure 3.28*.

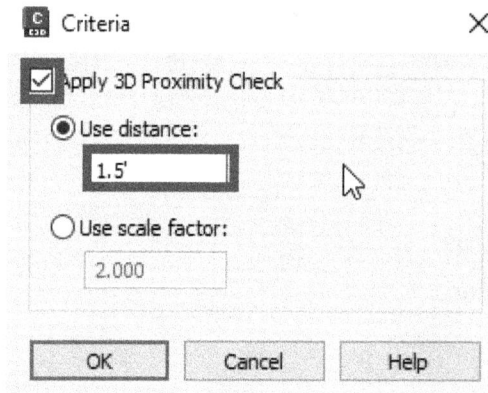

Figure 3.28 – Criteria Proximity Check Settings

After selecting **OK** to get out of the **Criteria** and **Create Interference Check** dialog boxes, another dialog box will appear letting us know that two interferences have been found.

9. For each of these interferences that have been located, it would be good to check these conflicts in a 3D view just to see what's going on and understand how we need to update our models. Let's navigate to one of the locations where interference has been located; select the Interference node along with the pipes and/or structures that are in conflict, right-click with our mouse in model space, and select **Object Viewer...** (as shown in *Figure 3.29*).

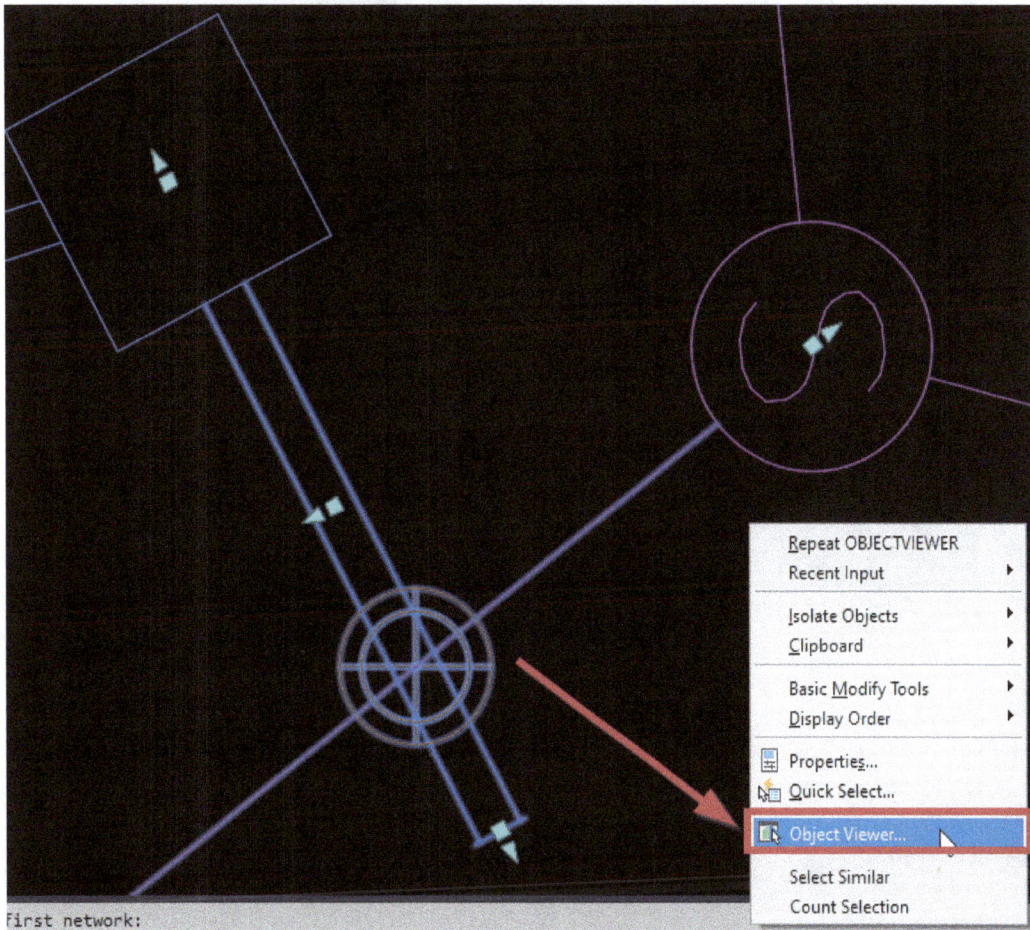

Figure 3.29 – Object Viewer of Selection Set

10. After selecting **Object Viewer...**, our **Object Viewer** dialog box will appear (refer to *Figure 3.30*), where we can apply different displays, views, and navigate around the interference to get a clear picture of how we need to update our model to ensure that it conforms to design requirements.

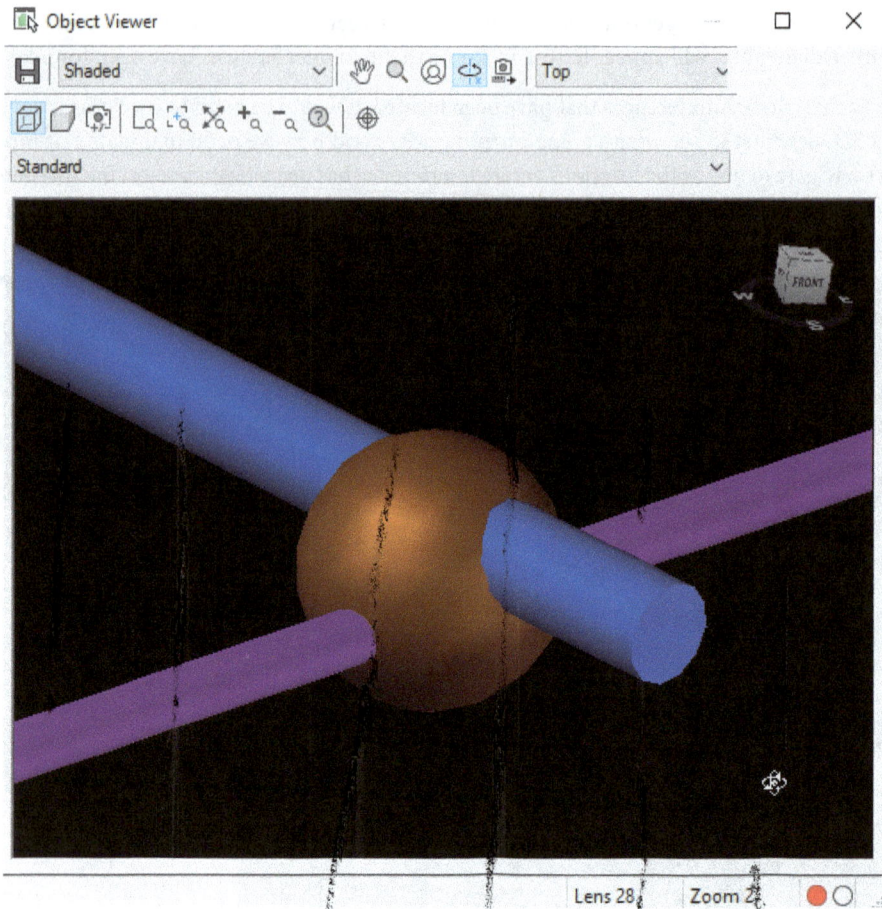

Figure 3.30 – Object Viewer of the interference

With this last step, we have successfully applied some learnings focusing on gravity networks within our advanced utility techniques. Civil 3D has much to offer beyond the basics of component modeling and after this section, we will apply similar principles to our next subsection of utilities, which is pressure networks.

Advanced utility analysis techniques and workflows for pressure networks

Now that we have an idea of what needs to be updated from a design standpoint related to our **gravity pipe networks (GPNs)**, let's move over to further analyze both our **pressure pipe networks (PPNs)**, properly named PPN – Proposed Sanitary Sewer, and PPN – Domestic Water Main Pressure Pipe Networks. The first analysis we're going to perform is the **design check**, followed by the **depth check**.

Performing a design check

To perform the design check, we'll start by using the following steps to access this tool (also displayed in *Figure 3.31*):

1. Right-click on the **PPN – Proposed Sanitary Sewer Pressure Pipe Network**.

2. Click on the **Select** option.

3. Select the **Design Check** tool located in the **Analyze** panel within the Pressure Pipe contextual ribbon.

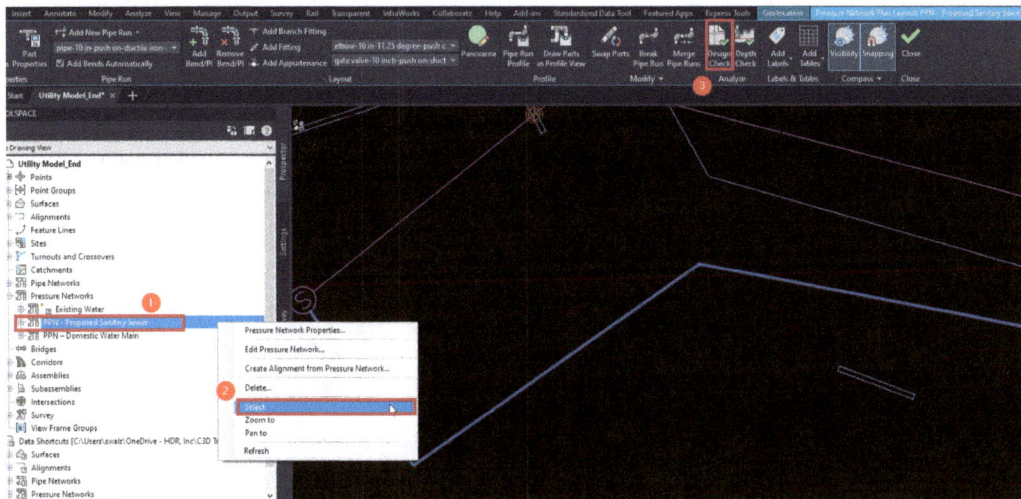

Figure 3.31 – Accessing the Pressure Pipe Network Design Check Tool

Upon selecting the **Design Check** tool, the **Run Design Check** dialog box is launched as follows:

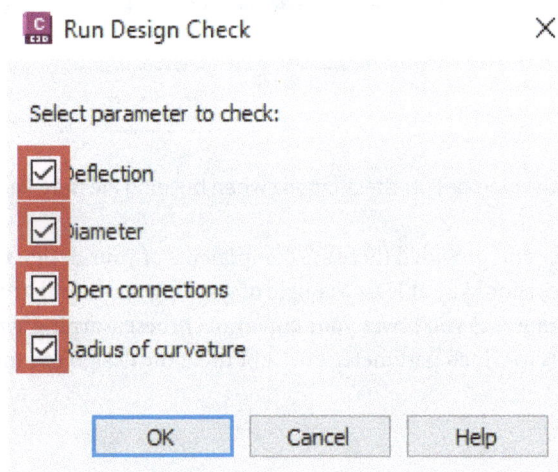

Figure 3.32 – Run Design Check dialog box

Now, let's run through the settings and check each of the boxes that we'd like to check for design conformance (as shown in *Figure 3.32*) and better understand how and where each criterion is being assessed:

- **Deflection**: Checking this parameter will check the deflection angle at all connection points between pipes and pipes, pipes and fittings, along with fittings and fittings. The **Allowable Deflection** angle value that this parameter is checking against is defined in the Pressure Network Parts List as Pipes and Fittings are added to the corresponding list (refer to *Figure 3.33*).

- **Diameter**: Checking this parameter will check that the diameters of pipes, fittings, and appurtenances match at all connection points.

- **Open Connections**: Checking this parameter will check that all pipes, fittings, and appurtenances are actually connected at all connection points.

- **Radius of Curvature**: Checking this parameter will check that all pipes meet the allowable **Minimum Flex Radius** as defined for each of the parts as they're added to the Pressure Network Parts List being used (refer to *Figure 3.33*).

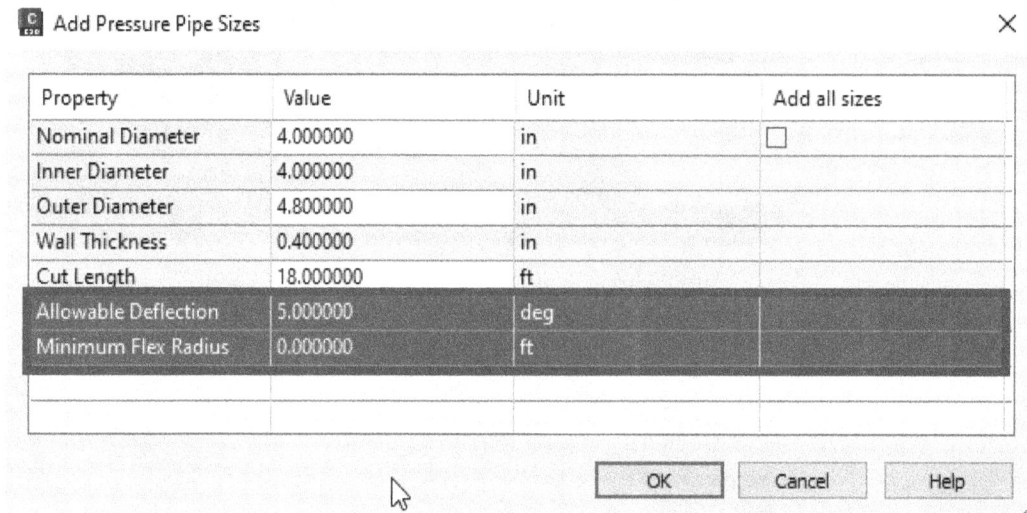

Figure 3.33 – Defining design check criteria when building Pressure Parts Lists

After checking all the boxes, and provided there are components of your design that do not fulfill the design check's criteria, there should be at least a couple of warning symbols that appear on top of our pressure network. Furthermore, as you hover your cursor over these warning symbols, they actually give you some indication as to which parameter does not meet the design requirements.

As can be seen in *Figure 3.34*, we have a pipe connection to a fitting that exceeds the maximum allowable deflection defined by 2.8947 degrees.

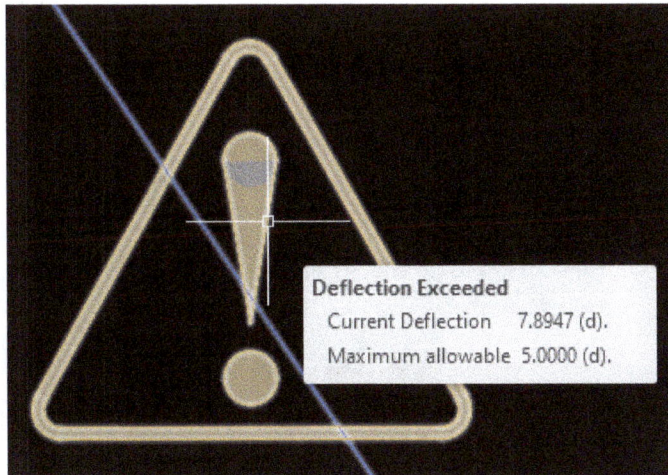

Figure 3.34 – Design check warning and details

It is worth noting that since half of the criteria that these design checks are checking our designs against, it's much more imperative that we have our model management hat on when setting up Pressure Network Parts Lists from the get-go. Otherwise, we may end up with false assessments and warnings throughout our design models.

Now that we have an idea of what needs to be updated to get our pressure networks in conformance with design requirements, let's go ahead and run a depth check on these networks as well.

> **Note**
>
> It is recommended that we run both design checks prior to actually updating our pressure network models so that we can update them all at once. It is also recommended to run both of these checks again after updating our models, just to ensure that our designs are still in conformance and we aren't updating one part of the network just to push another part of the network out of design conformance.

Performing a depth check

To perform the depth check, we'll start by using the following steps to access this tool (also displayed in *Figure 3.35*):

1. Right-click on the **PPN – Proposed Sanitary Sewer** pressure pipe network.
2. Click on the **Select** option.

3.   Select the **Depth Check** tool located in the **Analyze** panel within the Pressure Pipe Contextual Ribbon.

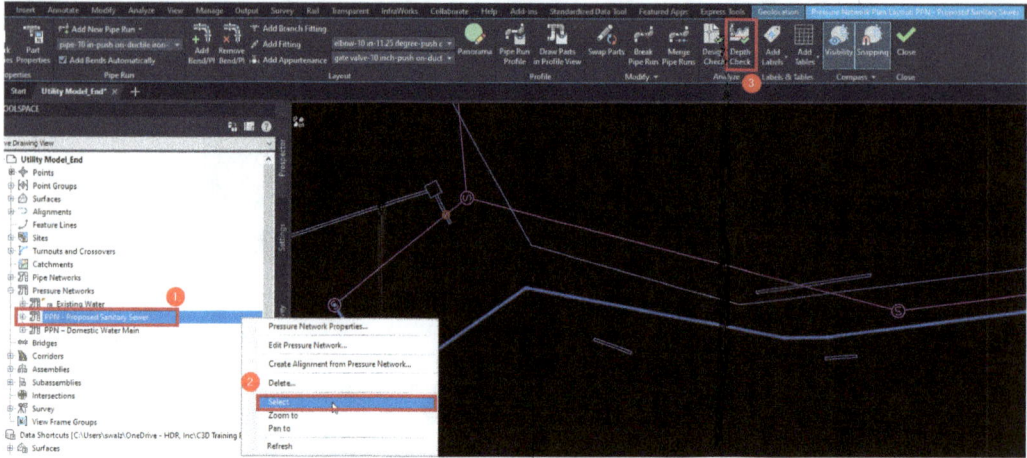

Figure 3.35 – Accessing the Pressure Pipe Network Depth Check Tool

Once selected, we'll be prompted at the command line to select a path along a pressure network in the plan or profile view, at which point we'll select the first pipe in the run followed by the last pipe or fitting in the run. After selecting this sequence, we should see the entire pressure pipe run highlight. Once the entire run has been highlighted as selected, click the *Enter* key on your keyboard to accept, and a **Run Depth Check** dialog box will appear.

Let's run through the settings and check each of the boxes that we'd like to check for design conformance (as shown in *Figure 3.36*) and better understand how and where each criterion is being assessed:

- **Minimum depth of cover**: Checking this parameter will check that all pipes, fittings, and appurtenances meet the minimum depth to the top of modeled objects in comparison to the referenced surface associated with, and defined in, the Pressure Network Properties.

- **Maximum depth of cover**: Checking this parameter will check that all pipes, fittings, and appurtenances meet the maximum depth to the top of modeled objects in comparison to the referenced surface associated with, and defined in, the Pressure Network Properties.

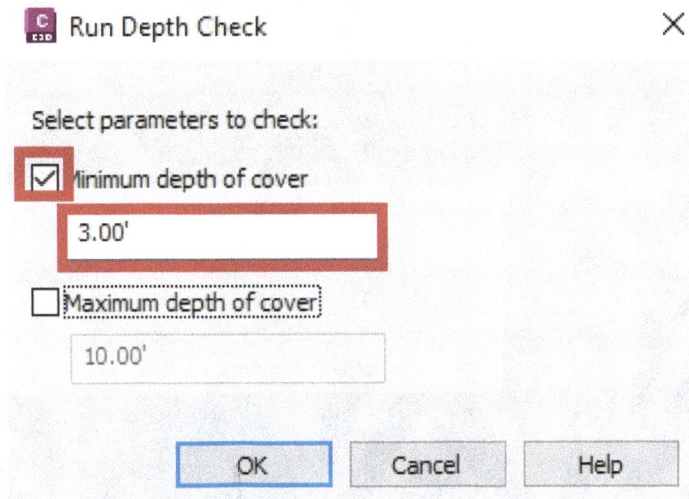

Figure 3.36 – The Run Depth Check dialog box

After checking the required boxes and setting the values, go ahead and click the **OK** button to run the depth check on our entire pressure network. Once completed, you'll likely see a series of warning symbols appear along the Pressure Network Pipe run (refer to *Figure 3.37*).

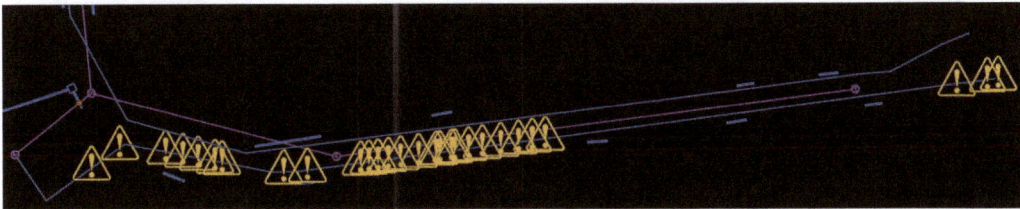

Figure 3.37 – Depth check warnings

Just as we did when we ran the design check earlier, if you hover your cursor over each of the warning symbols, you'll get some additional details that indicate the location and current actual coverage of your pressure pipe, along with the value that we defined as the minimum and/or maximum depth of cover in the **Run Depth Check** dialog box (refer to *Figure 3.38*).

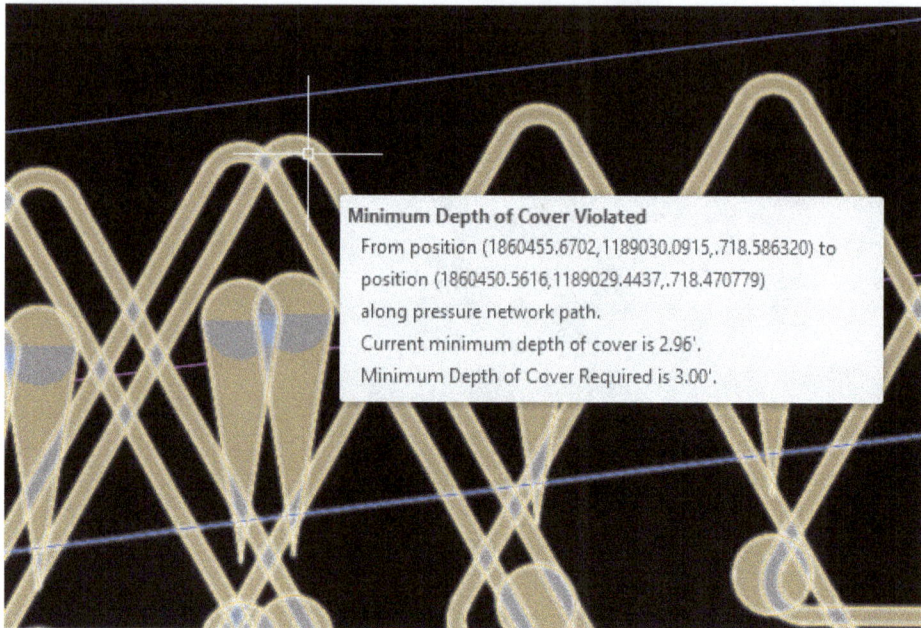

Figure 3.38 – Depth check warning and details

With this last step, we have officially completed our run of advanced functionality of pressure networks within Civil 3D. The detail and precision that comes with utility modeling in Civil 3D is very comprehensive and the applications do not end here.

Summary

As we worked through this chapter, we learned several techniques and workflows to further analyze and refine our subdivision design. This chapter took us through the advanced modeling and analysis capabilities of corridors and utilities, both gravity and pressure networks. These workflows are considered advanced and can get complicated as they layer upon each other in real-world projects. Tools covered are typically best handled by either a BIM model manager or senior-level designer who has intimate design application knowledge and is up to speed on design requirements that should be adhered to; with the chapters in this book, hopefully, you can make your way to that level much faster. We've also built on our comprehension and understanding of our foundational tool belts by truly elevating our skillsets to accommodate different situations we encounter throughout the design process, making us better prepared to update designs on the fly and improve our final product.

In the next chapter, we'll begin exploring how we can use our foundational knowledge of our Roadway Modeling Toolbelt and up our game a bit by understanding how we can use similar tools and workflows to develop custom railway designs. We'll also introduce a new dataset since the addition of a railway to our residential design doesn't make the most sense.

Rail Design Capabilities within Autodesk Civil 3D 2025

In the previous chapter, we reviewed and learned a few ways in which we can elevate our Model Management game as it relates to applying advanced design techniques and further analyzing our corridor and utility networks for both gravity and pressurized networks. In this chapter, we will take all our knowledge of developing roadway alignments, profiles, assemblies, and corridors and morph them slightly for a different type of design.

We'll begin to understand how we can continue building on our foundational knowledge of Civil 3D tools and workflows to develop a rail design. Since our initial site doesn't necessarily need to account for a rail design, we'll use a different dataset for this chapter to get more of a real-world scenario-type understanding in which we can utilize rail design tools and workflows. In this chapter, we'll cover the following topics:

- Setting up our rail alignments, assemblies, and corridor models
- Auxiliary track design workflows
- Viewing and editing rail models

Technical requirements

We will be using the same hardware requirements as discussed in the *Technical requirements* section of *Chapter 1*.

With that, let's go ahead and open up the `Survey Model_Start.dwg` file located within `Civil 3D 2025 Unleashed\Chapter 4\Model`. Once opened, you'll notice that the file has zoomed into our site, as displayed in *Figure 4.1*.

Figure 4.1 – Survey Model_Start.dwg

As I'm sure you've noticed, there really isn't much content in there except for a surface model. This dataset was actually created with the use of InfraWorks, which will be covered in *Chapter 14*, so we won't get into much of the details here, but InfraWorks is another civil application from Autodesk. However, it is good to note that whenever you need a quick existing conditions plan that doesn't necessarily need to be survey-grade accurate, InfraWorks is a great tool to use to get you a quick base model that incorporates open-source data.

Setting up our rail alignments, assemblies, and corridor models

In our previous book, *Autodesk Civil 3D 2024 from Start to Finish*, we covered many foundational tools to create surfaces, alignments, profiles, profile views, assemblies, and corridor models. Using similar workflows that we deployed to develop our roadway models throughout our previously designed subdivision, we can morph these, or make slight adjustments along the way, to generate a rail design that contains both a main track and an auxiliary track.

Since all we are starting with is a surface model, we'll want to start adding in our rail alignments manually as they are not typically recognized as corridor models in InfraWorks. That said, first, we'll want to turn on our aerial map by going to the **Geolocation** ribbon, then to the **Online Map** panel, and selecting the **Map Aerial** option as shown in *Figure 4.2*.

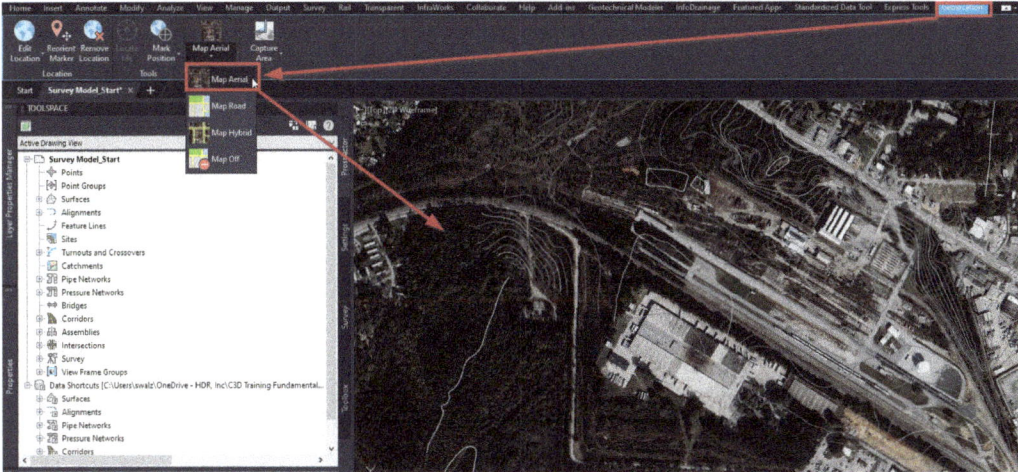

Figure 4.2 – Geolocation | Map Aerial

With the aerial imagery turned on, we can begin laying out our existing main railway line alignment.

Alignment creation tools

Using the foundational knowledge we've already gained, we have a few options in which we can develop our alignment geometry. In this case, I'd recommend we create our alignment by layout using **Alignment Creation Tools**, which is located in **Home | Create Design**, as shown in *Figure 4.3*.

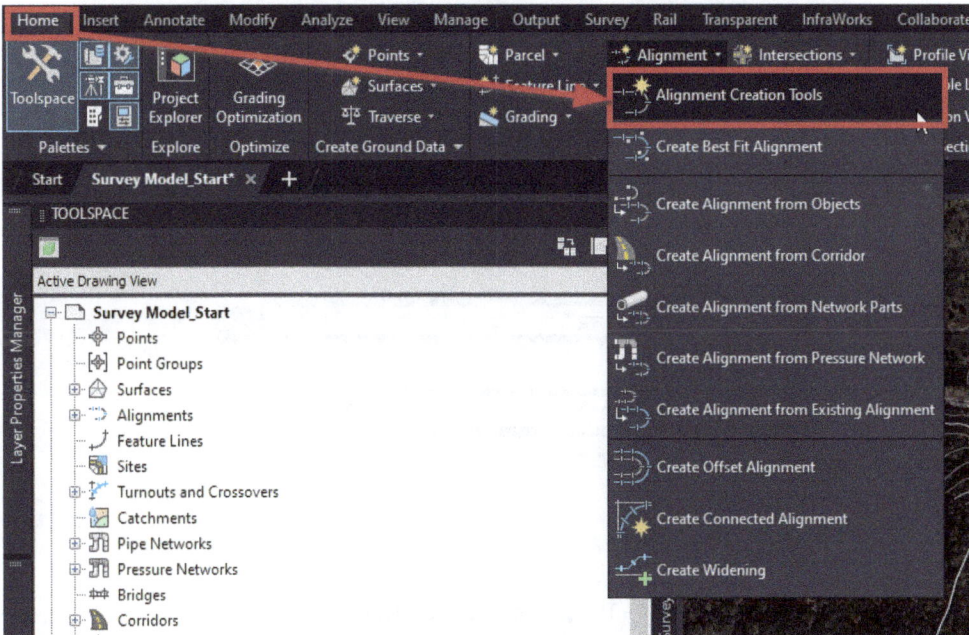

Figure 4.3 – Home | Alignment Creation Tools

Once selected, a **Create Alignment** dialog box will appear, which is where we'll input and make the following selections:

- **Name: ALG – Railroad – Main**
- **Type: Rail**
- **Description: Existing Main Rail Alignment**
- **Alignment style: Existing**
- **Alignment layer: C-RAIL-CANT-VIEW-CNTR**
- **Alignment label set: Major and Minor only**

Figure 4.4 shows the **Create Alignment** dialog box with these settings:

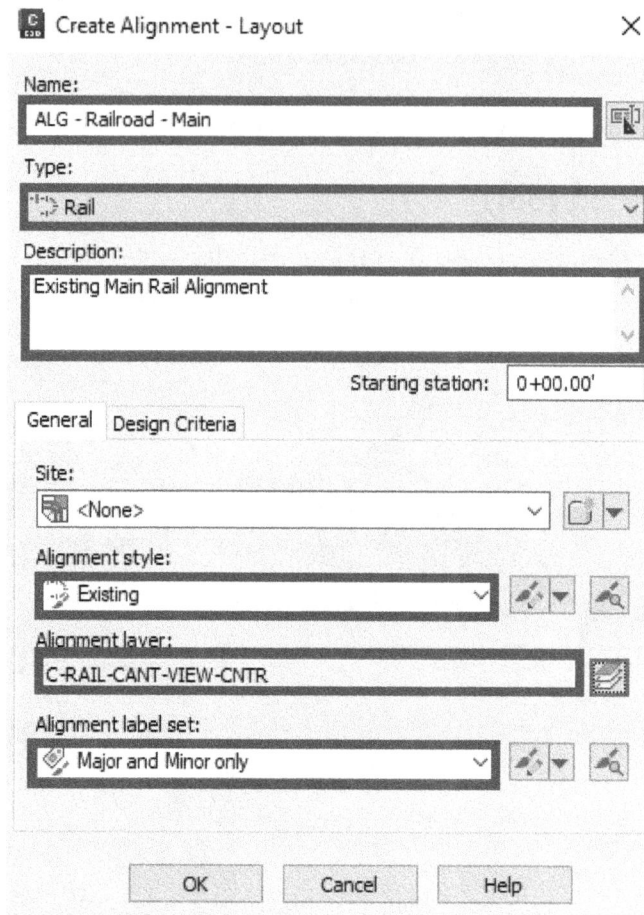

Figure 4.4 – The Create Alignment dialog box

Additionally, you'll notice that in our **Create Alignment** dialog box, we have some **General** and **Design Criteria** tabs. If we select the **Design Criteria** tab before selecting the **OK** button at the bottom of the **Create Alignment** dialog box, we gain access to – and have the ability to apply – design-based rules to our alignment.

That said, we'll want to make the following selections as well, and then we can select the **OK** button to exit out of the dialog box:

- **Starting design speed**: 50 mi/h
- **Use criteria-based design**: Check the box
- **Use design criteria file**: Check the box
- **Use design check set**: Check the box and apply a check next to the **Use design check set** checkbox and select the **Basic** design check set

Figure 4.5 shows the **Create Alignment** dialog box with these settings:

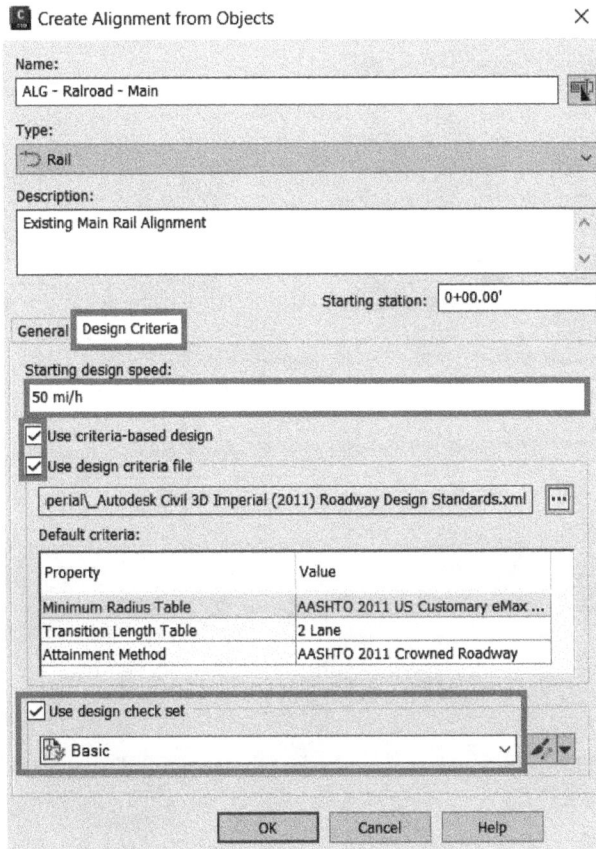

Figure 4.5 – The Create Alignment dialog box – Design Criteria

Next, an **Alignment Layout Tools** toolbar will appear (similar to that shown in *Figure 4.6*). From the start, we'll want to select the **Curve and Spiral Settings…** option so that we can specify the minimum allowable curve for rail designs.

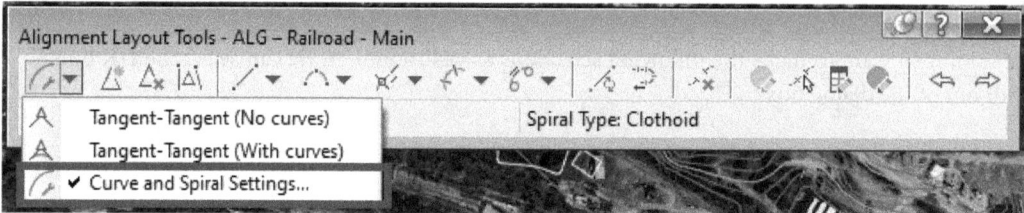

Figure 4.6 – Accessing Curve and Spiral Settings…

After selecting the **Curve and Spiral Settings…** option, we'll be presented with the **Curve and Spiral Settings** dialog box. When launched, we'll want to adjust the curve's **Default radius** value from 200.00' to 410.00' per typical design standards, as shown in *Figure 4.7*, and then click the **OK** button to exit out of the dialog box.

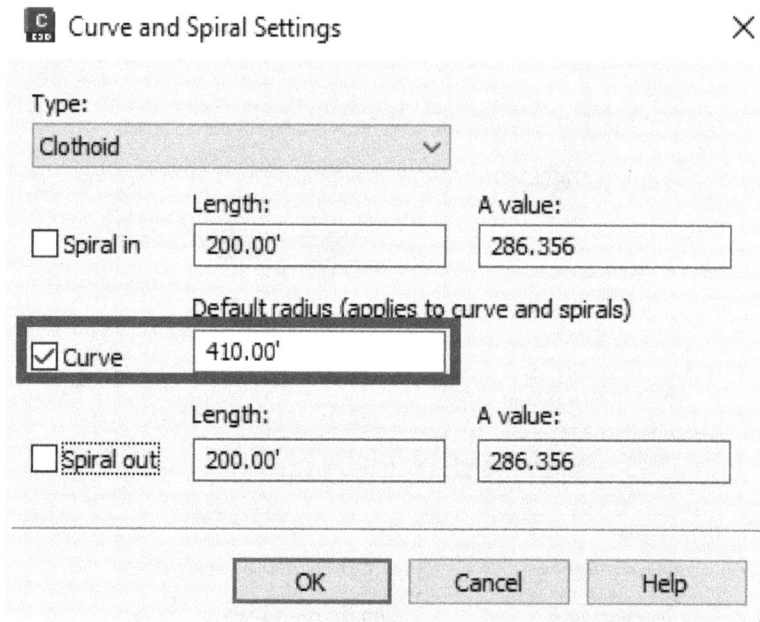

Figure 4.7 – Accessing the Curve and Spiral Settings dialog box

Next, we'll go back to the Draw Tangents icon and select the **Tangent-Tangent (With curves)** design tool, as shown in *Figure 4.8*.

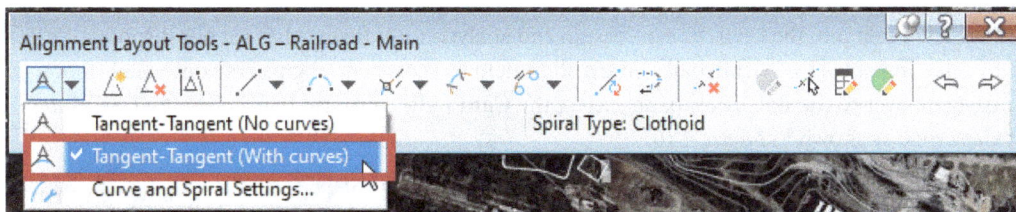

Figure 4.8 – Selecting the Tangent-Tangent (With curves) design tool

After selecting this tool, we'll notice that we're being prompted at the command line to specify a starting point for our alignment object. Let's go ahead and trace the existing **ALG – Railroad – Main** alignment displayed in the aerial as closely as possible by following the prompts at the command line. Be sure to lay out our alignment in a west-to-east direction. As with most alignment layouts, we can also select and grip edit the locations of our tangents and curves after we lay out our general alignment geometry. We should have an alignment that appears similar to that displayed in *Figure 4.9*.

Figure 4.9 – Main rail alignment

The next design aspect we'll want to consider along our rail alignment is **cants**, which we take a deeper look at in the next section.

Cant design and analysis tools

For those who may be unfamiliar with what a cant is within a rail design, the simplest way of explaining is to compare it to a superelevation within a roadway design. As trains travel through curves along a rail, it's important to note that greater force is transferred to the outer rail of the curve. To alleviate or balance this level of force, we'll need to raise the outer rail while lowering the inner rail to redistribute accordingly and ensure that the wheels on the train do not leave the track itself.

That said, let's jump into the Civil 3D cant design and analysis tools that we have at our disposal. We'll start by selecting our **ALG – Railroad – Main** alignment to access our alignment contextual ribbon, then navigate over to the Rail Tools panel, select the **Cant** pulldown, and then select the **Calculate/Edit Cant** tool, as shown in *Figure 4.10*.

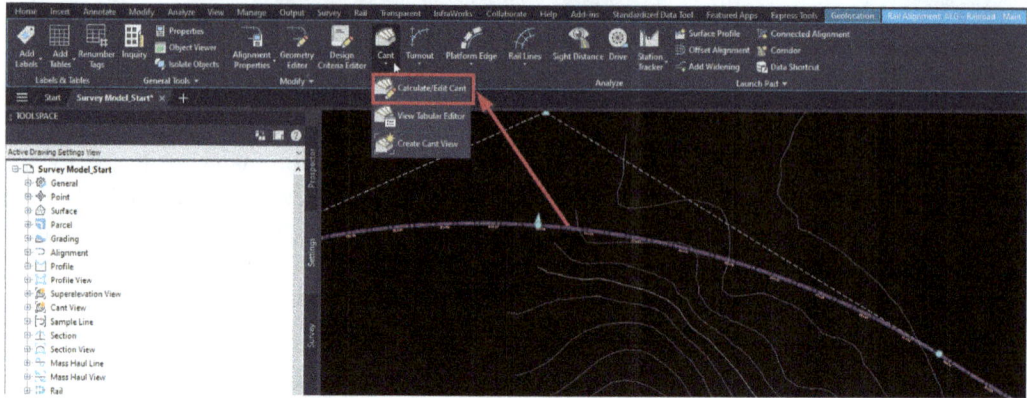

Figure 4.10 – Calculate/Edit Cant

Once selected, we'll be presented with a dialog box labeled **Cant – No Data Exists**, which has a message that says, **The alignment does not contain cant data. What do you want to do?** At this point, we have two options: **Calculate cant now** or **Open the cant curve manager**. Let's go ahead and select the **Calculate cant now** option, as shown in *Figure 4.11*.

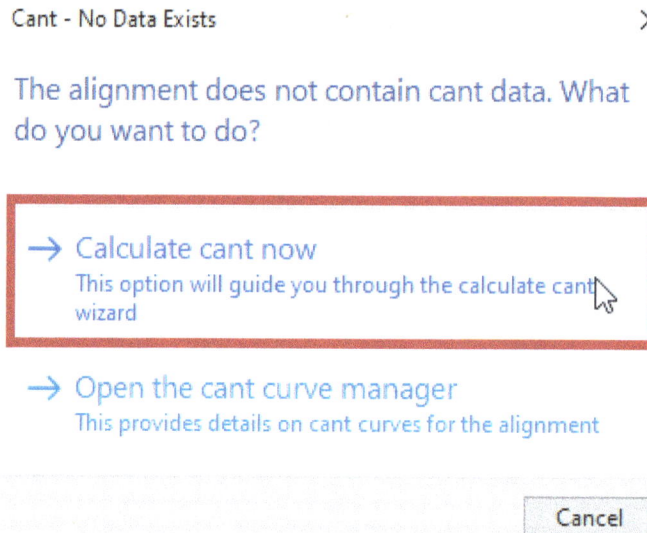

Cant - No Data Exists   ✕

The alignment does not contain cant data. What
do you want to do?

→ Calculate cant now
This option will guide you through the calculate cant
wizard

→ Open the cant curve manager
This provides details on cant curves for the alignment

Cancel

Figure 4.11 – The Cant – No Data Exists dialog box

Once the **Calculate Cant** dialog box appears, we'll make the following selections and inputs:

- **Railway Type:**

  - **Pivot Method: Low Side Rail**

  - **Track Width:** 4.92'

  Figure 4.12 shows these settings in the **Calculate Cant – Railway Type** dialog box:

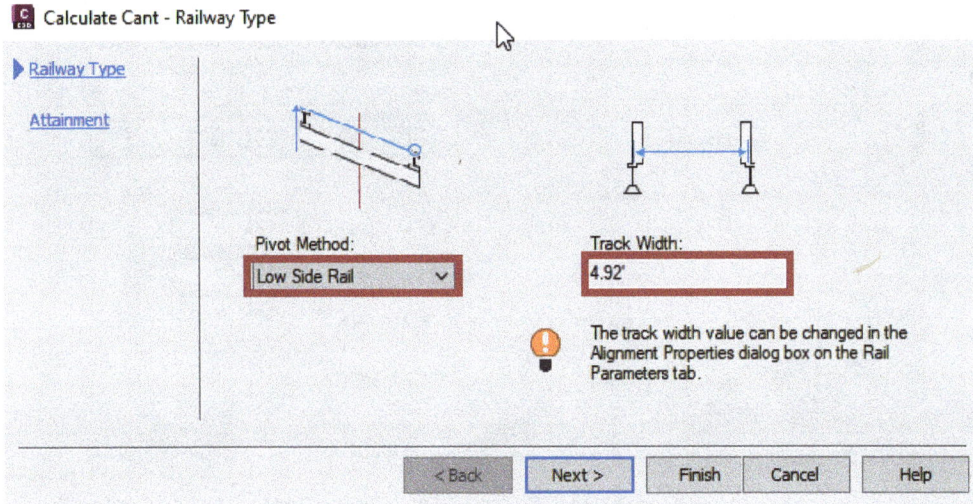

Figure 4.12 – Calculate Cant – Railway Type

- **Attainment:**

  - **Equilibrium cant:** 4.01 * {Design Speed}^2/Radius

  - **Maximum allowable cant deficiency:** 4.0

  - **Design criteria file:** C:\ProgramData\Autodesk\C3D 2025\enu\Data\Railway Design Standards\Imperial_Autodesk Civil 3D Imperial Rail Cant Design Standards.xml

  - **Applied cant table: Passenger train applied cant table**

  - **Transition length table: Minimum transition length(General)**

  - **Attainment method: by spiral lengths; and % on tangents if no spiral**

  - **% on tangent for tangent-curve:** 50.00%

  - **% on spiral for spiral-curve:** 100.00%

  - **Automatically resolve overlap:** Check the box

Figure 4.13 shows these settings in the **Calculate Cant - Attainment** dialog box:

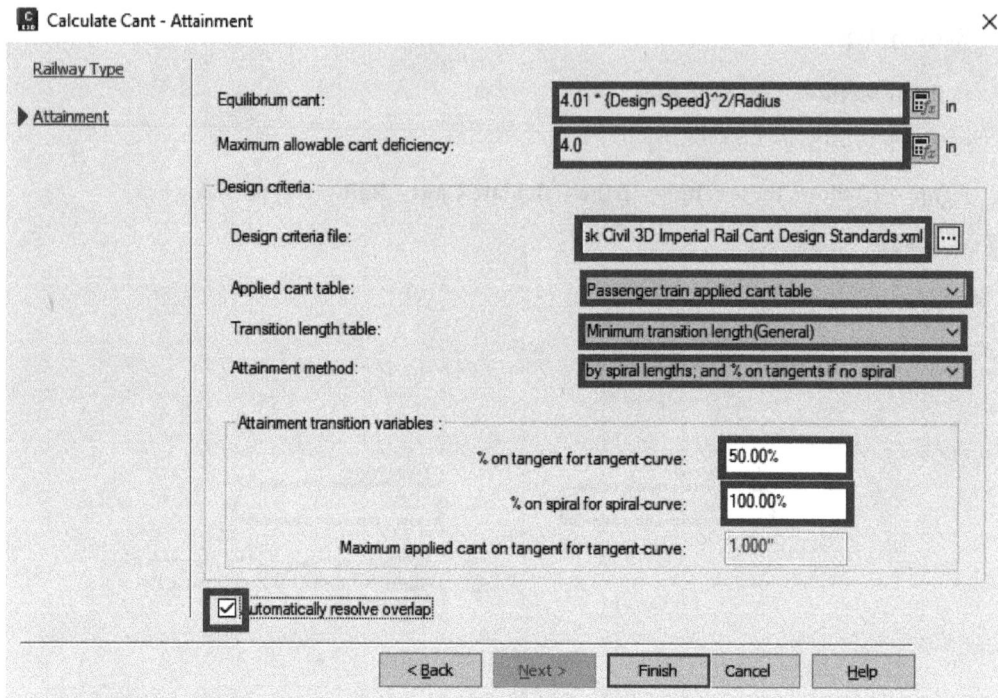

Figure 4.13 – Calculate Cant - Attainment

It's important to note that we selected many of the defaults to keep things simple in this case. It is recommended to select each of these pulldowns to understand what our full options are as each is a critical input that will impact our design, depending on the use of the rail.

Once all inputs have been defined, we can then click on the **Finish** button at the bottom of the **Calculate Cant** dialog box, at which point Civil 3D will run through the design criteria and identify locations and values of cants that will be required to conform to the design criteria defined previously.

To view the locations of cants, we'll need to turn on the cant labels for the alignment itself. You may recall that we actually set the label set for this particular alignment to **Minor and Major only** during the creation process. That said, let's quickly select our **ALG – Railroad – Main** alignment again, right-click, and select **Edit Alignment Labels…**. With the **Alignment Labels** dialog box open, we'll make the following selections:

- **Type**: **Cant Critical Points**
- **Cant Label Style**: **Cant Critical Points**
- Select the **Add>** > button

- **Cant Critical Points**: Check all options to ensure all critical points are being sampled and labeled

- Select the **OK** button

Figure 4.14 shows these settings in the **Alignment Labels** dialog box:

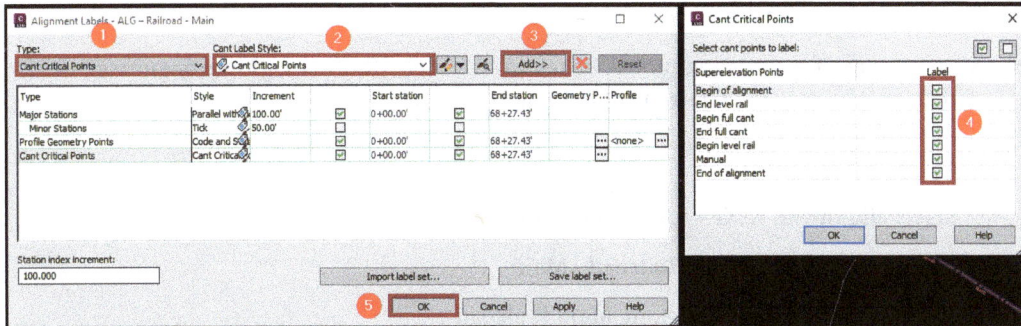

Figure 4.14 – Alignment Labels

After selecting the **OK** button in the **Alignment Labels** dialog box, we should have a layout that displays something similar to that shown in *Figure 4.15*.

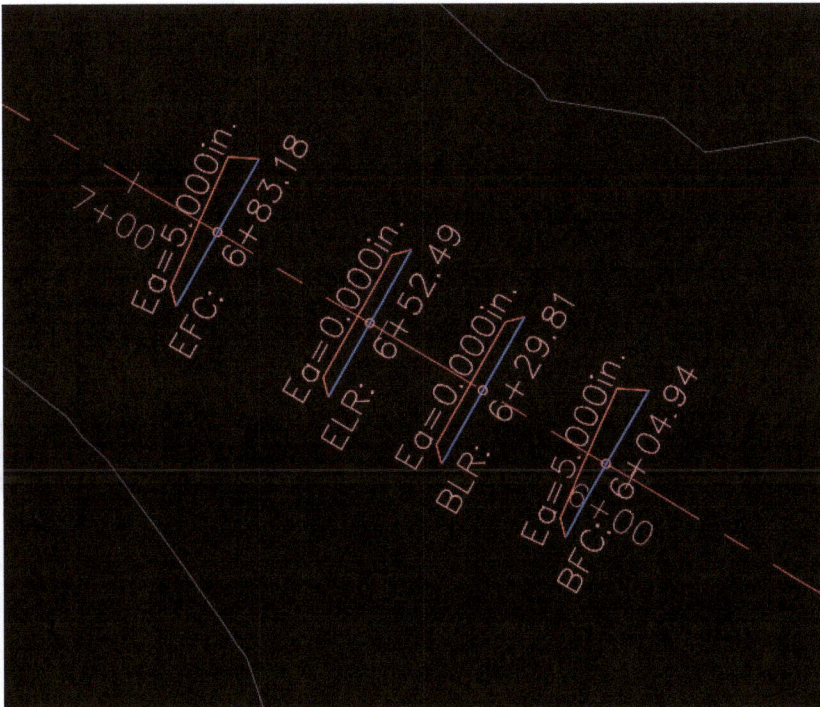

Figure 4.15 – Cant labels displayed

Using the same steps detailed in our previous book, *Autodesk Civil 3D 2024 from Start to Finish*, for creating a profile and profile views, let's go ahead and create another profile view for the **ALG – Railroad – Main** alignment.

Profile creation tools

First, create a profile of the existing surface linked to the **ALG – Railroad – Main** alignment, then use the following criteria in the **Create Profile View** dialog box:

- **General**:

    - **Select alignment**: ALG – Railroad – Main

    - **Profile view name**: PRV - Railroad - Main

    - **Description**: **Profile View along Main Railroad**

    - **Profile view style**: **Land Desktop Profile View**

- **Profile Station Range**:

    - **Station Range**: **Automatic**

- **Profile View Height**:

    - **Profile View Height**: **Automatic**

- **Profile Display Options**:

    - **Style**: **Existing Ground Profile**

- **Data Bands**:

    - **Select Band Set**: **EG-FG Elevations and Stations**

Figure 4.16 shows the final output:

Figure 4.16 – Placing our PRV – Railroad – Main Profile View

Moving over to the design side of our profile creation process, we'll need to go back up to the **Home** ribbon, and the Create Design panel to access **Profile Creation Tools** (refer to *Figure 4.17*).

Figure 4.17 – Accessing Profile Creation Tools

Once the **Profile Creation Tools** has been selected, we'll be prompted to select a profile view with which we'd like to place or associate our design profile.

Let's go ahead and select the **PRV – Railroad - Main** profile view, at which point the **Create Profile - Draw New** dialog box will appear. In the **Create Profile - Draw New** dialog box, we need to fill out the fields as follows:

- **Name**: PRF - Railroad - Main
- **Description**: Profile along Main Railroad Alignment
- **Profile style**: **Existing Ground Profile**
- **Profile label set**: **Complete Label Set**

Figure 4.18 shows these settings in the **Create Profile - Draw New** dialog box:

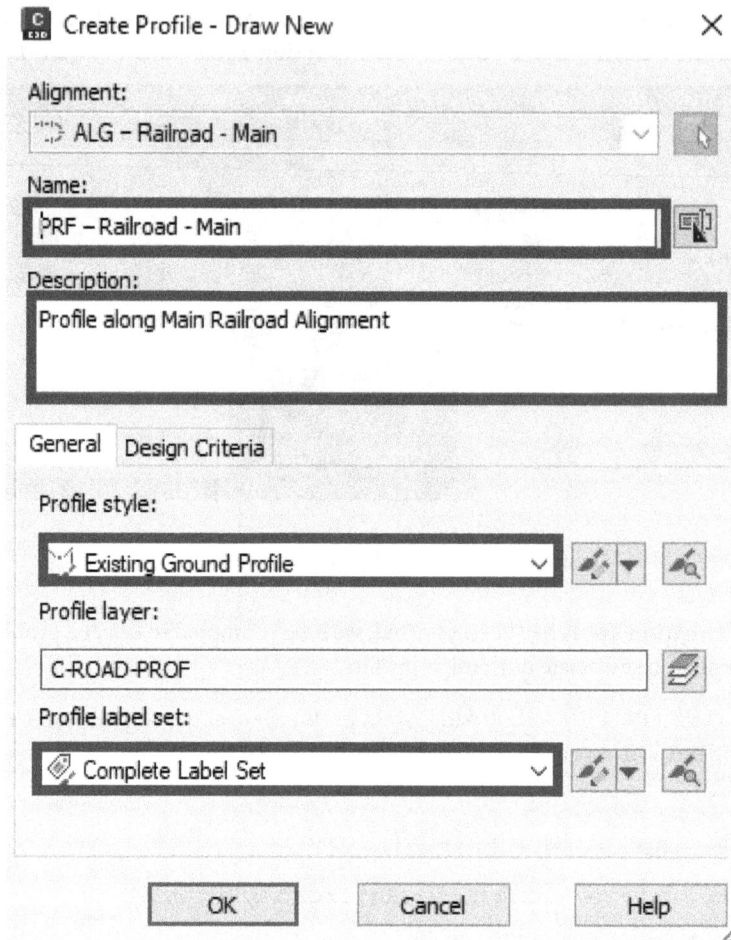

Figure 4.18 – The Create Profile - Draw New dialog box

Before closing out of the dialog box, we also need to switch over to the **Design Criteria** section and use the following selections (also displayed in *Figure 4.19*):

- **Use criteria-based design**: Check the box
- **Use design criteria file**: Check the box and navigate to `C:\ProgramData\Autodesk\ C3D 2024\enu\Data\Railway Design Standards\Imperial\_Autodesk Civil 3D Imperial Rail Cant Design Standards.xml`
- **Use design check set**: Check the box and select the **Basic** design check set

Figure 4.19 shows these settings in the **Create Profile** - **Draw New** dialog box:

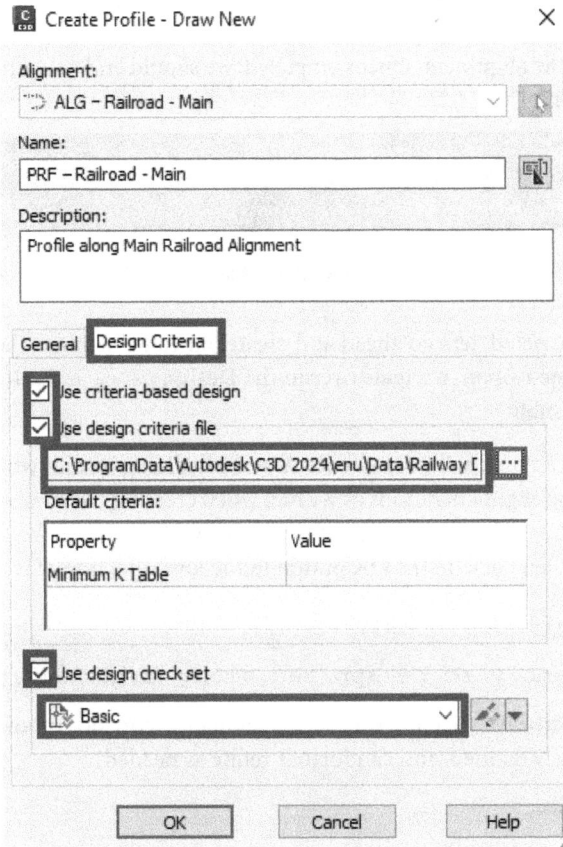

Figure 4.19 – Create Profile - Draw New

After all the selections have been made, click the **OK** button at the bottom of the **Create Profile - Draw New** dialog box. The **Profile Layout Tools** toolbar will then appear. Let's go ahead and select the down arrow at the first icon, the Draw Tangents icon, and select **Draw Tangents with Curves**, as shown in *Figure 4.20*.

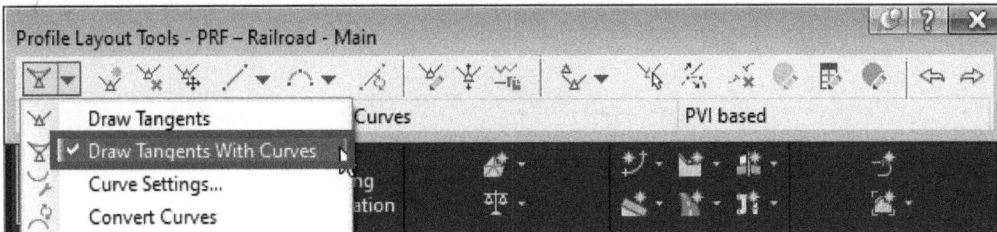

Figure 4.20 – The Profile Layout Tools toolbar

Let's now go ahead and lay out our existing vertical rail alignment, following the existing grade already displayed in our profile view depicting the existing surface. We'll want to make sure we space out our **PVIs (points of vertical intersection)**, so they're shown at high and low points to ensure smooth vertical transitions along the alignment. Once completed, we should end up with a vertical alignment similar to that displayed in *Figure 4.21*.

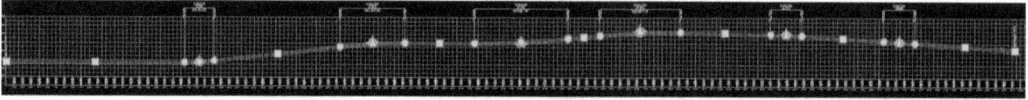

Figure 4.21 – PRF – Railroad – Main Profile View

With our vertical profile created, let's go ahead and create our rail assembly. That said, we'll want to jump back up to our **Home** ribbon, navigate over to the **Design** panel, and select **Create Assembly** to begin creating our assembly.

Once selected, we'll be presented with the **Create Assembly** dialog box, at which point we'll fill in the fields available, making slight alterations as we had when creating our roadway assemblies in the previous book, *Autodesk Civil 3D 2024 from Start to Finish*. We'll want to make selections and fill out the fields as follows, and then click on the **OK** button in the lower portion of the dialog box:

- **Name**: `ASM - Railroad - Main`
- **Description**: `Assembly to be applied along Main Railroad Alignment`
- **Assembly Type: Railway** (selecting this option allows the corridor model to account for cant as we have previously defined, and can further refine as needed)
- **Assembly Style: Basic**
- **Code Set Style: All Codes with No Shading** (selecting this option allows us to quickly visualize only the points and shapes included in our assembly without getting clouded with additional hatching for ease of understanding)

Figure 4.22 shows these settings in the **Create Assembly** dialog box:

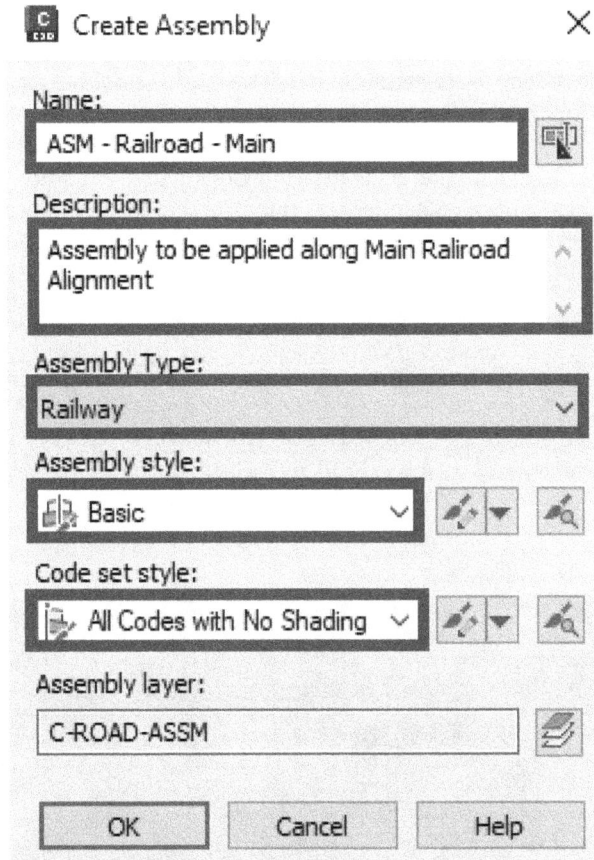

Figure 4.22 – The Create Assembly dialog box

After selecting the **OK** button, we must go back to the command line, where we'll be asked to specify a location for the **Assembly Baseline Location** setting. What we tend to do in these cases is specify the assembly baseline location near the profile views so that we can group horizontal and vertical components relatively close to each other for better ease of use.

That said, choose a location that you prefer (above, below, besides, and so on) next to your current profile views to place your assembly baseline. Once a location has been identified, Civil 3D will automatically zoom into that location so that only the assembly baseline is visible in your view.

Next, we'll take the following steps to create our rail assembly:

1. Navigate up to our **Rail** ribbon.
2. Select the **Subassemblies** option in our **Content** panel.
3. Select the **RailSingleTrackCANT_w_ExtraLayers** subassembly.

> **Important note**
>
> We are selecting **RailSingleTrackCANT_w_ExtraLayers** for this situation because we generated our horizontal alignment along the centerline of one of the tracks only. If we toggle **Geolocation Aerial Imagery** back on, we'll notice that there are two tracks side by side. To model both, we could have generated our horizontal alignment between the two tracks and then used the **RailDoubleTrackCANT_w_ExtraLayers** subassembly to knock them both out at once. Feel free to run through this exercise again once you complete this chapter to better understand the differences and workflows required.

4. After selecting the subassembly we wish to import, we'll move over to our **Properties** dialog box and update the parameters as we had done with our roadway subassemblies as seen in *Figure 4.23*.

5. You'll also notice that we're being prompted at the command line to select a marker point within the assembly, at which point we'll go ahead and select the center marker in our previously inserted assembly.

Figure 4.23 shows all the preceding steps:

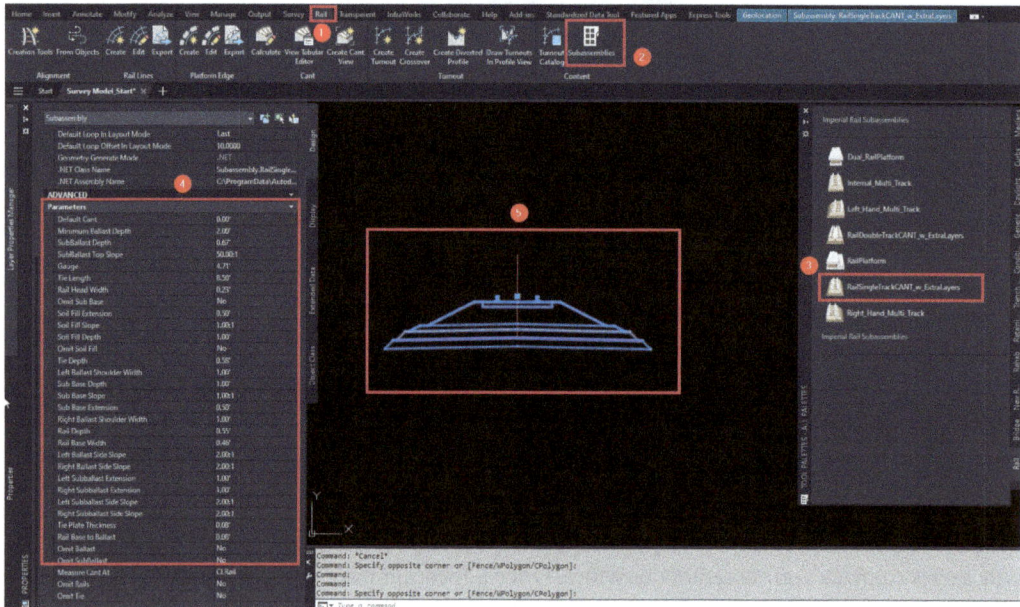

Figure 4.23 – Steps to create the main rail assembly

Finally, the next steps for us now are to run through the corridor creation process, taking the same steps we took in *Autodesk Civil 3D from Start to Finish* to generate each of our roadway corridor models. These steps will be identical, so no need to republish them here, but be sure to give the corridor a unique name (i.e., COR – Railroad – Main). We should have a final corridor model that looks similar to that displayed in *Figure 4.24.*

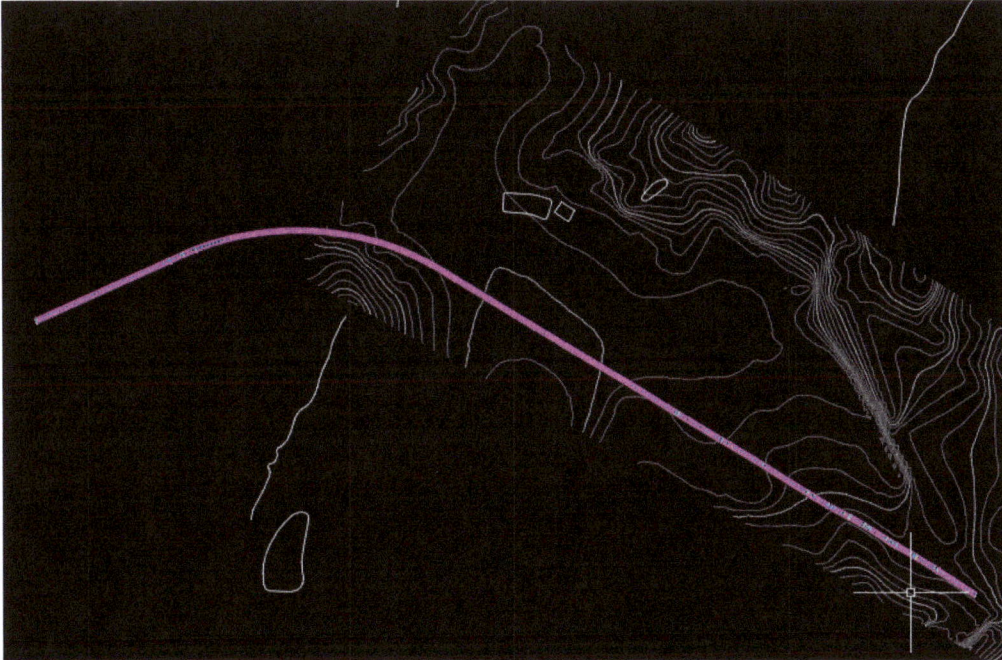

Figure 4.24 – Main rail corridor plan view

If your main rail alignment and corresponding corridor model follow ours, you'll notice that when you look at the resulting corridor model in a 3D view, or even in the Object Viewer, only the portion of the corridor model where we have an existing surface displayed is shown at true elevation, whereas the remainder drops to elevation 0. Without having a surface or profile for the corridor to sample and follow, Civil 3D simply assumes a 0.00' elevation. Rest assured; we'll adjust this later on in the chapter.

Auxiliary track design workflows

With our ALG – Railroad – Main alignment created and passing through our existing rail yard, we'll want to add at least one additional rail line to allow for trains to load/unload materials and pass other trains as needed. This concept is known as an **auxiliary track**.

Developing an auxiliary track consists of several different components and steps, as outlined through the following:

1. Create an offset alignment that will be dynamically linked to our ALG – Railroad – Main alignment.

2. Identify two locations where we'll branch off our ALG – Railroad – Main alignment to meet our newly generated offset alignment. This is known as a **crossover alignment**.

3. Create a new assembly that contains the **RailDoubleTrackCANT_w_ExtraLayers** subassembly.

Update our COR – Railroad – Main to incorporate the new assembly throughout the auxiliary track.

With that, let's go ahead and create our offset alignment.

Offset alignments

To do so, we'll simply select **ALG – Railroad – Main** alignment in the model space to activate our **Alignment Contextual** ribbon. Once activated, we'll select the **Offset Alignment** option within our **Launch Pad** panel and fill out fields as follows:

- **Alignment to offset from: ALG – Railroad – Main**

- **Offset name template:** `ALG - Railroad - Auxiliary`

- **Station range:** Check the **From start** and **To end** boxes

    Note that you can also uncheck these boxes and specify start and end stations to be closer to the range of where the auxiliary track design will be located.

- **No. of Offsets on left:** `0`

- **No. of Offsets on right:** `1`

- **Incremental offset on left:** `10.00'`

- **Incremental offset on right:** `14.50'`

    Note that since we applied `0` as the **No. of Offsets on left**, the `10.00'` value will be ignored.

- **General:**

    - **Alignment style: Offsets**

    - **Alignment label set: Major and Minor only**

- **Create Offset Profile:**

    - **Create profile for offset alignment:** Check this option

    - **Parent profile: PRF – Railroad – Main**

    - **Cross slope from parent profile:** `0.00%`

- **Profile name**: `PRF - Railroad - Auxiliary`
- **Profile style**: **Existing Ground Profile**

Figure 4.25 shows some of these settings:

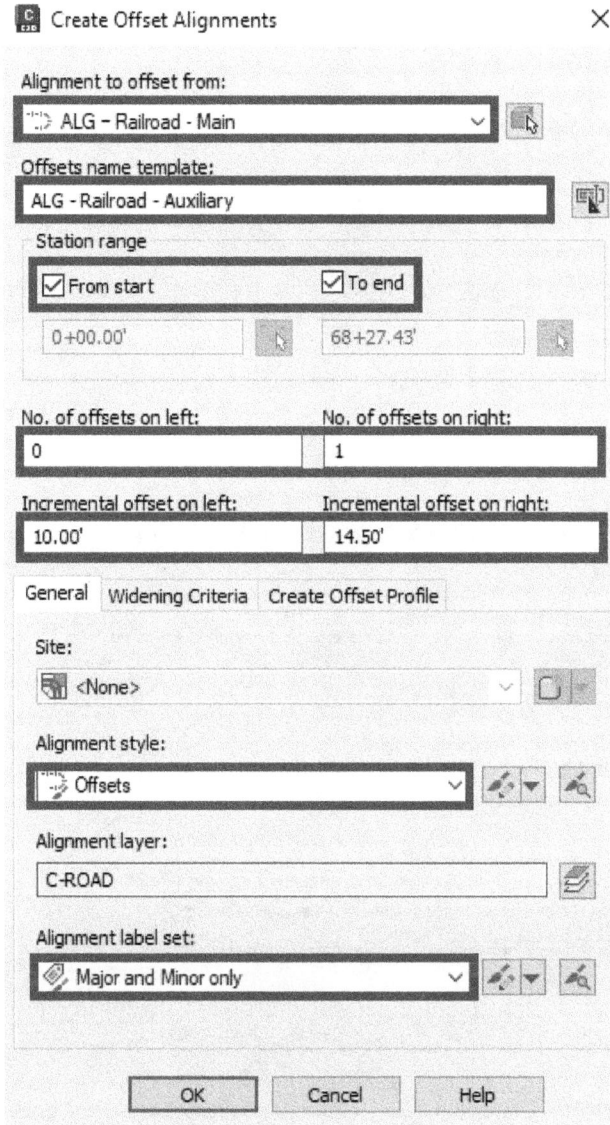

Figure 4.25 – The Create Offset Alignments dialog box

Figure 4.26 shows the settings for the **Create Offset Profile** tab:

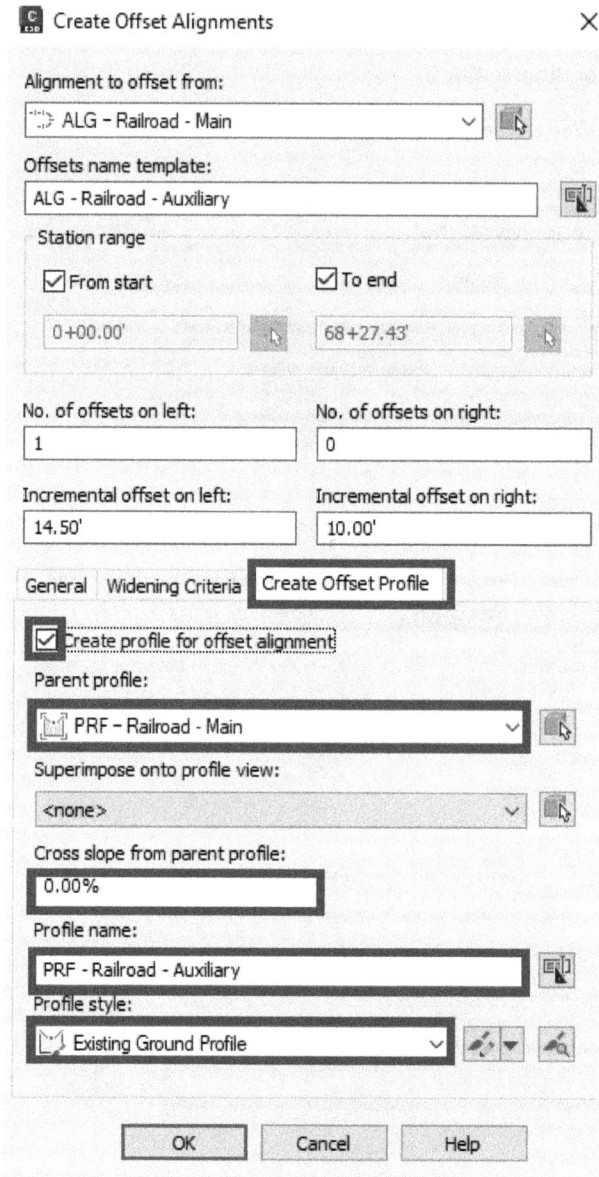

Figure 4.26 – The Create Offset Alignments | Create Offset Profile tab

With our offset alignment and profile created, let's go ahead and identify two locations where we'll branch off our ALG – Railroad – Main alignment to create our crossover alignments.

Crossover alignments

For this workflow, it may be beneficial to toggle the **Map Aerial** option back on to locate our crossover alignments a bit more accurately. Since we're not exactly designing anything at this point, and really just running through the paces to understand the overall workflow, we're going to pick two random locations on site.

Using the following steps, we'll go ahead and create our first crossover alignment:

1. Navigate to and Activate the **Rail** ribbon.

2. Select the **Create Crossover** tool in the **Turnout** panel.

3. We'll be prompted at the command line with the `Select an existing turnout or a rail alignment` prompt, at which point we'll select **ALG – Railroad – Main** alignment.

4. Next, we'll be prompted with the `Specify the reference station for the turnout` prompt. Here, we can either manually type in `at Station` or click in the model space with our mouse.

5. Next, we'll be prompted with the `Specify the next rail alignment` prompt, at which point we'll select our newly created **ALG – Railroad – Auxiliary Offset Alignment**.

6. If not already previously defined, our **Turnout Catalog** dialog box will appear on the screen where we'll need to load our rail catalog (as shown in *Figure 4.27*). If our file already has this defined, the **Turnout Catalog** dialog box will not appear.

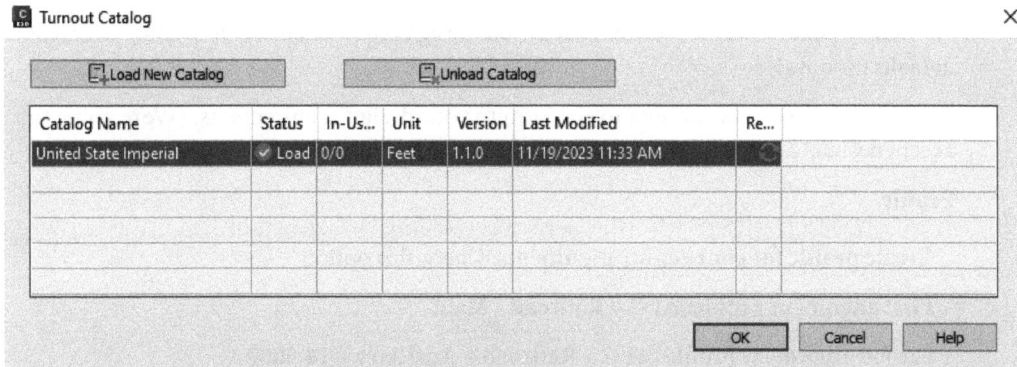

Catalog Name	Status	In-Us...	Unit	Version	Last Modified	Re...
United State Imperial	Load	0/0	Feet	1.1.0	11/19/2023 11:33 AM	

Figure 4.27 – The Turnout Catalog dialog box

Note that Civil 3D has several catalogs available in Metric and as part of Country Kit installation packages; if a different catalog is needed to conform to your design requirements, you can select that catalog by clicking on the **Load New Catalog** option in the **Turnout Catalog** dialog box (may need to click on the **Unload Catalog** option prior to loading a new one).

7. Next, our **Create Crossover** dialog box will appear, which is where we'll fill out fields and make selections, and then click on the **OK** button at the bottom, as follows:

- **Crossover group name**: `CSG - Railroad - Auxiliary`

- **Main alignment: ALG – Railroad – Main**

- **Main station**: `56.97.76'` (or whichever station you had previously defined)

- **Second alignment: ALG – Railroad – Auxiliary**.

- **General**:

  - **Turnout name (on main alignment)**: `Crossover - Turnout - (<[Next Counter(CP)]>0`

  - **Turnout style: Standard**

    Feel free to update, create, and/or select a different **Turnout style** display that matches the display standards required for the project; and remember to save your adjustments into your `Company Template File.dwt` file.

  - **Turnout label set: Standard**

    Feel free to update, create, and/or select a different **Turnout label set** display that matches the display standards required for the project; and remember to save your modifications into your `Company Template File.dwt` file.

- **Content**: Update as needed to conform to your design requirements. (We're going to accept defaults for now.)

- **Connection**: Update as needed to conform to your design requirements. (We're going to accept defaults for now.)

- **Profile**:

  - **Create profile for connection alignments**: Check this option

  - **First alignment profile: ALG – Railroad - Main**

  - **Second alignment profile: ALG – Railroad – Auxiliary – 14.5000**

  - **Diverted profile name**: `CSD - (<[Next Counter(CP)]>)`

  - **Profile style: Existing Ground Profile**

Figure 4.28 shows some of these settings:

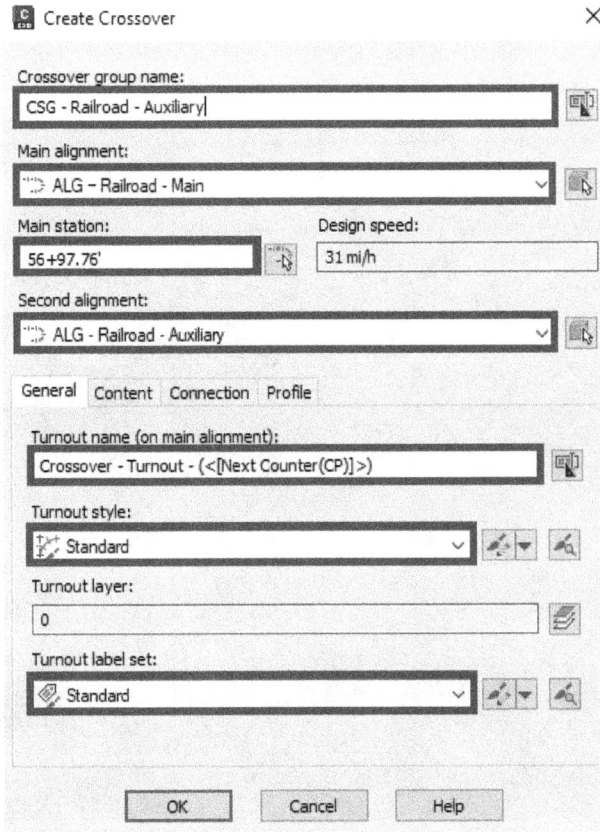

Figure 4.28 – The Create Crossover dialog box

Figure 4.29 shows the settings of the **Profile** tab:

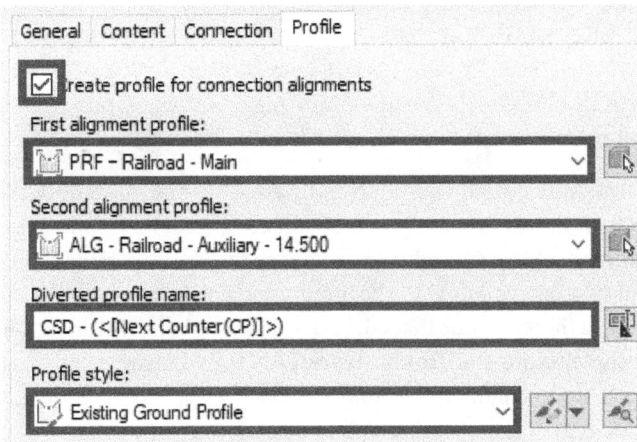

Figure 4.29 – Create Crossover | Profile

We should now have a crossover added to our design that looks similar to that shown in *Figure 4.30*. Note that it may not be exactly based on selections and values filled out in your fields to conform to your specific design requirements.

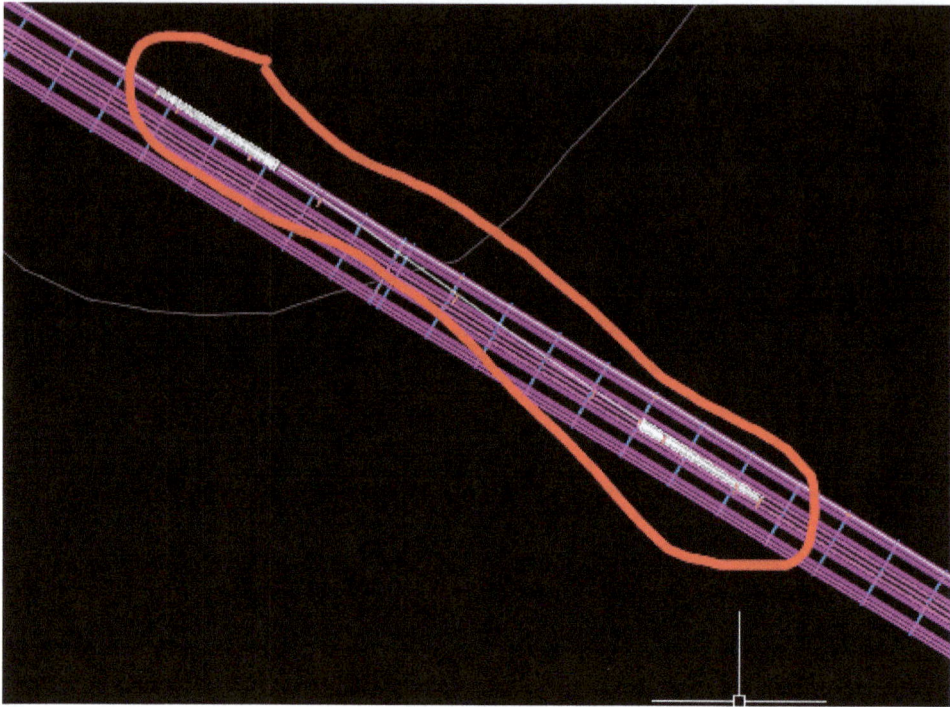

Figure 4.30 – Crossover display

Next, we'll want to tie our ALG – Railroad – Auxiliary offset alignment back into our ALG – Railroad – Main alignment by adding another crossover. We'll follow similar steps to create our second crossover. However, we'll want to select **ALG – Railroad – Auxiliary Offset Alignment** first when prompted at the command line with the `Select an existing turnout or a rail alignment` prompt, followed by selecting **ALG – Railroad – Main** alignment when prompted with the **Specify the next rail alignment** prompt.

With our crossovers established, all that's left is to create a new assembly that contains the **RailDoubleTrackCANT_w_ExtraLayers** subassembly. Once created, we'll update our corridor by first selecting our **COR – Railroad – Main Corridor** model. Using the Modify Region tools in our right-click menu, let's go ahead and split the regions at the beginning and end locations of our crossovers, update our Region Properties in all three regions that cover the crossovers and Offset Rail in between to use the new assembly that contains our **RailDoubleTrackCANT_w_ExtraLayers** subassembly. Finally, update the targets in each of these new regions to point to each respective alignment and profile. We should end up with a design that looks similar to that displayed in *Figure 4.31*.

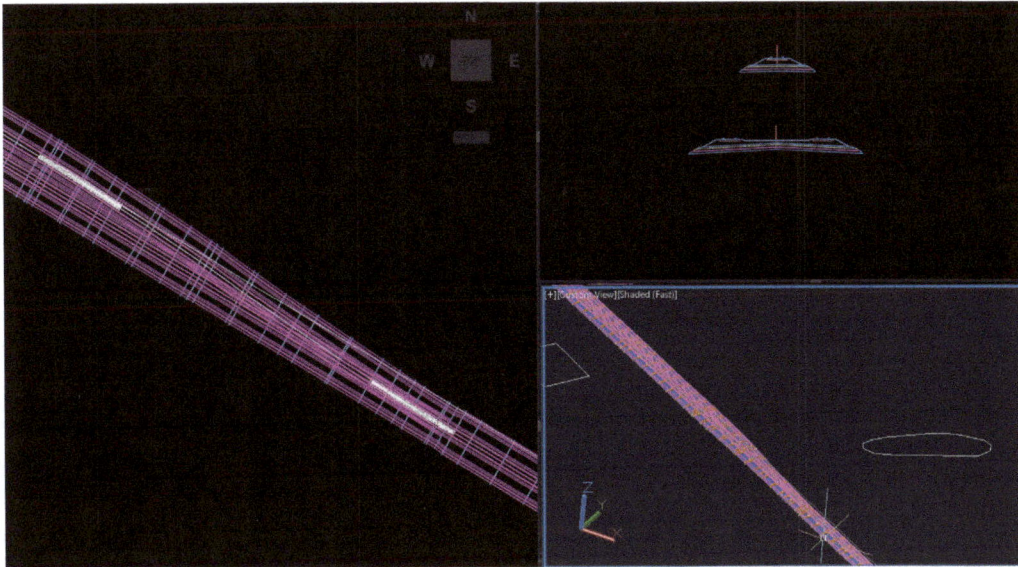

Figure 4.31 – Corridor update

With our rail corridor model updated, let's go ahead and learn how we can further analyze and modify it to ensure that we are in conformance with the design standards and requirements.

Viewing and editing rail models

As mentioned earlier, our full corridor model may appear to abruptly change elevation down to 0.00', where it extends beyond our SRF – Existing Grade – From OpenSource Surface model. We'll want to revisit this and make sure we update our COR – Railroad – Main Corridor model.

To do so, let's navigate over to our PRV – Railroad – Main Profile and select the **PRF – Railroad – Main Profile** that our COR – Railroad – Main Corridor model is referencing for elevation purposes. Looking over at our **Properties** dialog box with **PRF – Railroad – Main Profile** selected, we can quickly identify the start (0+00.00') and end (159+18.36') station points.

Knowing these start and end station points, let's navigate back to and select our **COR - Railroad – Main Corridor** model. Right-click in the model space and select the **Corridor Properties...** option to pull up our **Corridor Properties** dialog box. Once opened, move over to the **Parameters** tab and change the **End Station** for our last region to 159+18.36 as shown in *Figure 4.32*. Go ahead and select **OK** and **Rebuild Corridor** when prompted.

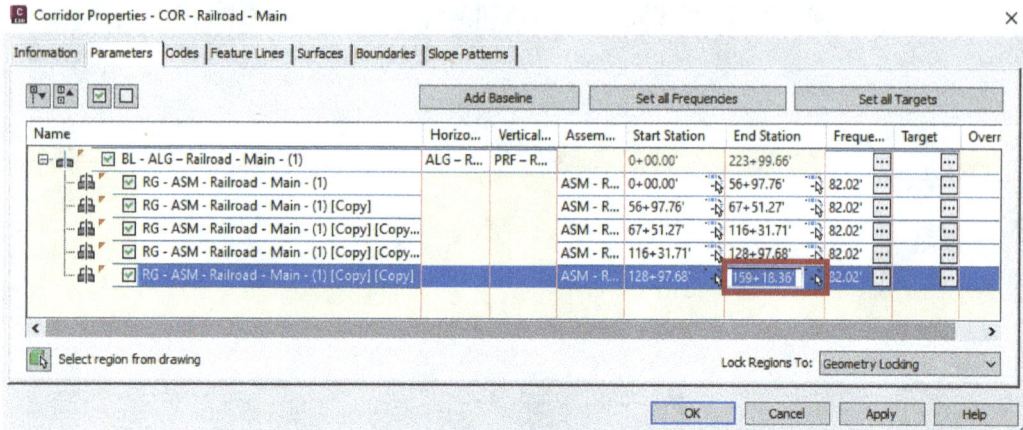

Figure 4.32 – Corridor Properties | Parameters

Next, we'll select our corridor model again. Right-click with your mouse in the model space, select the **Modify Corridor Sections** option, and then the **Corridor Section Editor** tool. Once the **Corridor Section Editor** tool has been selected, our model space view will be split up into four viewports by default as follows:

- Plan view

- Profile view

- Section view

- Assembly view

Figure 4.33 shows these viewports:

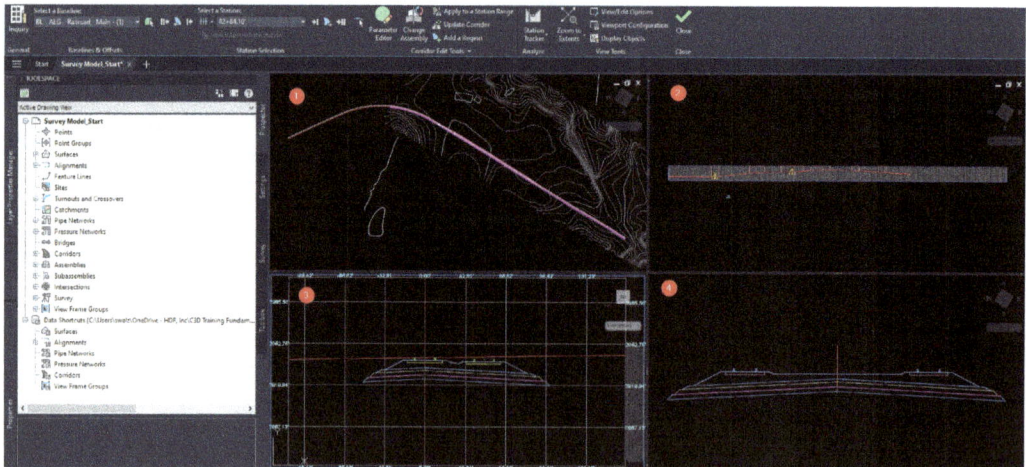

Figure 4.33 – The corridor Section Editor viewports

If we take a look at the **Section Editor** contextual ribbon along the top, we'll see that we're presented with a whole slew of tools available to us to further analyze our corridor model. Running from left to right, we have the following panels with tools available to us:

- **Inquiry**: This provides various tools to interrogate and retrieve information related to our surfaces, alignments, profiles, sections, and so on

- **Baseline & Offsets**: This allows us to switch between baseline alignments for our corridors to view

- **Station Selection**: This allows us to navigate through our various views at specified stations

- **Corridor Edit Tools**: This provides us with the ability to refine our corridor models at specific stations

- **Analyze**: This provides various **Station Tracker** controls to quickly identify locations in multiple views as we analyze our corridor models

- **View Tools**: This provides quick access to basic view and display tools

- **Close**: This allows us to close out of our contextual ribbon after all analysis and refinement edits have been performed

Figure 4.34 shows these panels:

Figure 4.34 – The corridor Section Editor contextual ribbon

Once all refinement edits have been performed and we select the **Close** option within our contextual ribbon, our view will go back to a singular viewport showing just our plan view. The next analysis tool that can come in handy is the **Drive** tool. To access this one, we'll select our COR - Railroad – Main Corridor model again, right-click with our mouse in the model space, and select **Drive**.

After selecting **Drive**, we'll be prompted at the command line with the `Select corridor feature line` prompt. Here, we'll want to select the centerline of our COR – Railroad – Main Corridor model so that we are centered in our assembly while running a drive simulation within Civil 3D.

If we take a look at the **Drive** contextual ribbon along the top, we'll see that we're presented with a few tools available to us to further analyze and run simulations on our corridor model. Running from left to right, we have the following panels with tools available to us:

- **Path**: This provides various tools to specify and/or change our path and view

- **Eye**: This allows us to modify our viewpoint in which we are running our simulation along the main path

- **Navigate**: This allows us to specify station, speed, and displays of our simulation to gain quick access and analysis to specified locations

- **Target**: This allows us to modify our focal point, which we are looking at throughout the simulation along the main path

- **Close**: This allows us to close out of our contextual ribbon after all analysis and refinement edits have been performed

Figure 4.35 shows these panels:

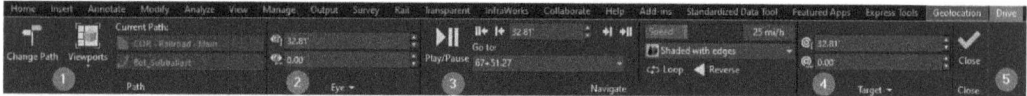

Figure 4.35 – The Drive contextual ribbon

Common practice is to switch back and forth between the **Drive** and **Section Editor** tools to get a more holistic understanding of how our rail corridor design is coming along. We also gain further insights as to how our rail corridor model will be integrated into the existing built environment and validate our design from a constructability standpoint.

Summary

As we worked through this chapter, we learned several ways through which we can adapt our roadway corridor modeling techniques and workflows and apply them to our rail design, all while introducing a new dataset entirely. We learned how we can define and configure our alignments, profiles, assemblies, and corridor models for both main railroad and auxiliary track alignments and how we can specify industry standards for our modeled objects to ensure that we are following design requirements for our projects.

In the next chapter, we'll begin exploring how we can introduce LiDAR and Reality Capture to supplement any additional existing conditions data received and used to develop our existing conditions model. We'll also introduce **Autodesk ReCap** and learn how we can use it to our benefit to essentially complement our Civil 3D designs.

Part 2: Improving Our Civil BIM Designs with Civil 3D Extensions and Customized Design Workflows

Autodesk Civil 3D 2025 has many extensions that will broaden our design capabilities well beyond what's available directly within the software. With greater access to additional extensions and tools comes greater responsibility. Accessing and applying advanced workflows are two completely different discussions, but we'll begin to familiarize ourselves with more of the extensions and tools that are available to us and how we can use them to develop custom design solutions that are directly integrated into our civil BIM designs.

This part contains the following chapters:

- *Chapter 5, Harnessing Reality Capture to Enhance Civil Projects within Autodesk Civil 3D 2025*
- *Chapter 6, Streamlining Design with Grading Optimization*
- *Chapter 7, Exploring Content Catalog Editor*
- *Chapter 8, Empowering Utility Modeling with Infrastructure Parts Editor*
- *Chapter 9, Custom Roadway Design with Subassembly Composer*

Harnessing Reality Capture to Enhance Civil Projects within Autodesk Civil 3D 2025

So far, we have gone through new and advanced workflows of Civil 3D to build upon the foundational workflows within the program. In this chapter, we will continue to take our workflows to the next level, but instead of building on them, we will back up and reexamine our existing conditions and how we can leverage them and their capture methods for incredible project accuracy and stronger foundational design from the beginning of the project through to project delivery.

The main topic of this chapter is, of course, reality capture, and we will dive into how it can build a truly strong level-up in any BIM model manager's career and skill set. Reality capture is a whole industry by itself, but it is immensely helpful to the civil engineering industry. We will investigate what reality capture actually means and includes, as well as what we should look for when implementing it and how we can integrate it into our existing conditions, project communication, and more.

In this chapter, we will focus on several key topics, including the following:

- Understanding reality capture and its function within civil projects
- Highlighting key data for incredible accuracy in civil design projects
- Leveraging reality capture workflows in civil design projects

With that, let's start with a broad introduction to the topic of reality capture before we get specific and begin integrating this new data into our designs. As we start our journey, it may be helpful to picture what we mean by reality capture with an image. You can see in *Figure 5.1* an image of a point cloud, which is the typical output of a reality capture practice (which we will dive further into in the upcoming sections).

Figure 5.1 – Example of a point cloud file

Technical requirements

Here are the technical requirements for this chapter:

- **Software**: Autodesk Civil 3D 2024 and Autodesk ReCap 2024

- **Operating system**: 64-bit Microsoft Windows 10

- **Processor**: 4+ GHz

- **Memory**: 32 GB RAM (we suggest going with either 64 GB or 128 GB)

- **Graphics card**: 8 GB

- **Display resolution**: 1,980 x 1,080 with true color

- **Disk space**: 20 GB

- **Pointing device**: MS-Mouse compliant

Understanding reality capture and its function within civil projects

Let's begin with an understanding of reality capture. The term **reality capture** seems a bit ethereal when you first encounter it, but it is meant to cover many applications of capturing existing conditions. Another term for reality capture is **remote sensing**, the practice of sensing or measuring an object or place of interest remotely. The industry has essentially broadened to include many methods of measuring projects, whether measuring a road, arena, cave system, buildings, or mines, you name it.

There are many methods for reality capture, including laser scanning, photogrammetry, and even surveying. We will not dive into all of them as we could write several books on the differences, nuances, and applications. In this book, we will focus primarily on laser scanning, also known as LiDAR.

Understanding LiDAR

LiDAR is an acronym for **Light Detection and Ranging**. LiDAR uses light to measure an object remotely from a point source, typically known as a laser scanner; see *Figure 5.2* for an idea of what a laser scanner looks like. Think of a laser scanner as similar to a laser measurer or laser distance measurer, typically used in residential applications to measure distances between walls or other objects, but measuring many times over and rotating as it is measuring.

Figure 5.2 – Example of a laser scanner (image credit: https://flickr.com/photos/armyengineersnorfolk/)

So, a laser scanner uses the same principles but amplifies them and spins while simultaneously rotating, capturing points of measurement in a sphere from the source of the laser. While a laser distance measurer captures one *point* as it measures, a laser scanner can measure thousands, if not millions, of points per second. This is a tremendous amount of data, but often creates a photorealistic representation of the area captured, called a *point cloud*. A point cloud is a collection of points captured during a *scan*.

Now, laser scanning and all other types of reality capture are typically only capable of capturing points that they can physically see from the source. In most cases, it will be from the laser scanner. So, if we were to capture a roadway that has several turns and bends in it, if the laser scanner could not physically "see" an area of the project site, it would not be able to capture it. With this limitation on the capture method, projects are typically captured with multiple scans. As you can see in *Figure 5.3*, the image shows multiple red dots, and each dot represents its own scan.

Figure 5.3 – Example of multiple scans

When we merge all of those scans together, we are able to combine the data into one holistic dataset that can be used for our analysis and corresponding project design.

Terrestrial or static scanning

To add another layer of complexity, the examples we have been referencing in this chapter show a laser scanner that is on a tripod. This method of scanning is typically known as **terrestrial scanning** or **static scanning** because the scanning is in a static position when it is capturing. The process typically involves an operator setting up the scanner in a predetermined location and then letting the scanner conduct its scan, which can take from 20 minutes all the way down to 20 seconds. Once the scan has completed, the operator moves the scanner to another location and repeats the process again. Once the entire area is scanned in the field, the data captured must be brought back to the office. Once captured, there is office work to be done to combine the scans into one holistic scan file.

Other methods of laser scanning can be considered mobile scanning and vary in many ways.

Mobile scanning

Mobile scanning can be conducted through a variety of means, from handheld devices to wearable backpacks, to car-mounted or even plane-mounted setups that can capture project areas from thousands of feet away. There are plenty of methods of scanning, but they all have the overall goal of capturing reality.

Understanding flown LiDAR

A common method of reality capture is called flown LiDAR. **Flown LiDAR** is typically associated with very large laser-scanning apparatuses mounted within a plane or UAV that integrate **Global Positioning System (GPS)** and **Inertial Measurement Units (IMUs)** to help correctly position the captured data in a real-world coordinate system, as well as adjusting the point locations as the plane is moving while capturing. These systems can be very complex and have only in recent years been adapted to smaller capture devices that can now similarly capture from a device the size of a water bottle.

Now, with the broadening of reality capture techniques and tools, there has been a growing understanding of the accuracy and precision of the data captured. We could fill several books if we were to branch into this topic, but there are many already available and we simply want to give a broad understanding of this concept.

Navigating the accuracy, cost, and speed triangle

For the sake of brevity, let's consider the word *accuracy* as a general term and not a specific one as there can be many interpretations of accuracy and even how "accuracy" itself can be measured. But we don't just want to consider accuracy; we need to consider the golden triangle of reality capture: accuracy, cost, and speed.

Think of all these terms with respect to the capture process, that is, the accuracy of the capture, the cost of the capture, and the speed of the capture. In this industry today, typically users get to pick two of the three. If you hope to get a fast and accurate point cloud, expect it to be expensive. If you hope to get a fast and cheap point cloud, expect it to be less accurate. If you hope to get an accurate and cheap point cloud, expect it to take a longer time.

We've looked at static scanners, which we explained take one scan at a time and are a bit more manual than mobile scanning. With static scanning, there is less of a need to compensate for the movement of the scanner, so the accuracy is incredible. With mobile scanning, the movement of the scanner must be taken into consideration when computing the measurements and stitching together the final point cloud, which causes less accuracy. However, since the scanner is capturing while it is moving, you are able to capture faster and you also get more vantage points for the scanner.

So, with every advantage of a capture method, there are trade-offs that users and buyers must be aware of. With surveyors being early adopters of this technology, they tend to be biased toward accuracy because they are liable for the deliverable; so, they lean more toward static for that similar accuracy they expect when being in the field.

Streamlining workflows

In traditional workflows for civil engineering, the existing conditions data was received by surveyors, and they traditionally used a survey rod to capture similar points. These points are X, Y, Z coordinates located in real-world coordinate systems. This workflow consisted of walking around the site and using a standard survey rod to touch a point of interest to be measured, such as a manhole cover or the corner of a curb and gutter transition. They would then mark it with a specific point code to indicate what the point captured represented when the technician began processing the data back in the office.

This workflow was very manual, having to walk to each point and wait to measure and record it, and on a typical day, you could capture maybe several hundred points. With modern applications, that number is increasing, but at a linear rate. With the introduction of laser scanning, current equipment can capture millions of points per second. This is a tremendous leap in the amount of data captured for the surveying industry, but with this solution, a new problem was born.

The datasets captured were incredibly accurate and almost photorealistic, but with the immense amounts of data, technicians couldn't actually utilize it to extract the specific information needed for the project design. Extracting the ground points for surface creation was a very difficult process, and since then, many technologies have been created to alleviate this and automate the creation and segmentation of these datasets for efficient workflows.

Now, surveyors were early adopters of laser-scanning technology for tremendously more efficient workflows, but they still needed to process the data and deliver a stamped survey file for legal processing when handing over to a civil engineer for design. In this chapter, we will walk through this workflow to give a basic understanding of what occurs in the process and to help provide an understanding of the value of reality capture and why it is becoming a standard operating procedure in today's world.

This book primarily teaches you about workflows within Civil 3D, but in this chapter, we will take our first detour from that and introduce **Autodesk ReCap**, a point cloud processing software that is essential to visualizing point clouds as well as modifying them to fit our specific needs when importing into Civil 3D. ReCap is utilized first as the primary importer of point cloud data. It is capable of handling the many millions of points that come with a point cloud and allows for the basic functionality of viewing, cropping, editing, and exporting. With that, let's dive into some workflows and put our knowledge into action.

Delving into reality capture workflows

Reality capture is becoming more and more of a standard operating procedure in not only the civil engineering world but also many others, such as industrial facilities, architecture, and construction. It is not uncommon nowadays for civil engineers to encounter a point cloud and be expected to understand how to navigate and utilize such a dataset.

With that, let's begin by opening Autodesk ReCap and exploring the interface. In *Figure 5.4*, we can see the home screen of Autodesk ReCap. The home screen allows for basic project creation or the reopening of previous projects for further development.

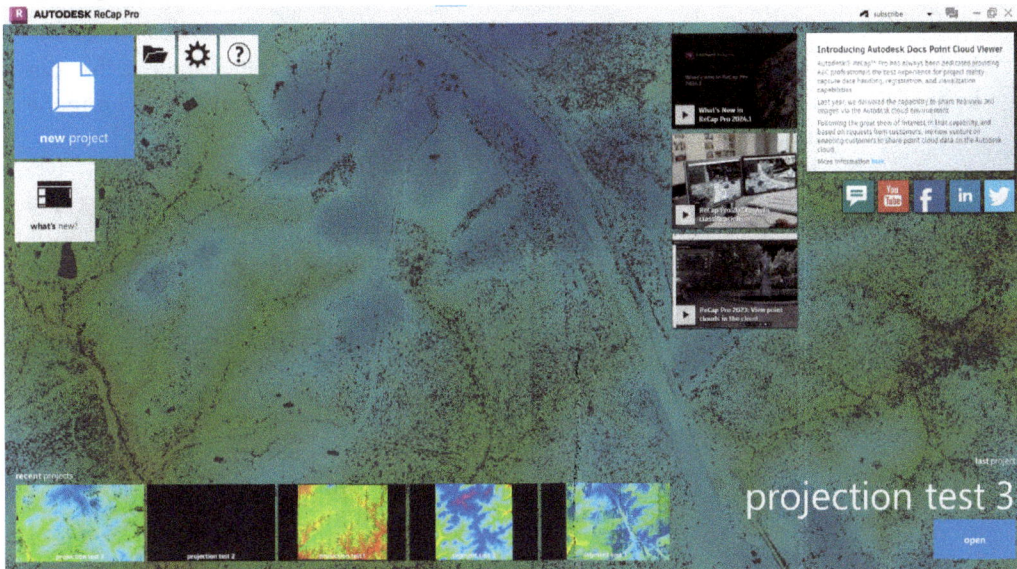

Figure 5.4 – Autodesk ReCap home screen

For our first exercise, we will create our own project and import a dataset we received from a surveyor, using the following steps:

1. Click the *open a previous project* button in the top-left corner of the home screen.

2. Navigate in the file explorer to the file named `laserscan.rcp`. In *Figure 5.5*, you can see `laserscan.rcp` and a similarly named folder, `laserscan Support`. ReCap works with a main file, with the `.RCP` extension, with the corresponding raw data files located in the `Support` folder. You'll notice the size of the `.RCP` file is only 44 KB, while the support folder is 77 MB.

Figure 5.5 – Autodesk ReCap file and folder structure

3. Click on **Open**.

4.   In *Figure 5.6*, we now see the user interface of a project. We should see a square conglomeration of dots in the middle of the screen. This is the point cloud, and we are looking at its visualization style, which always defaults to **RGB**, meaning the red, green, and blue colors are assigned to the points if they exist.

In this dataset, they do not, but in some point clouds, when they are being captured, they can be overlaid with matching photos to project color on them and assign an RGB color to the point in addition to its typical *X*, *Y*, *Z* coordinate.

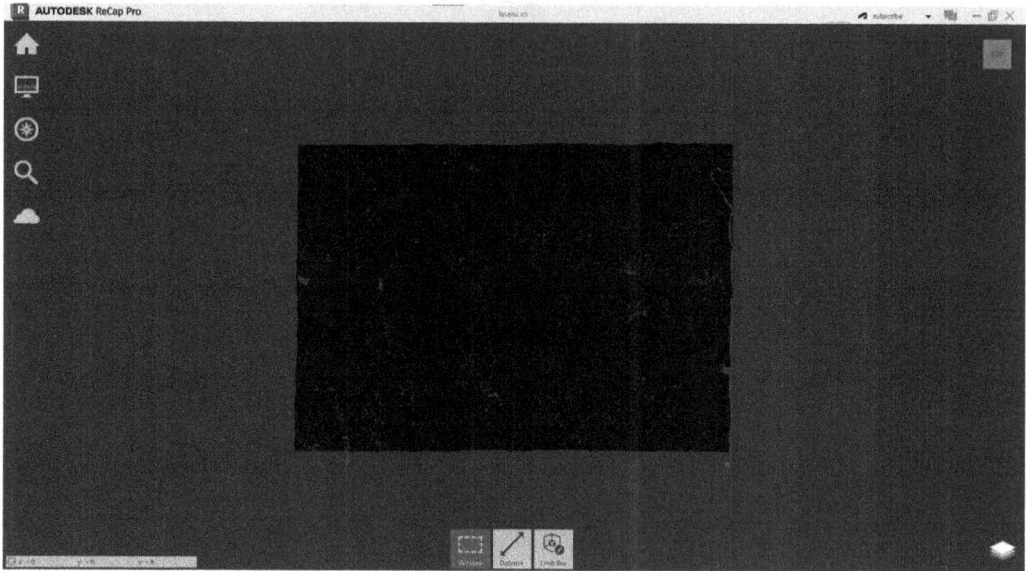

Figure 5.6 – Autodesk ReCap user interface

5.   Let's first begin by giving some color to the point cloud and showing the data in a more visual way for interpretation. Begin by hovering over the monitor icon, as shown in *Figure 5.7*. Then, while hovering over it, move your cursor over to the color wheel and hover. From here, you should see many options to select. Each of these represents a different method of visualizing the point cloud and each can help to show different information about the point cloud.

Let's look at these options:

I.   We begin with **RGB**, and since there are no values assigned for this point cloud, it defaults all points to the same color.

II.   Next, we have **Elevation**, which will display a color gradient based on the varying elevations or *Z* values of each point.

III.   We will not go into **Intensity** and **Normal** very much, but they can be quite useful when looking at static laser scan data.

IV.   Earlier, we talked about individual scans, which are merged together to create a larger scan. The **Scan Location** option will help differentiate each scan in this visualization style.

V.   Lastly, there is **Classification**, which is another property that can be mapped onto points for differentiating between points classified as ground points versus vegetation.

6.   Select **Elevation**.

Figure 5.7 – Visualization options

7.   With **Elevation** selected, we should see something more similar to *Figure 5.8*. We can see almost a colorized version of contours that we may be used to. We can see land variations with high and low spots and even a roadway running through the middle. In the top-right corner of your screen, you should see a gray cube that says **Top**. This is a view cube to help us see different perspectives of the dataset.

Right now, we are looking from a top-down view, or a plan view similar to Civil 3D. Let's hover over the bottom-right corner of the view cube and wait until it turns light blue. Once light blue, click that corner.

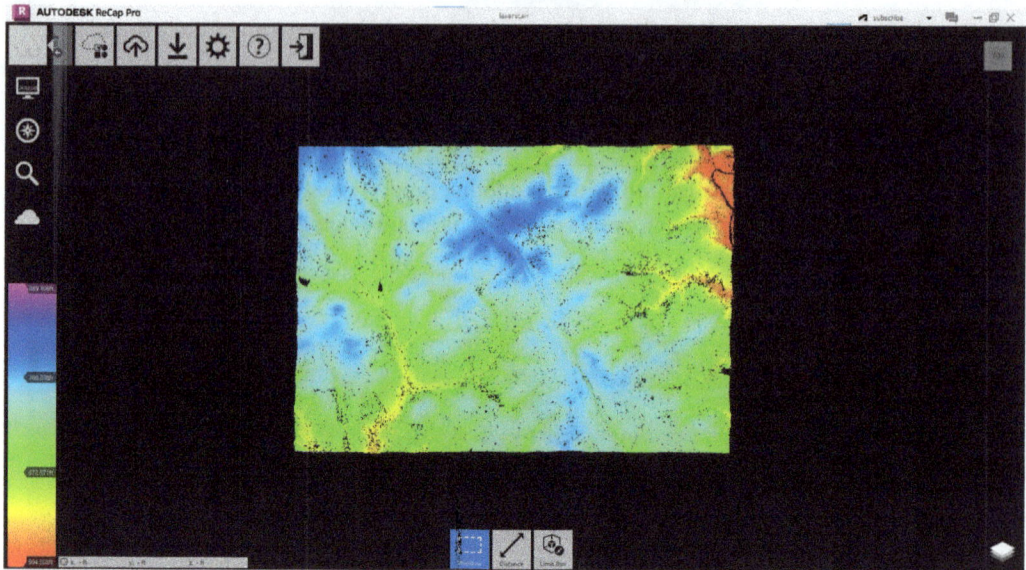

Figure 5.8 – Elevation color mode

8. Now that we can see the data and its variations, let's get into the navigation of the data itself. By clicking on the corner of the view cube, we have changed the view to a perspective. This is just one way to do this, but let's first run through all the functions for orbiting, zooming, and panning around in our dataset:

    I.   To zoom, use your mouse wheel to scroll up or down. This will allow you to zoom in or out of the dataset for more or less detail. This is perfect for zooming into the details or a certain area, or for a holistic view of the entire project site. You should notice that when you zoom in with your cursor on the dataset or point cloud, a small gray sphere appears where your cursor was when you began zooming. This is important as it centers the view on that area as you orbit and navigate in other areas as well. So, if you are having a hard time zooming into a specific area, watch where your gray sphere is and then adjust it if needed for a better vantage point. See *Figure 5.9* for a better idea of this.

    II. To pan in ReCap, click and hold on the mouse wheel and this will lock the perspective of the screen and allow you to strife or pan in the viewing area for better perspective of any areas of interest while exploring the model.

    III. To orbit, you can right-click and hold on the point cloud and move your mouse up or down or left or right, to rotate and orbit in many different ways. You should see the same gray sphere appear wherever you right-click on the point cloud. This will help you understand where you are orbiting around, and then you can adjust it if needed to focus attention on a different area of the point cloud.

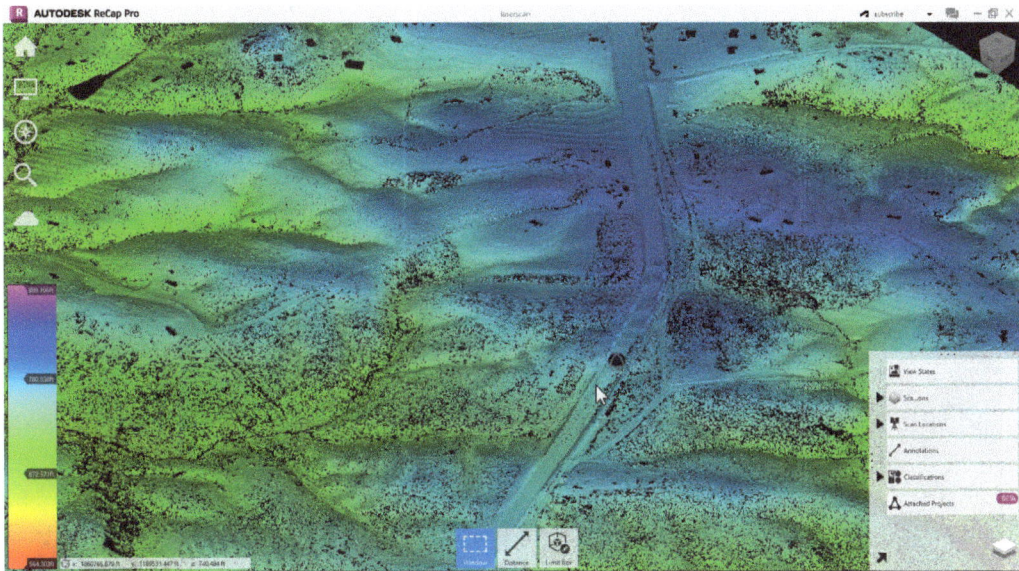

Figure 5.9 – Orbit sphere

9.   Practice orbiting, panning, and zooming around the model for a moment and get familiar with the navigation methods as well as the dataset itself. Once you have a good handle on the basics there, go back to the view cube in the top-right corner and hover over it until you see the top highlighted in light blue. Click that and this will reorient our dataset to the view we first came to when we began.

10.   Now, as we went through earlier, there are a few other visualization styles we can use in ReCap, but if the values are not assigned to the points, then they will essentially have the same value. This means they will not look different when looked at under that style. So, this dataset does not have **RGB**, **Normal**, **Scan Location**, or **Classification** values. What it does have is **Elevation** values and **Intensity** values.

Let's go back to the monitor, hover over the color wheel, and then change the visual style to **Intensity**, as seen in *Figure 5.10*. Here, you should see a lot of colors similar to the elevation rainbow we saw previously, but the many colors are oriented completely differently. The **Intensity** values are simply that, the intensity with which the laser was returned when it measured different parts of the project site.

One thing you can notice is how red certain areas and specific lines look in this view. These solid red lines represent roads, and the other greener areas show softer ground areas. So, this is an interesting way for us to better understand our dataset and highlight different areas based on what we are looking for.

If we want to get an understanding of the ground and overall contour movement, we can select **Elevation**. If we want to investigate the varying elements captured, such as roads, we can look through the **Intensity** style.

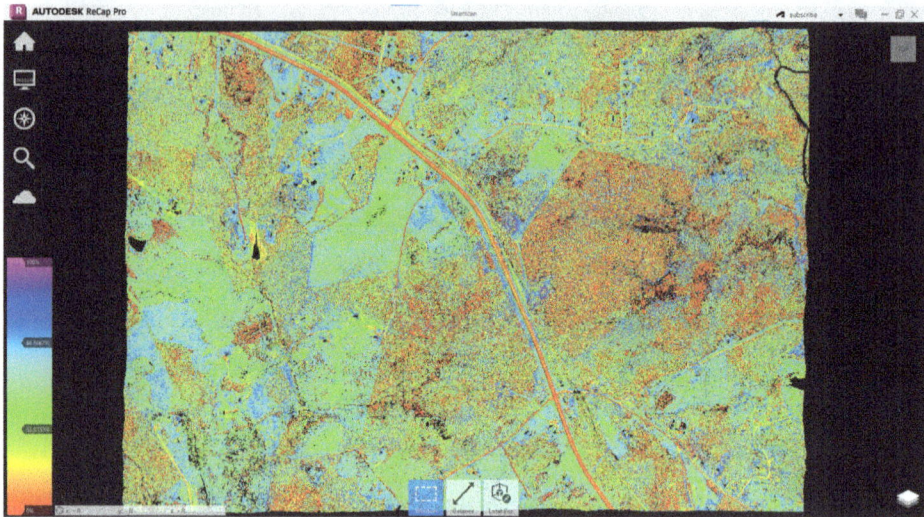

Figure 5.10 – Intensity color mode

11. As we look at this dataset, we can see it is covering a large area. It is much larger than our specific project site, so let's dive into reducing the amount of points visible and better zoom into our project site. If we look at *Figure 5.11*, we can get a rough idea of where our project site is.

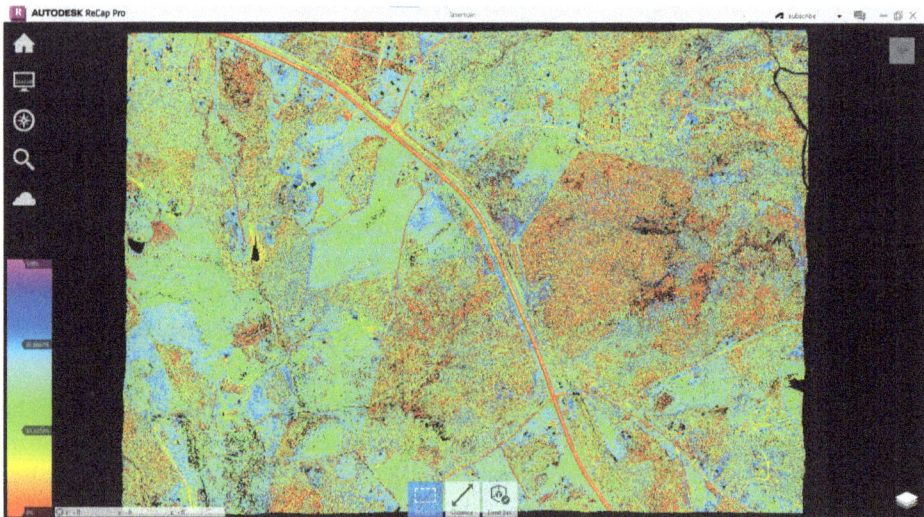

Figure 5.11 – Approximate project boundary location

12. Now, let's direct our attention to the three buttons in the bottom middle of our screen. You can see the **Window** button highlighted in blue. Up until now, we haven't really left-clicked on the model space of the program. That is because when we left-click, we are in some way "selecting" points. That is dictated by the **Window** button. If you hover over the **Window** button, you will see several other options appear. Our options are **Window**, **Fence**, and **Plane**, and each allows us to select points within our dataset for different purposes.

Let's click on **Window**. Then, left-click and hold on the approximate left corner of our project site. Then, drag to the bottom right of our project site and release. This should select our project area, as shown highlighted in gray in *Figure 5.12*. This does not need to be exact – as long as we capture enough of our project site and do not miss selecting the main portion. You can use the roads in red as guides to help your selection.

Figure 5.12 – Selected project location

13. With our project site selected, new buttons have appeared at the bottom of our screen. We will use them to hide unnecessary points in our dataset to reduce the file size and make it easier to work with. Click the **Region** button next to the **Window** button and then click **New Region**. We will name it Project Site. Then, hit the *Esc* key. Our point cloud should look no different, but what we have just done is create a "region" similar to a selection or group of points that can function similarly to AutoCAD layers where we can turn them on or off for display purposes.

14. Now we can use ReCap to dissect and manipulate our point clouds to extreme detail and complexity by creating different "regions" and deleting unnecessary points. But for the purposes of this exercise, we simply want to reduce the file size and focus on our project site.

So now, go to the bottom-right corner of your screen and hover over the button that looks like three sheets of paper, shown in *Figure 5.13*. This is your Project Navigator, and it helps you turn different regions of your drawing on or off, show annotations and classifications, and much more.

15. For now, hover over your Project Navigator and click the black arrow next to **Scan Regions**. You may have to expand the Project Navigator to the left a bit to read each option. But in it, you should now see two regions, one being **Project site** and the other **Unassigned Points**.

Figure 5.13 – Regions in the project navigator

16. When we hover over the different regions, we can see them highlighted in green. Next to each region is a *lock*, which allows you as a user to protect certain regions so they cannot be modified unless "unlocked." Next to the lock is an *eye*, which will turn the regions on and off depending on what you need to do. Next to that is an *X*, which will delete points in that region. Hover over the **Unassigned Points** region and click the *eye* to hide those points outside of our project site.

17. Now, window-select the `Project site` points again by left-clicking and holding with your mouse and dragging to fully select the visible points. They should be the only points visible now.

18. Next, hover over the home icon in the top-left corner of the screen. Move your cursor over to the icon that looks like a large down arrow and hover over it. Then, click the arrow below it that is pointing up, as shown in *Figure 5.14*. This is the **Export** button, which will make us a whole new project with just the project site data within it.

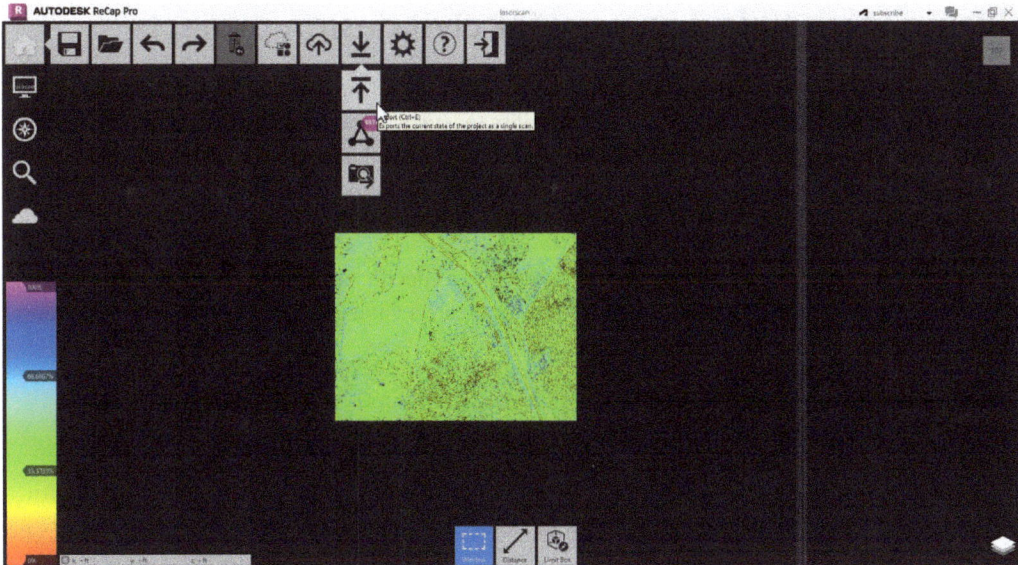

Figure 5.14 – Exporting a new condensed project dataset

19. Let's name it `laserscan project site` and hit **Save**. We will get a warning because we are unifying our project data as well as removing scan data from our previous project. This is okay and what we want to make the dataset usable for our purposes in Civil 3D. Too much data in Civil 3D and the program could crash on us mid-project. So, the more refined our dataset, the better. Click **Let's go! Unify my project scans**. We have just created a new condensed file for our project work, but you will notice we are currently in our previous project. Go back to the Project Navigator and turn on the **Unassigned Points** region again. Nothing has changed in our old project, so let's open our new project to see what has changed.

20. Hover over the home button and click the folder icon to open our new project, `laserscan project site.rcp`. Now, adjust the visual style to your liking and orient how you wish, and then hover over the Project Navigator and expand your regions. You should notice none of your previous regions are there and the project dataset is much smaller now.

This is the perfect place to end this section and move on to the next one, where we will learn how to move this condensed dataset into Civil 3D to create a surface from a point cloud. In this section, we learned the basics of Autodesk ReCap and how to navigate around our dataset and view it differently to fully understand how our project area is behaving. We also learned how to edit and specify what we want to utilize as our working dataset for migrating to Civil 3D. Next, we will learn how to leverage this data in Civil 3D and take advantage of the incredible accuracy provided to us with this dataset.

Utilizing incredible accuracy for civil design projects

Now that we have developed a basic understanding of Autodesk ReCap, you can imagine the potential for larger datasets with more complexity and the ability to leverage them for further use within either Autodesk Civil 3D or other programs. Reality capture in general has many use cases and nuances, and being able to manipulate and handle this type of data can help a civil engineer of BIM model manager stand out

Now that we know how to use point clouds, let's dive into a specific use case in Civil 3D and use this data to develop our existing conditions and ultimately create a Civil 3D surface from it to act as the foundational layer of the rest of our project design.

Let's begin by opening our `Survey Model.dwg` file located within `Civil 3D 2025 Unleashed\ Chapter 5\Model`, then take these steps:

1. First, let's understand how Civil 3D refers to point clouds. Civil 3D is a detailed program with layers of functionality within it that prohibits it from working with large files, similar to Autodesk ReCap. So, what Civil 3D does is look at point clouds similar to XREFs where they are attached to the drawing, but not actually in the drawing, acting as a reference and sampling as you work with it. So, let's navigate on the ribbon to the **Insert** tab and look for the **Attach** option under **Point Cloud**, as shown in *Figure 5.15*.

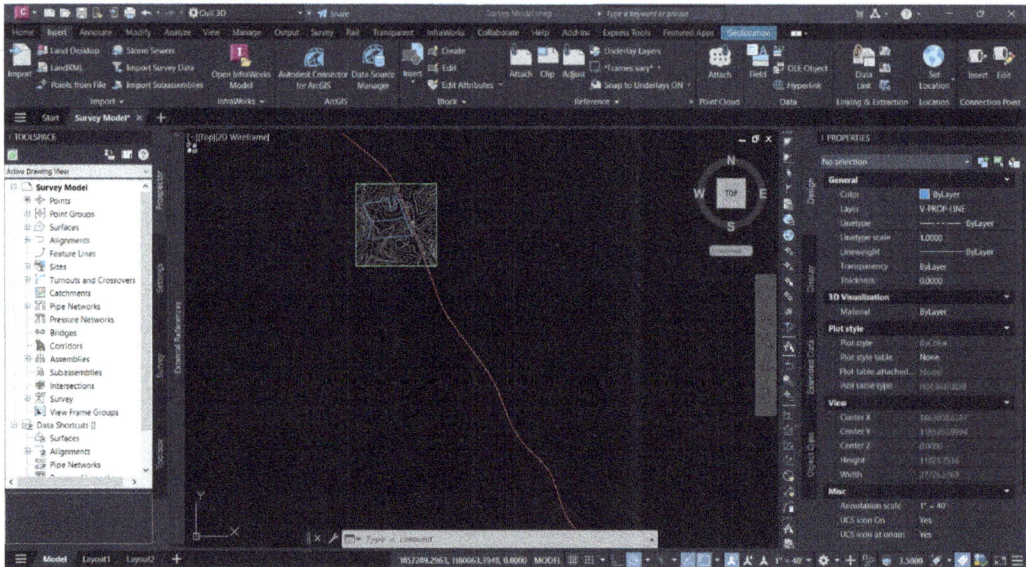

Figure 5.15 – Attach a point cloud to Civil 3D

2. Navigate in your file explorer to the file named `laserscan project site.rcp` and click **Open**. We will get a new dialog box pop up here asking how we would like to insert the point cloud and what location, scale, and so on we would like. With existing conditions data

in civil engineering, we want to rely on the surveyor or whoever captured the data to set up the proper scale, units, and coordinate system. So, if we assume everything is correct and taken care of already, the dataset should simply drop right into place and we would not need to modify anything.

Our drawing is already set up for NAD83 South Carolina State Planes with US Survey feet as the units, so we should assume the data collector has set them up the same way. So, we should ensure the **Insertion point** is unchecked. The path type should be **Relative path**, and **Rotation** and **Scale** should be unchecked. See *Figure 5.16* for how the parameters look altogether.

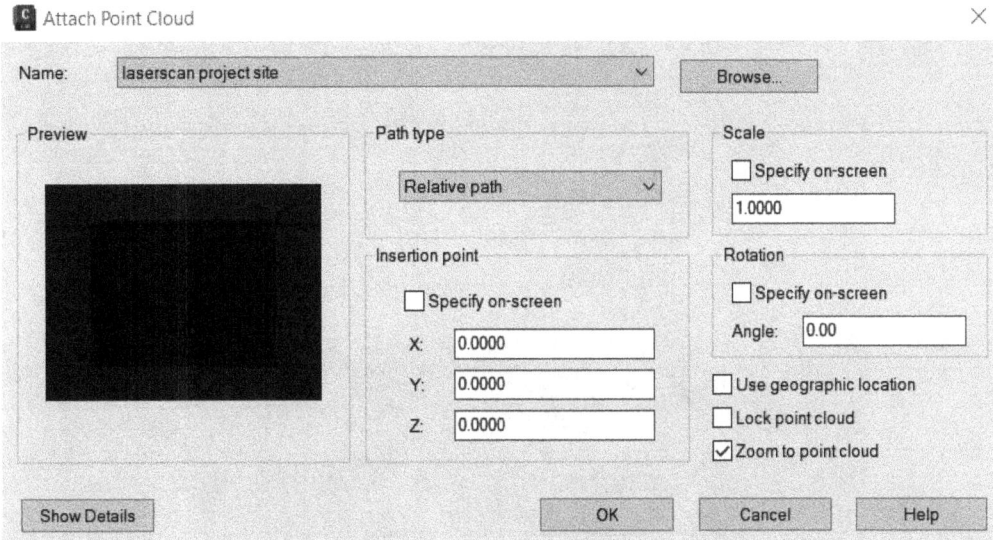

Figure 5.16 – Point cloud attachment parameters

3. When the point cloud is attached, we should see it drop behind our existing topography file to check our accuracy, see *Figure 5.17*. Ensure you have the **PROPERTIES** tab on and displayed. Hover over the point cloud, and when it highlights, click it. You should see in the **PROPERTIES** tab that a point cloud is selected with many other parameters Civil 3D is reading from the dataset.

Similarly to when we inserted our dataset into ReCap, in Civil 3D, the point cloud is a bunch of points represented in black, which is difficult to orient when working. In **PROPERTIES**, you should see a **3D Visualization** parameter. Under that is **Stylization**; change it from **Scan Colors** to **Elevation**. Then, change **Color scheme** to **Spectrum** and we should see a similar rainbow style applied to the point cloud that we saw in ReCap. Depending on your preference, you can leave the visualization style to show contours like this or adjust **Stylization** to **Intensity** as well to highlight the different materials scanned in the point cloud.

4. As we get our stylization how we prefer, notice the contextual ribbon of Civil 3D has changed since we have our point cloud selected. We now have many options to adjust how our point cloud looks. Starting from the left, you will see options to adjust the size of the points within the point cloud as well as the level of detail of the point cloud and several more.

    Again, Civil 3D does not like to work with too many points at a time, so the more points you have visible and the smaller the points are, the more Civil 3D will have to render each time you maneuver within your working space. We recommend leaving the parameters as is with **Point Size** at 2 and **Level of Detail** at 9 or 10 depending on the dataset. You can adjust as you go to see which works better, but be sure to save your work as you progress.

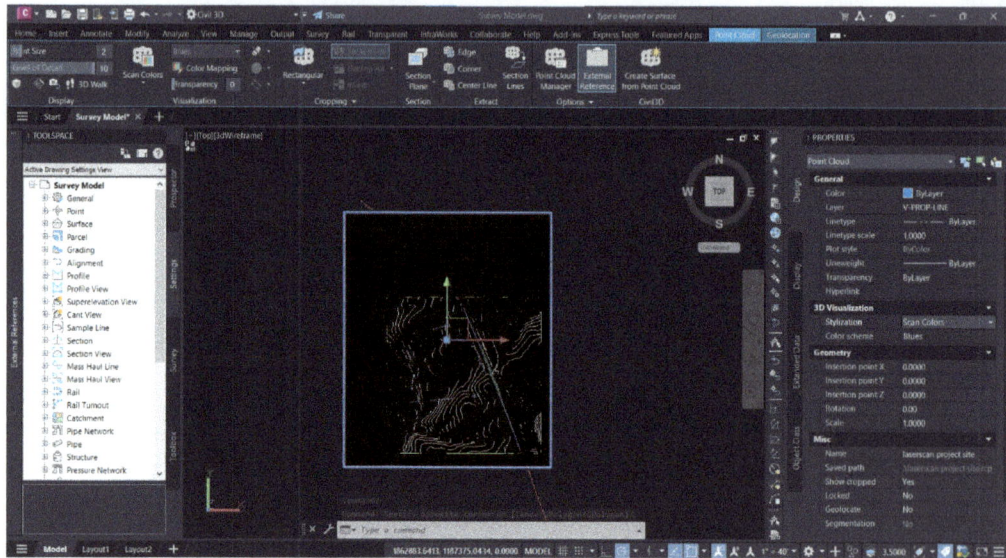

Figure 5.17 – Point cloud attached within Civil 3D

5. Moving to the right now, we have similar options to what we just modified for **Scan Colors**. If you click the drop-down arrow, you should see all the same options for elevation, intensity, and so on. You can also adjust the transparency of the point cloud, which is very useful when coordinating between the point cloud and the existing Civil 3D model. Let's adjust **Transparency** to 10.

6. Continuing to move right, we can adjust the lighting, but we will not modify this for now. This is useful when looking at different rendering options of the point cloud.

7. From there, we see tools to allow for cropping the data even further. Remember how large point cloud datasets can be, so there are always many options for filtering and minimizing the actual data to be worked on when processing these workflows. We will not utilize these in this book, but they can be very handy to create saved crop states for specific views when working on detailed areas of a point cloud. Similarly, next to the cropping options is the option to utilize **Section Planes**. We will pass over these options and the next ones for **Extract** tools as well.

8. The second-to-last option moving to the right is **Point Cloud Manager**. Let's click that to expand its options. As shown in *Figure 5.18*, **Point Cloud Manager** gives us options similar to the *Project Navigator* within ReCap, but these same controls have been carried through to Civil 3D, which can be incredibly useful in maintaining the work done previously. In **Point Cloud Manager**, we can see the same regions we have created as well as different scan locations if we had these in our point cloud project. We do not, however, have these options now because we "unified" our scans and deleted the other data to allow for a smaller, more agile dataset while working in Civil 3D.

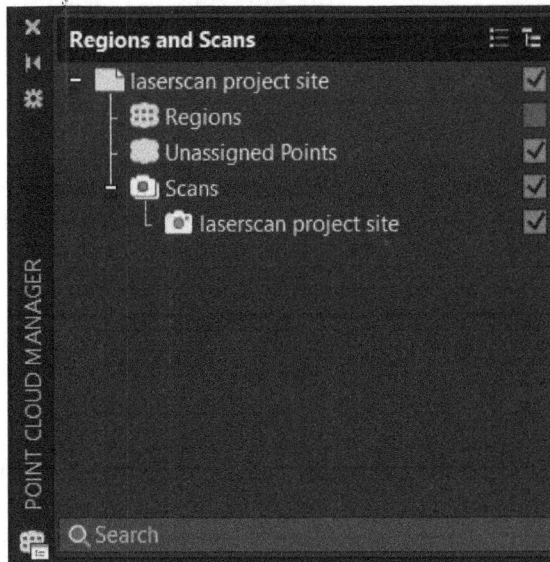

Figure 5.18 – Point Cloud Manager

9. Finally, as we move from left to right, as we can see in *Figure 5.19*, on the **Point Cloud** contextual ribbon, we have the **Create Surface from Point Cloud** button. With our point cloud selected, let's click that button.

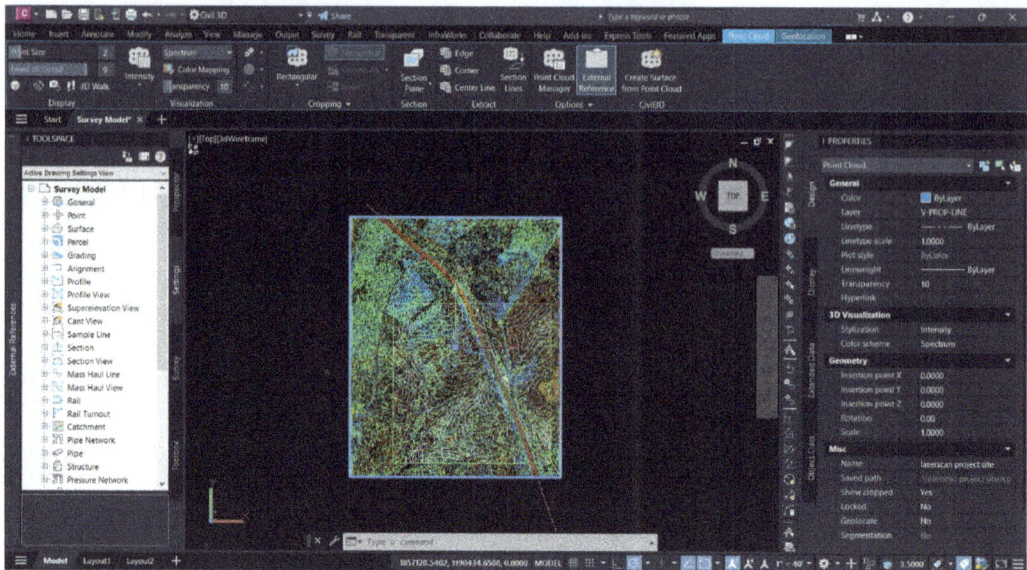

Figure 5.19 – Create Surface from Point Cloud

10. Now, we get a new dialog box pop up titled **Create TIN Surface from Point Cloud – Point Cloud Selection**. We have three tabs that we need to go through to fully stylize and adjust how we want our surface to be created. We begin in the **General** tab, where we can name our surface, add a description for it, and choose what style we would like, how we hope to render it, and finally, what layer to put it on. Let's begin by naming our surface `Point Cloud Surface`. Then, hit the **Next** button.

11. The second tab we have is the **Point Cloud Selection** tab, which will help us narrow down exactly where and how we want to create our surface and decide which point to include as well as how many samples of points we want. You can see in *Figure 5.20* that if we left this with its defaults, we would create a surface that included almost 600,000 points. Most surfaces are typically made of hundreds of points, not hundreds of thousands of points. It is vital to understand this concept because if we create a surface from too many points, Civil 3D could struggle and crash, or at least make our production process lag tremendously throughout the project.

    We have *three icons* at the top, which allow us to add entire point clouds at a time or select and exclude regions of points for our surface creation. Let's click the *middle button* to select only the areas we wish. When we click this button, Civil 3D will put us into model space again and ask us to window-select the areas we want to add. Simply window around the border of the existing surface; there is no need to be exact.

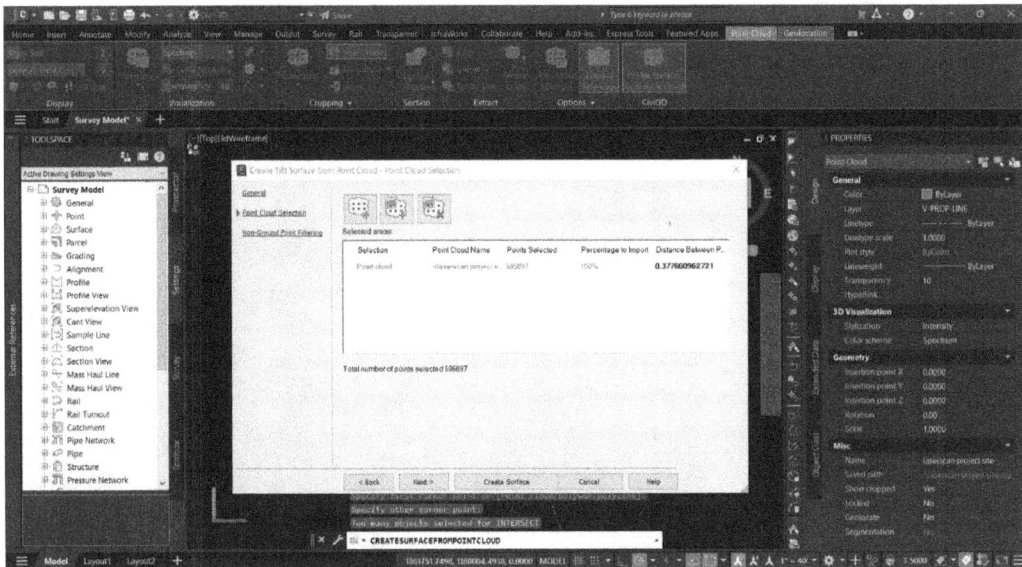

Figure 5.20 – Point Cloud Selection

12. Once we window-select, we should have two "selections" for what to create for a surface. We will delete the first selection, which is our entire point cloud, by using the third button from the top, **Remove a selection from the list**. We can do this by clicking the point cloud named **Selection** in our list and then clicking the **Remove a selection from the list** button. Now, this is not deleting any points from our point cloud; it is simply isolating the points we wish to create for a surface. Our previous total number of points selected was around 600,000 and should now be under 300,000.

13. Now that we have found our surface extents, we need to further chisel down what we want to create our surface from. We can do this by clicking on our only point cloud selection and then clicking on the **Distance Between Points** cell. It should be highlighted in blue, as seen in *Figure 5.21*. By default, it should show the current distance between points as **0.377**, which means the distance between points from the point cloud is 0.377 feet, which is around 4 inches. So, the surface would have calculated a triangulation every 4 inches to create the surface. This kind of accuracy and detail is not typical for traditional survey methods. So, we can experiment with this number and adjust it as we like to see how the subsampling affects our point count. Let's adjust the number to 10, which would give us a point every 10 feet. So, click into the cell, type in 10, and then click out of it. Now, our total number of points should be around 54,000.

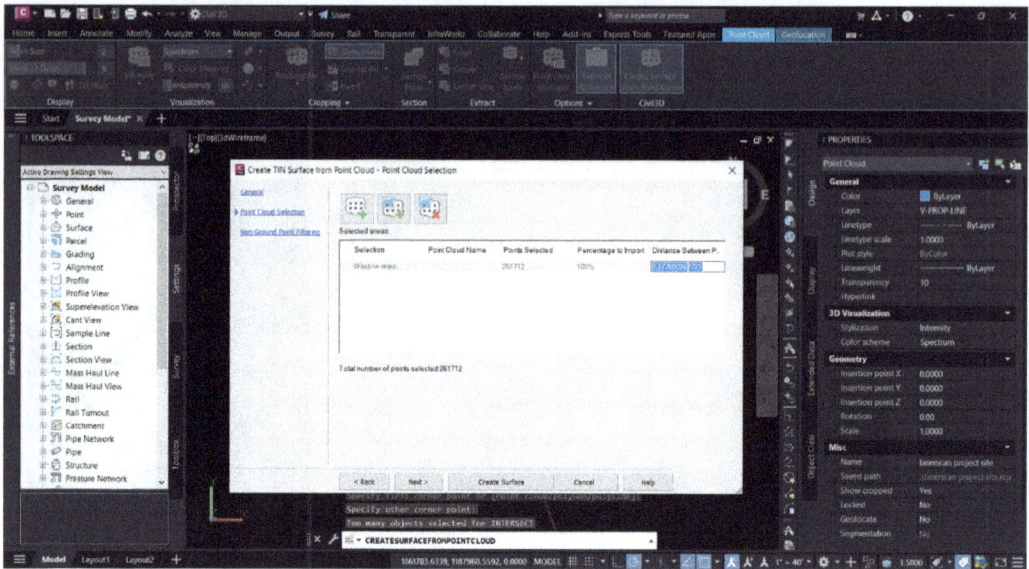

Figure 5.21 – Point cloud subsampling

14. With these much more manageable numbers from where we started at 600,000, we can proceed to the next tab, **Non-Ground Point Filtering**. This tab allows us to begin averaging or adjusting our interpolation of points. We have three options, **Planar Average**, **Kriging Interpolation**, and **No filter**. The first two options allow us to average or adjust the approximation between points if we think it is required, but in our example, the dataset provided includes only ground points, so we will select **No Filter**.

15. Once we select **No Filter**, we are ready to click the **Create Surface** button and let Civil 3D churn and process our request. We should get a dialog box informing us that Civil 3D will be processing this in the background and will notify us when it is complete.

16. Once our point cloud surface has been created, we can zoom into it. First, we should hide the point cloud, since it has done its job for now. Select it, right-click, and select **Isolate Objects**, then click **Hide selected objects**. This should hide the point cloud temporarily so we can see how it turned out. Our previous surface should still be visible, as well as what looks like a new surface with very similar features, which is what we created from the point cloud. We can select each one separately and look at the properties to notice slight differences. One such difference is that the number of points within each varies, where one is composed of 35,000 points and our new one is composed of around 54,000 points.

This shows how many different parameters there are for creating surfaces as well as the need for filtering and isolating only the points you need from point clouds. If we recall, we began with large swathes of data captured and had a point count in the millions. We first began cutting out non-necessary data for our surface within ReCap and created our own point file to import into Civil 3D.

Then, we attached it to Civil 3D and used the tools within to select even further down into that point cloud to get us down to a point count of around 600,000. From there, we were able to use the point cloud surface tools to adjust our subsampling of the point cloud to only give us points every 10 feet, which finally reduced our point cloud to a manageable 54,000 points.

Now, this is still a large number of points to create a surface from. Remember a traditional road survey would have points in the hundreds, maybe in the low thousands, through the traditional capture process with a survey rod. This is a huge advantage of utilizing reality capture data and workflows.

The more points you include when creating a surface or any conversion of point cloud data, the more accurate the surface or object will be. But there needs to be a middle ground because the more points, the more triangulation and the more difficult the object will be to work with. So, it is vital to understand the end goal of this data, and then you can better know how to filter and trim out what is needed from what is unnecessary for your purposes.

For surfaces specifically, there are ways to filter data less on areas that we care more about, such as roadways, and then also filter more out on areas we do not need as much, such as open field areas. These are ways to preserve the accuracy of the data and create a more informed surface than simply applying the same level of filtering to the entire dataset that we performed.

Summary

In this chapter, we took a broad look at the reality capture industry in general to learn how it works as well as what its uses are. With many possibilities and many data points, reality capture offers great workflow enhancements, but also must be utilized with the proper knowledge and end goal. We saw how complicated some of the data can be, and if not utilized properly, it can overwhelm our projects and actually make us less efficient. This new technology is not going away anytime soon; if anything, it is moving at a faster pace than ever. With this basic understanding of reality capture, a BIM model manager can truly propel to the next level in their careers and become an even stronger asset for their teams and projects.

We hope this was helpful to understand the basics of reality capture as well as how to approach and apply it to your projects for incredible value and accuracy. With our exercises, we filtered out almost 90% of the data captured. To some, this could seem incredibly wasteful, but the data we used brought incredible accuracy for our purposes. The remaining data can be leveraged in other programs we will cover later on in this book for many more uses and disciplines.

This is a broad characteristic of reality capture: understanding a large dataset and being able to drill into the nuances and find exactly what is needed for your project. We took a broad approach to reality capture in this chapter.

In the next chapter, we will begin diving into the details of a Civil 3D-specific function: grading optimization. We will build on the basics of grading, introduce some new tools and workflows for edge cases, and equip you with the tools you need to be able to handle everything around grading.

6

Streamlining Design with Grading Optimization

In the previous chapter, we provided a comprehensive overview of how we can incorporate reality capture into our projects from the get-go. Using proper workflows, we can increase the efficiency and overall accuracy of our design models moving forward. We highlighted some of the complexities experienced while managing reality capture data and the risk of inefficiencies if not utilized correctly.

Continuing the efficiency-gaining trends when it comes to generating our surface models developed within Civil 3D, we'll shift our focus from generating existing conditions to proposed ones. In this chapter, we'll cover a relatively newer tool that has been made available over the past several releases of Civil 3D, called Grading Optimization.

In this chapter, we'll jump into the world of advanced cloud-computing grading tools that, when used properly, can increase design efficiencies and save teams valuable time earlier on in the project as alternative designs are being considered. This tool, when utilized properly, can provide design teams tremendous advantages at a conceptual design phase/level to quickly realize optimal grading slopes and elevations for their proposed site design. It's recommended to use these results as a starting point, from which we can further refine as the remainder of our site design progresses.

With that said, in this chapter, we'll cover the following topics:

- Getting started with Grading Optimization
- Defining grading criteria
- Optimizing our grading models

Technical requirements

We will be using the same hardware and software requirements as discussed in the *Technical requirements* section of *Chapter 1*.

With that, let's go ahead and open up the `Grading Model_Start.dwg` file located within the `Civil 3D 2025 Unleashed\Chapter 4\Model` location. Once you've opened it, you'll notice that the file has been zoomed into our site, as displayed in *Figure 6.1*.

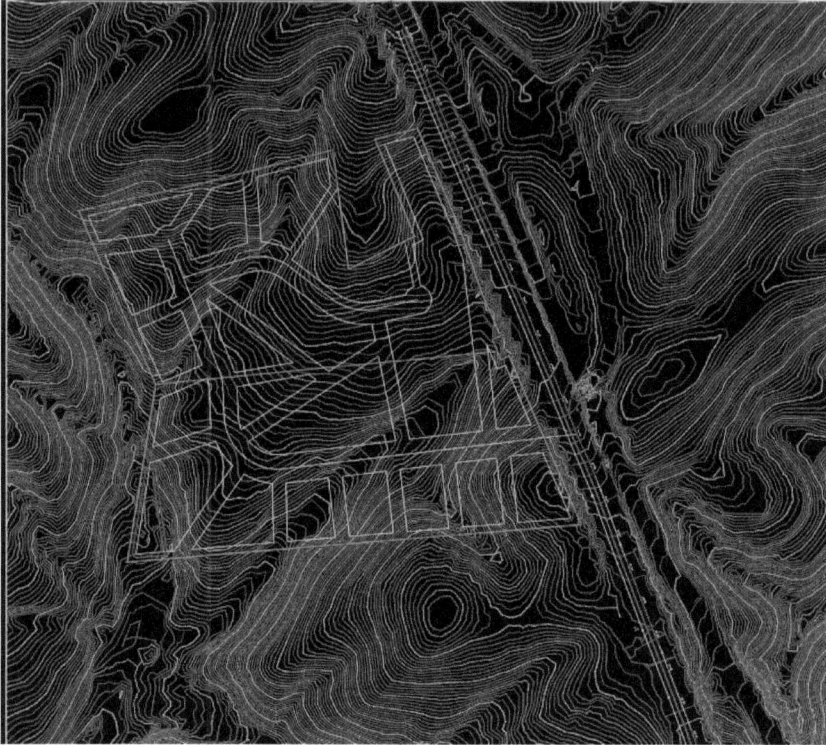

Figure 6.1 – Grading Model_Start.dwg

As you can see, we currently have our `Survey Model.dwg` externally referenced and our **SRF - Existing Grade – FromSurveyPoints** surface model data referenced in our file. We also have a few polylines depicting our overall proposed parcel, composed of two existing parcels, along with our building pads for each individual lot within our subdivision design. We also have created a Cogo Point, which we'll want to apply as a low point in our surface model to define a drainage location. We'll be utilizing these to define different grading conditions throughout our site shortly.

Getting started with Grading Optimization

Autodesk Civil 3D's **Grading Optimization tool** is a user-friendly tool designed to automatically grade the land in various settings, such as construction sites, road interchanges, and around buildings. Using advanced cloud-based optimization algorithms, the tool focuses on creating a smooth surface while considering user-defined constraints, including grading and drainage elements, high/low points, roads, and pathways.

Grading Optimization also offers real-time visualization, allowing users to observe the optimization process, make adjustments as needed, or let it run until an optimal solution is reached. This is particularly beneficial for civil designers, as it automates the identification of grading objects and the execution of analyses, reducing the time and effort typically spent on grading processes. Additionally, by facilitating efficient monitoring, the tool helps us identify potential issues early on, leading to production and construction cost reduction through minimized material and effort wasted.

To access Autodesk Civil 3D's Grading Optimization tools, we'll jump up to our **Analyze** ribbon and focus on the **Grading Optimization** panel at the end (right), as shown in *Figure 6.2*:

Figure 6.2 – The Analyze ribbon | the Grading Optimization panel

Taking a closer look at our **Grading Optimization** panel, you'll notice that we have five different selection options available to us. These tools are as follows (refer to *Figure 6.3*):

1. **Grading Objects**: Selecting this tool will launch both the **Grading Objects** browser and the **Grading Objects Tools** palettes (additional info on each of these follows).

2. The **Grading Objects** browser: Selecting this tool will enable a tree view palette to appear, providing us with some details of each type of grading object currently defined within our file.

3. **Grading Objects Tools**: Selecting this tool will enable a tool palette to appear, where we can specify which grading object types we'd like to assign to various objects within our current file.

4. **Optimize**: Selecting this tool will enable the **Grading Optimization** viewer to appear, where we can run through different grading optimization scenarios based on the criteria and grading objects defined previously.

5. **Help & Learning**: Selecting this tool will give us quick access within our current Civil 3D session to guided workflows, videos, tutorials, and resources to quickly understand how best we can utilize Autodesk Civil 3D's Grading Optimization tools.

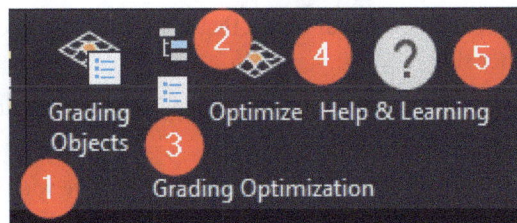

Figure 6.3 – The Grading Optimization panel

For now, let's go ahead and select **Grading Objects** (**1**) so that both the **Grading Objects** browser (**2**) and **Grading Objects Tools** (**3**) appear. If we first focus on the **Grading Objects Tool** palette that appears, we'll notice that we have many different types of grading objects that we can classify the 2D and 3D objects in our design as. From the top down, we have the following grading object classifications (also displayed in *Figure 6.4*):

1. **Building Pad**: Creates a flat area for potential siting of buildings to be constructed

2. **Reveal**: Can be applied to specified **Building Pad** grading objects to add additional elevation criteria for smoother transitions with the surrounding surface

3. **Pathway**: Identifies linear drainage paths within a proposed site

4. **Pond**: Creates areas where either wet or dry ponds will be placed within a proposed site

5. **Retaining Wall**: Allows for stepped grading conditions to be applied within a proposed site

6. **Curb**: Similar to **Retaining Wall**, **Curb** grading objects will apply a stepped grading condition within a proposed site

7. **Parking Lot**: An enclosed area where we can specify sloped grading conditions within a proposed site

8. **Zone**: Similar to **Parking Lot**, we can apply various grading conditions within a proposed site

9. **Exclusion Zone**: Area in which we would like to maintain existing grade conditions within a proposed site

10. **Grading Limit**: Allows us to define the surface boundary where we will tie proposed grading conditions into our existing built environment

11. **Aligned Edge**: Similar to a Breakline, we can apply a consistent slope along a specified linear object

12. **Bend Line**: Similar to **Aligned Edge**, **Bend Line** grading objects are a bit more relaxed, where we can specify minimum and maximum slope conditions

13. **Drain Line**: Allows us to define a linear object that we wish to drain surface water within our proposed site to

14. **Ridge Line**: Allows us to define a linear object that will act as an apex so that surface water will drain away on either side

15. **Low Point**: Allows us to define a specific location where we wish to drain surface water within our proposed site to

16. **Bounded Point**: Allows us to define a specific location where we wish to maintain a specific elevation within our proposed site

17. **Offset Points**: Allows us to define an elevation difference between the beginning and end points of a linear object or segment

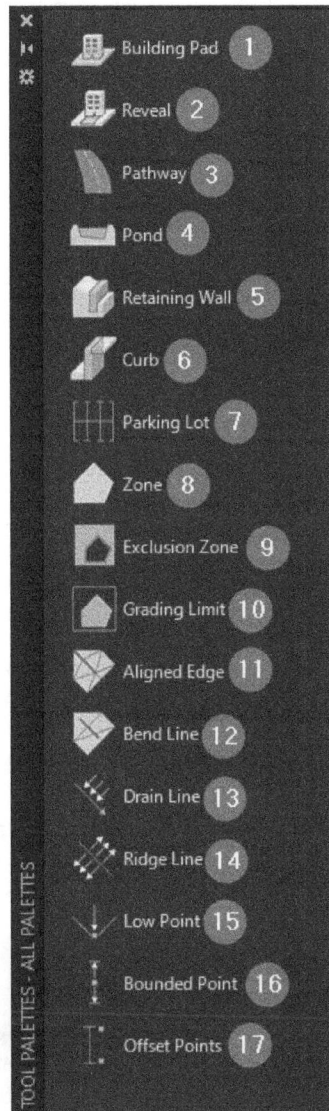

Figure 6.4 – The Grading Objects Tool palette

Moving our focus to the **Grading Objects** browser, we'll can see that we have the same categorizations that we have available to us within the **Grading Objects Tool** palette, with the exception of the **Reveal** grading object, which will be classified and associated with the **Building Pad** grading object (refer to *Figure 6.5*).

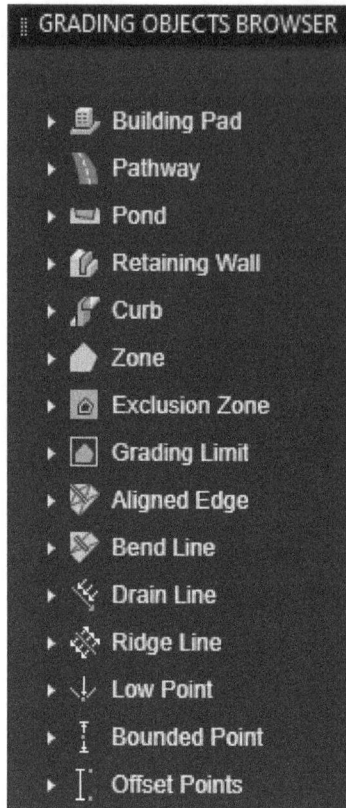

Figure 6.5 – The Grading Objects browser

Now that we have a decent idea of what potential the Grading Optimization tool has as we design our grading models, let's jump into how we can utilize this great tool to our benefit as we design our Residential Subdivision design.

Defining grading criteria

With our Grading Model_Start.dwg file still open, we can begin assigning different grading objects to our 2D polylines depicting our overall proposed parcel, composed of two existing parcels, along with our building pads for each individual lot within our Subdivision design and our Cogo Point, which we'll identify as a low point to drain to in our surface.

Jumping back to our **Grading Objects Tool** palette, let's identify our Grading Limit Object first. To do so, we'll simply click on the **Grading Limit** option within the **Grading Objects Tool** palette, and then select our overall site parcel line when prompted at the command line to select objects set as grading limit. After selecting the overall 2D polyline representing our proposed site parcel line, we'll then press *Enter* on our keyboard to accept that as our Grading Limit Object.

If we shift focus over to the **GRADING OBJECTS BROWSER** window, we can see that we now have a Grading Limit Object listed within as we click on the *arrow* icon to expand that particular category. If we select that particular object within the **GRADING OBJECTS BROWSER** window now, we can see that the object is selected in Model Space, and a **Grading Limit Properties** dialog box appears, where we can further define additional parameters and controls associated with this particular object, as shown in *Figure 6.6*.

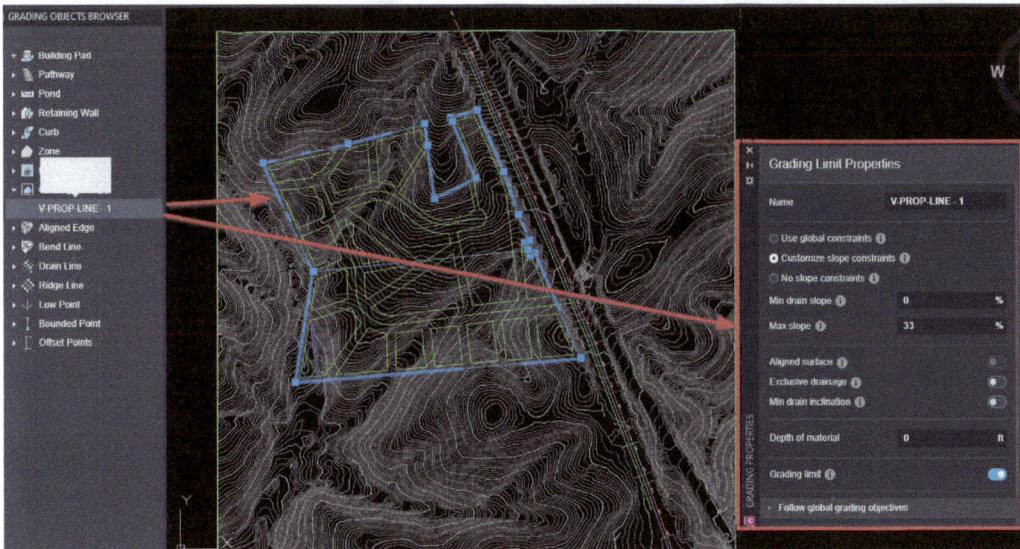

Figure 6.6 – The Grading Limit Properties dialog box

Furthermore, in the event we made a mistake and accidentally assigned an object in our file to the wrong grading object, we have the ability to correct this. To do so, we select the grading object within the **GRADING OBJECTS BROWSER** window, as we just did to access the **Grading Limit Properties** dialog box. Once selected, we can then right-click with our mouse in Model Space and select the **Remove Grading Limit** option, as shown in *Figure 6.7*.

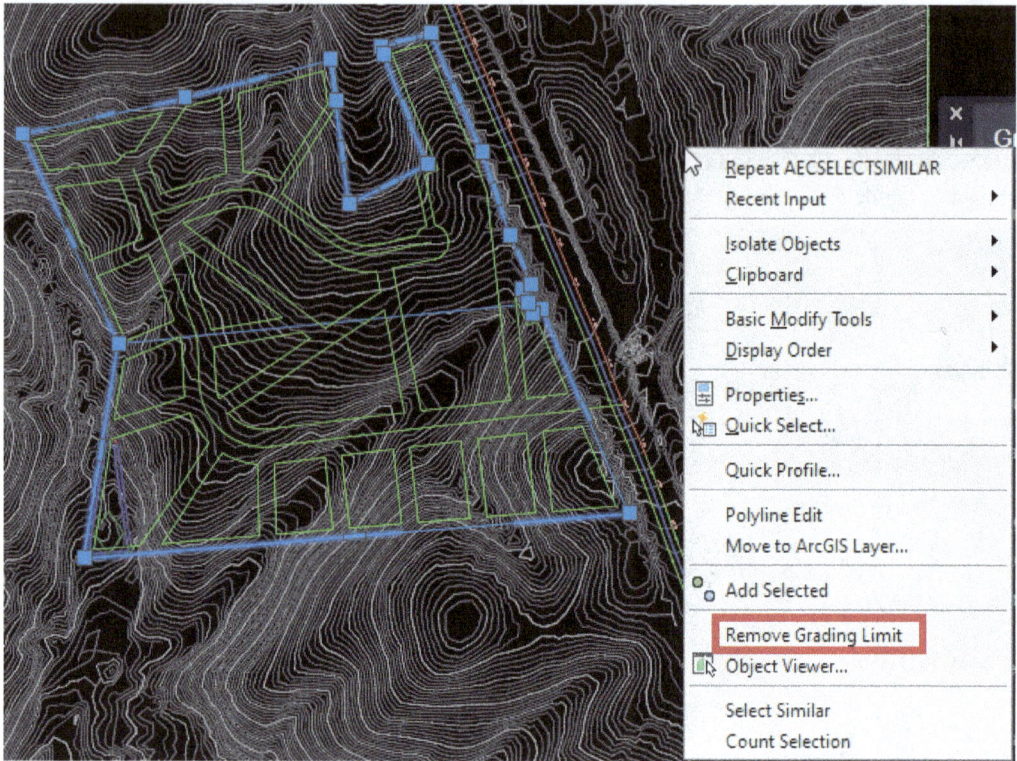

Figure 6.7 – Remove Grading Limit

With that, let's go ahead and assign the remaining grading objects to the additional polylines and Cogo Point objects mentioned earlier. This time, instead of selecting our grading object in the **Grading Object Tool** palette first, we will go ahead and select all of the individual lot lines representing green polylines by going to **Layer | C-TOPO-FEAT**, as shown in *Figure 6.8*.

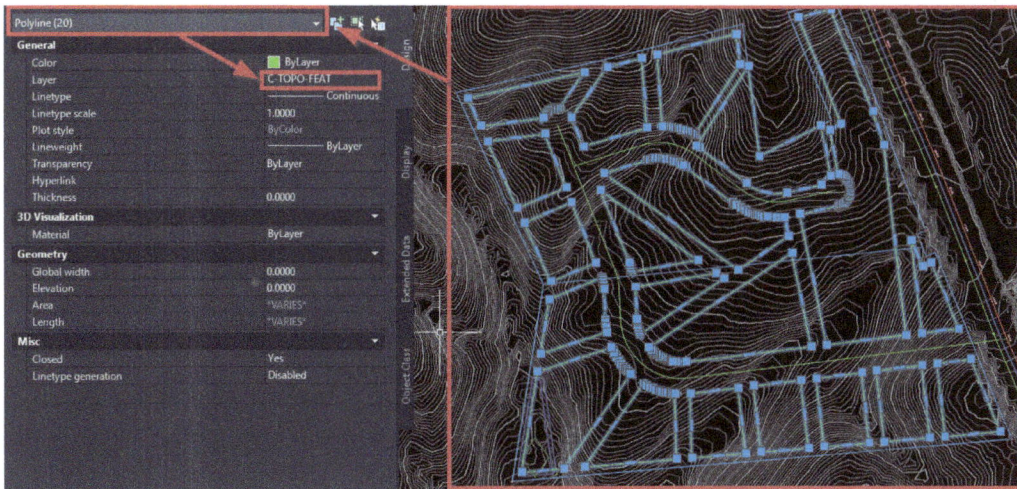

Figure 6.8 – Selecting all the individual parcel polylines

After all have been selected, we can now go back to our **Grading Object Tool** palette and select the **Building Pad** grading object. Once selected, we can see that we now have all of the grading objects listed within our **Building Pad** category within the **GRADING OBJECTS BROWSER** window. Also, the **GRADING PROPERTIES** dialog has automatically appeared as well (refer to *Figure 6.9*). That said, if you have multiple objects to select and categorize as a particular grading object, I'd recommend selecting them, possibly by using the **Select Similar** command in the right-click menu, prior to assigning to speed things up a bit.

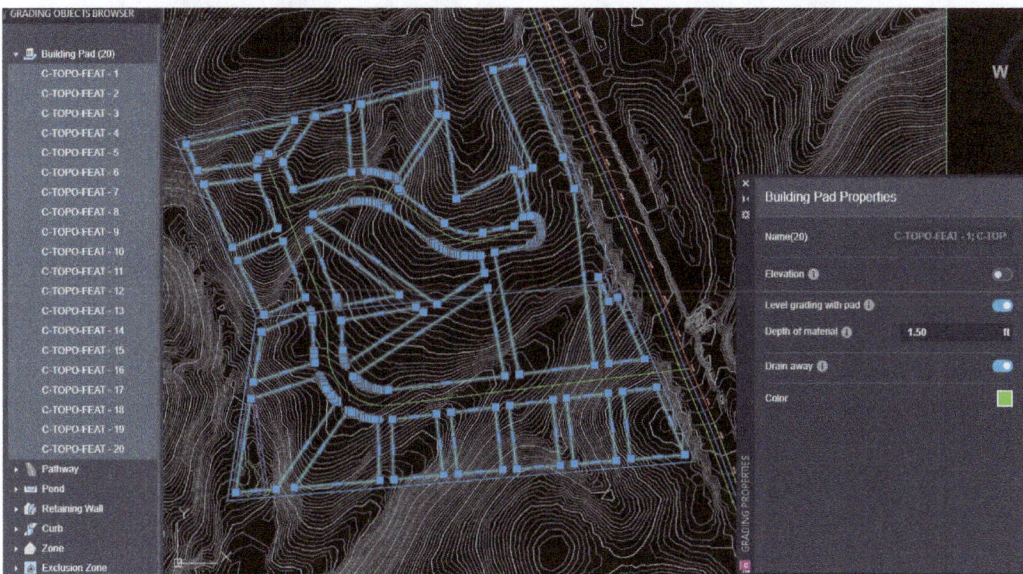

Figure 6.9 – Assigning all Building Pad grading objects

Finally, let's go ahead and select our Cogo Point, located in the southwestern corner of our proposed site, and classify it as a **Low Point** grading object by selecting this option in the **Grading Object Tool** palette, as shown in *Figure 6.10*.

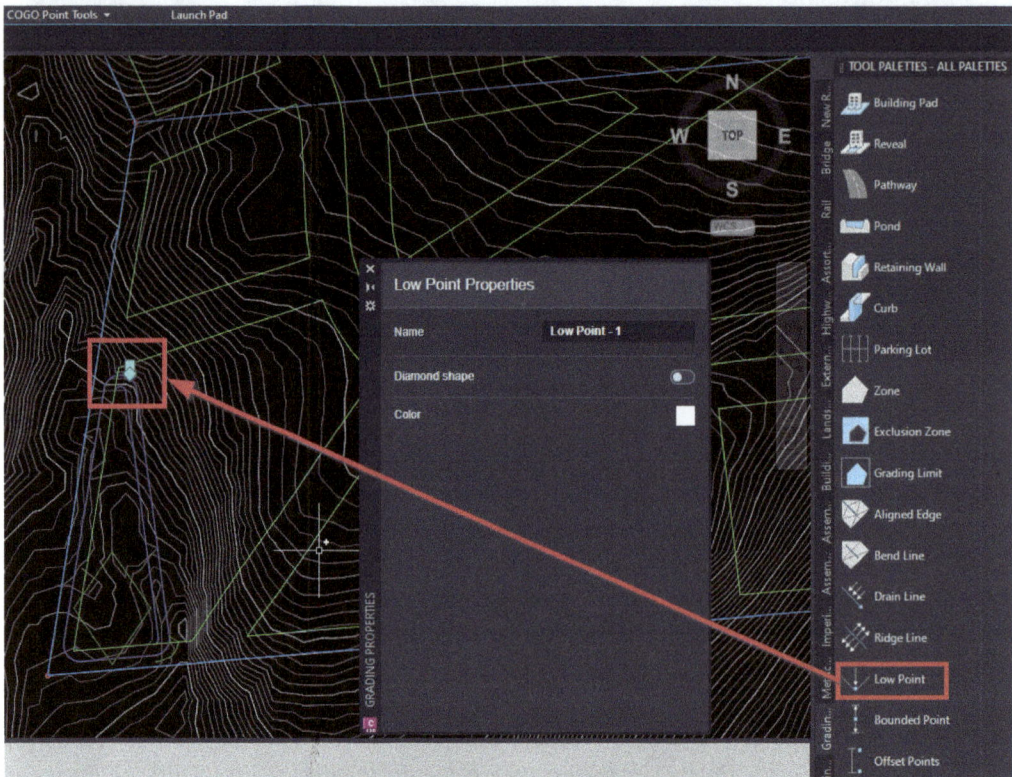

Figure 6.10 – Assigning a Low Point grading object

At this point, we have simply assigned several grading objects that we can further define and apply additional parameters to, using our **GRADING PROPERTIES** dialog box. This process will certainly come in handy as we want to further refine our Grading Optimization results later on, but there's no need to do this now.

You also may be thinking, how this is going to work without defining any of these additional parameters or applying any specific elevation data to the polylines and Cogo Point we've selected? This is the beauty of the Grading Optimization tool. We don't have to think too much from the get-go, specifying some very basic geometry to achieve some really good grading concepts and a more thorough understanding of what design options are available to us.

That said, let's go ahead and jump into our full Grading Optimization tool, take a peek at what's under the hood, and see how well this advanced tool can help us out quickly with relative ease.

Optimizing our grading models

Returning to the **Analyze** ribbon and **Grading Optimization** panel, let's go ahead and select the **Optimize** icon to launch the full Grading Optimization tool. In our current situation, we only have one surface model in our file, **SRF - Existing Grade – FromSurveyPoints**. If we had multiple surfaces in our file, we would be prompted to select a surface in our file first that we wish to optimize or integrate our design conditions into. Also note that we can move the Grading Optimization tool anywhere on our screen and still be able to interact directly with our file.

With our Grading Optimization tool launched, let's quickly familiarize ourselves with the different panels and tools we have available to us. Right off the bat, we have the following groupings available to us (refer to *Figure 6.11*):

1. **Object Browser**: Provides quick access to details of grading objects already assigned within our file. As we expand each category, we can select individual items and apply additional values and parameters directly within this interface. We can also zoom into each of our grading objects in our model display (**2**) by right-clicking on each grading object and selecting **Zoom To**. We can also activate or deactivate grading objects in our optimization results.

2. **Model display**: Provides a thematic view of our modeled conditions.

3. **Help & Learning**: Provides quick access to online help articles and tutorials.

4. **Optimization toolbar**: Provides quick access to apply various options that we'd like to apply to our optimization results and displays. We can also run our Optimization tool here, view progress, and push our results back into our current file.

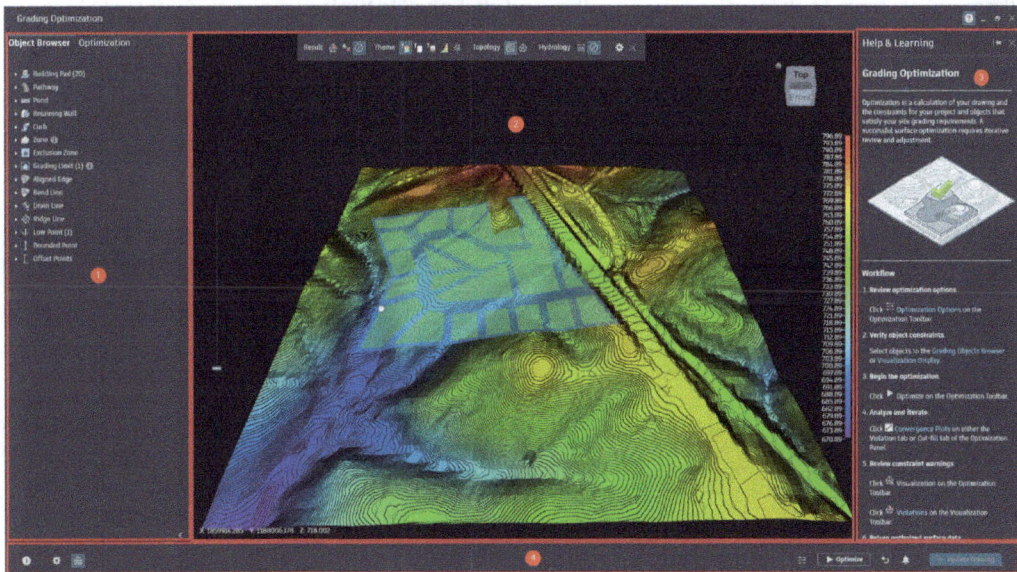

Figure 6.11 – Assigning a Low Point grading object

For our first pass, let's simplify our Grading Optimization results a bit just so that we can get a feel for what we can do. That said, go ahead and jump over to the **Object Browser** window and deactivate and turn off the display of the **Building Pad** objects (as shown in *Figure 6.12*).

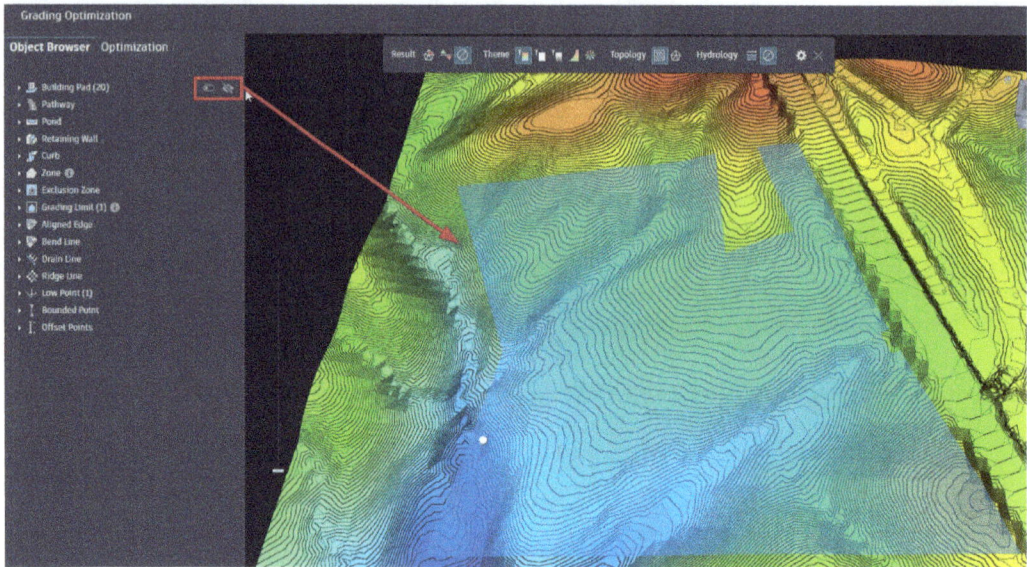

Figure 6.12 – Deactivating and turning off grading objects

With only our Grading Limits and Low Point grading objects applied in our current design scenario, we'll now jump down to the Optimization toolbar and click the **Optimize** icon to run our tool. As the Optimization tool is running, you'll notice the display of contours within the grading limits continues to change. The tool itself runs through a generative process that will take all the parameters that we've applied into consideration and ultimately provide us with optimal surface conditions. The longer we let this tool run, the more precise it will be.

When the tool has considered everything, a notification will appear in the lower right-hand corner, letting us know that the tool is just about complete and has found optimal results (as shown in *Figure 6.13*).

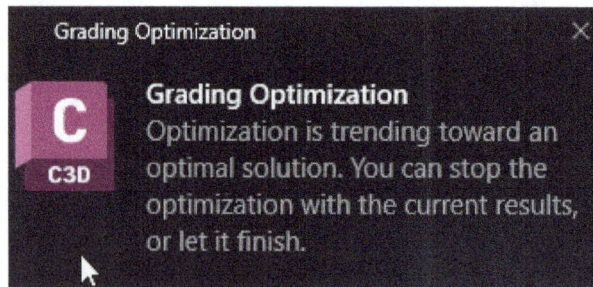

Figure 6.13 – A Grading Optimization notification

When this notification appears, we can either let it run to completion (which could take a considerable amount of time, depending on the complexities of the surface and parameters applied), or we can simply stop it when we're satisfied with the results displayed. To end the optimization process, in the same location where we selected **Optimize** within our optimize toolbar, we can now select **Stop**, with the final results in the model display area looking similar to that shown in *Figure 6.14*.

Figure 6.14 – The Grading Optimization results from our first pass

We also can see a few different options on the right-hand side of our Grading Optimization interface in place of where we originally viewed our **Object Browser** window. As shown in *Figure 6.15*, we can apply visual violations that are color-coded and can be seen within the model display space as the Optimization tool is running. We can also watch the optimization **Convergence** graph as the Optimization tool is running, enabling us to quickly see how likely we are to find an optimal grading solution with the given parameters applied.

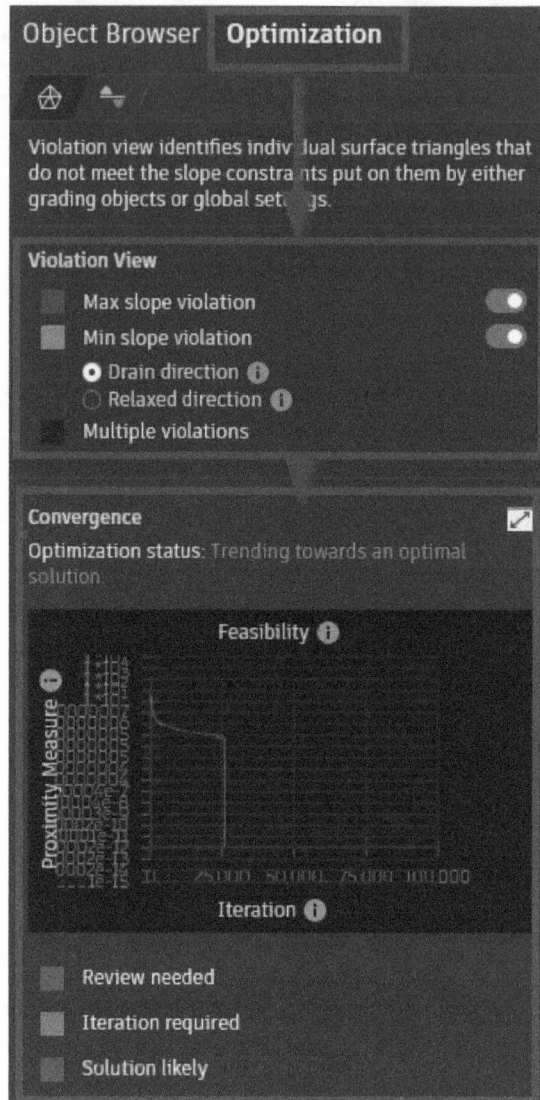

Figure 6.15 – The Optimization tab

Now that we have a decent understanding of what we can expect and areas that we should focus on as our Optimization tool is running, let's go ahead and activate our **Building Pad** grading objects, change our model display back to the *Elevation* theme, and run the Optimization tool again until we are satisfied with the results (the steps are shown in *Figure 6.16*).

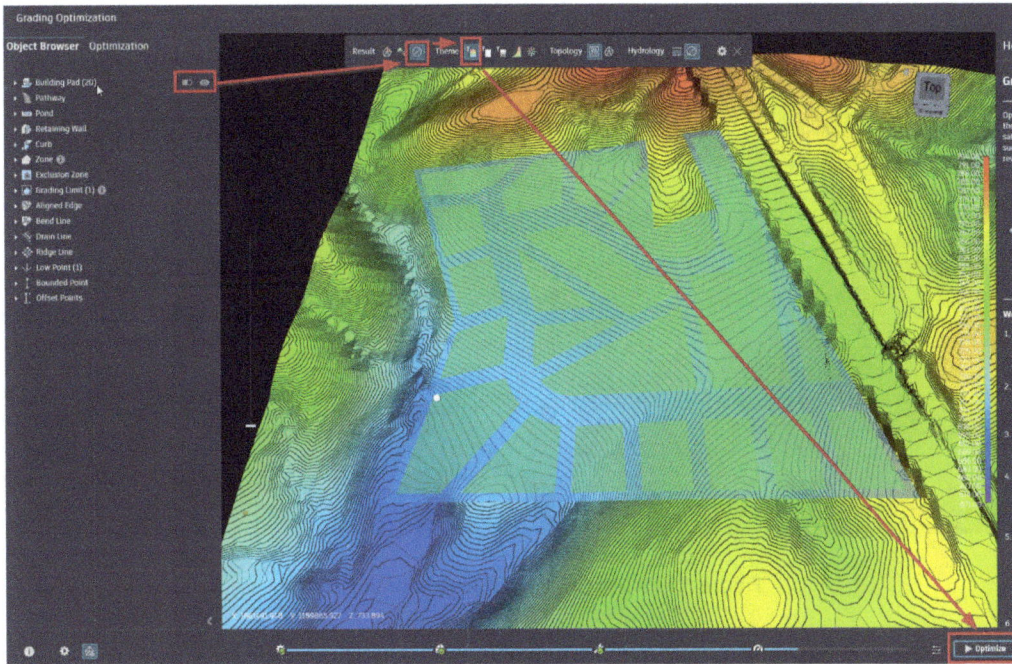

Figure 6.16 – The steps to run Grading Optimization with all the grading objects and parameters applied

Note that this optimization may take a little while to start and a lot longer to finish, as there are way more parameters and criteria to consider when attempting to identify an optimal grading solution. Also, when we first initiate the **Optimize** tool, the majority of our site will be displayed as maroon, indicating that there are multiple violations located in our site. The longer we let the tool run though, the more these areas will mostly correct themselves to provide us with a cleaner result, based on the parameters and criteria we previously defined.

It's worth noting that there will be a few small areas that will continue to display as maroon, which lets us know that there are some violations still located within our Residential Subdivision design. We can clean these areas up after we transfer results back into our current file. After letting the **Optimize** tool run for a bit, we should end up with a result that looks similar to *Figure 6.17*.

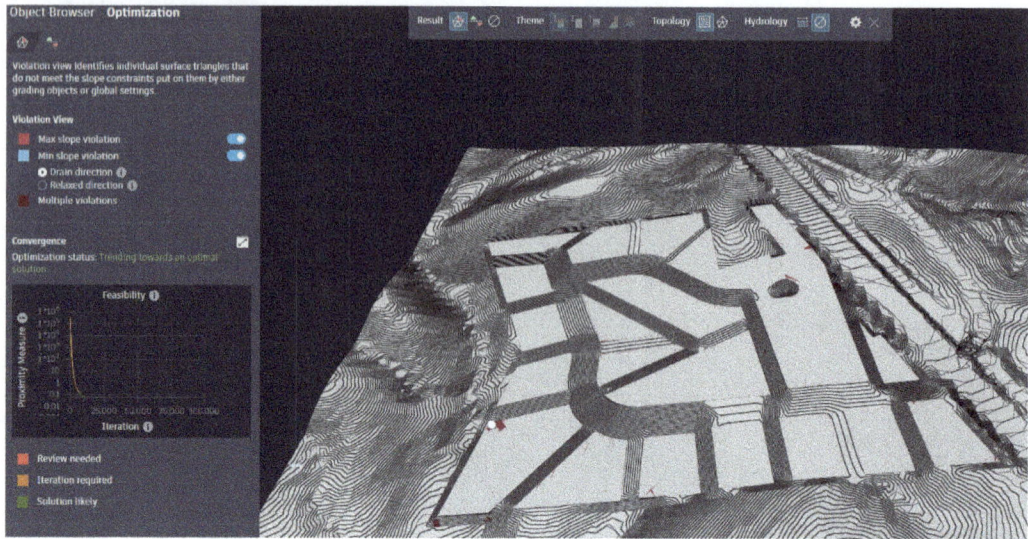

Figure 6.17 – The Grading Optimization results

Occasionally, the site can struggle to visualize changes in grades, elevations and slopes, especially since there aren't a lot of interrogation tools available directly within Grading Optimization.

That said, I'd recommend changing the display back to *elevations* (or any other display that you prefer), using the toolbar at the top of the model display area, and then adjust the *vertical distortion slider* on the left-hand side so that we can see a little easier how the Residential Subdivision site will drain from the northeastern corner to the southwestern corner, as shown in *Figure 6.18*.

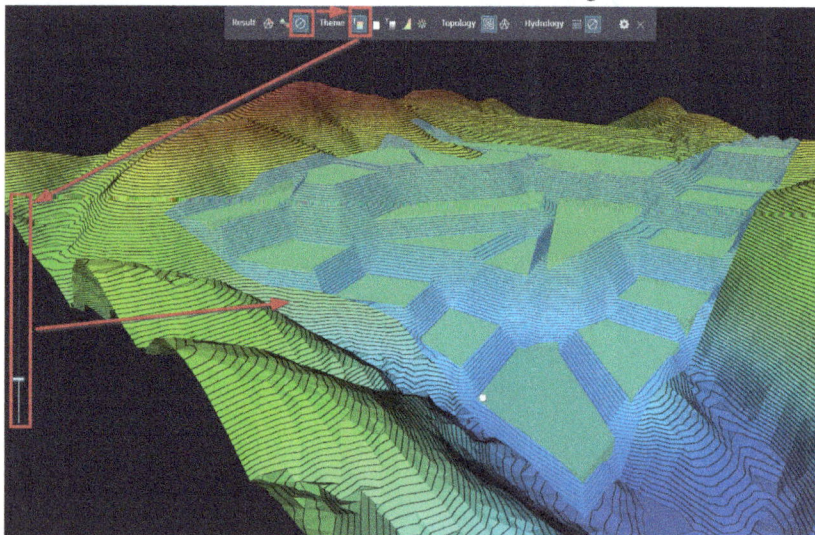

Figure 6.18 – Changing the vertical distortion of the grading models

Once we're satisfied with the results, we can finally go ahead and select the *Update Drawing* icon in the lower right-hand corner of our Grading Optimization tool to add to our current file. Once selected, a **Save Optimization Result** dialog box will appear where we'll fill out the fields and make the following selections (also displayed in *Figure 6.19*) and then click the **Finish** button at the bottom of the dialog box:

- **Create new surface**: Select this option
- **Name**: SRF - Proposed Grade - From Grading Optimization
- **Style**: Contours 1' and 5' (Design)

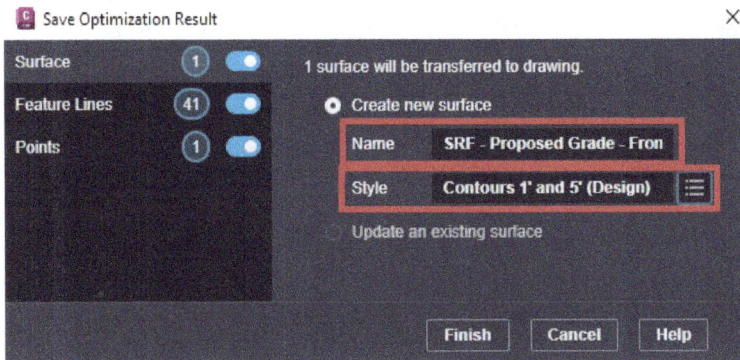

Figure 6.19 – The Save Optimization Result dialog box

With our new grading model now pushed back into our file, we can interrogate and refine it next. If we select our newly created surface, **SRF – Proposed Grade – From Grading Optimization**, now, right-click with our mouse, and select **Object Viewer**, we should see results similar to those displayed in *Figure 6.20*.

Figure 6.20 – The Object Viewer display of our Grading Optimization results

As mentioned early on in this chapter, the results that come out of the Grading Optimization are a great way of running through some quick alternative grading designs to understand the overall feasibility, based on the various parameters and criteria applied. In the majority of cases, we will want to refine our results to get our grading models to remain at a true constructability level of detail.

Summary

As we worked through this chapter, we shifted focus on streamlining conceptual designs and generating alternative grading designs with the Grading Optimization toolset, available in Autodesk Civil 3D 2025. This tool enhances design efficiencies and saves a lot of time at the conceptual design phase, by automating land grading rapidly. We can also quickly see real-time visualizations of our results as it considers the various user-defined constraints that are applied. As we walked through, we can simply start with some very basic AutoCAD geometry and Cogo Points.

In the next chapter, we will switch focus from grading to utility design tools. Here, we'll begin exploring how best we can utilize both Civil 3D and the Content Catalog Editor, generating custom utility fittings that we can apply directly to our designs. Civil 3D comes with a lot of great foundational utility components in its out-of-the-box part catalogs, but it isn't always a one-size-fits-all solution. Occasionally, we'll need to use some advanced methods to generate custom solutions, which we'll begin understanding in the next chapter.

7

Exploring Content Catalog Editor

In the previous chapter, we explored new ways of developing grading models that make our work more efficient as we tackle various site constraints in the early conceptual design phase. We used a special tool called Grading Optimization in **Autodesk Civil 3D 2025** to rapidly generate our designs and run through multiple design scenarios as required. It also allowed us to see our designs changing in real time and adjust them based on the different rules and parameters that we set. The tool proved it can work with simple AutoCAD shapes and COGO Points, showing how adaptable and versatile it is for creating grading models.

As we move to this chapter, our continued exploration of simple and effective designs leads us to the forefront of Autodesk Civil 3D innovation. The Grading Optimization toolset, more than just a regular feature, is a powerful force that changes how we think about technology and creativity. Now, in this chapter, we shift our focus to **utility modeling**, using a tool called **Content Catalog Editor**. This tool lets us create custom solutions, going beyond the standard Civil 3D utility parts. Here, we'll learn about the capabilities of Content Catalog Editor, giving design teams the freedom to make unique solutions that take our utility modeling to new levels. That said, in this chapter, we'll cover the following topics:

- Creating custom pressure parts in Civil 3D
- Defining parts in Content Catalog Editor
- Integrating custom parts into our BIM designs

Technical requirements

We will be using the same hardware and software requirements as discussed in the *Technical requirements* section of *Chapter 1*.

With that, let's go ahead and launch Civil 3D 2025 and start a new drawing using our `Company Template File.dwt` file located within the `Civil 3D 2025 Unleashed\Chapter 7` location. Once opened, we'll want to start by navigating to our local `C:\ProgramData\Autodesk\ C3D 2024\enu\Pressure Pipes Catalog\Imperial` location using Windows Explorer and create a new custom catalog subfolder; in our case, we'll call it `Imperial_Custom_Pressure_ Parts`. Once created, we'll also want to add a `DWG` subfolder and an `IMG` subfolder within the `Imperial_Custom_Pressure_Parts` folder to ensure that Civil 3D and Content Catalog Editor recognize the supporting custom content that we'll be creating.

> **Important note**
>
> If you intend to have multiple design team members working on a utility design using the custom pressure network catalog, the entire directory will either need to be copied over to each additional user's local at the same location or it can be stored on a shared network location that all users can access, eliminating the need to replicate for each user that needs access. Also, it is a lot easier to manage in the end.

After all folders and subfolders have been created, go ahead and perform a `SAVEAS` function with our current file, navigate to the newly created `DWG` subfolder, and give it a descriptive name; in our case, we'll call it `12x8 Reducing Wye.dwg`.

Creating custom pressure parts in Civil 3D

With our `12x8 Reducing Wye.dwg` file created and saved, let's go ahead and start creating the bases of our custom part. If not done so already, it's recommended that we first establish the **level of detail** (**LOD**) needed for our part based on design requirements along with familiarizing ourselves with the part specifications from a manufacturer's website for the part that will be installed.

In the case of this demonstration, we're going to refer to the U.S. Pipe fitting catalog to pull dimensions for our part from: `USP_Flanged_Fittings_C110_Submittal_digital.pdf` (uspipe. com). As we navigate this specification document, we'll see a table that details dimensions specific to `12x8 Reducing Wye`, as shown in *Figure 7.1*.

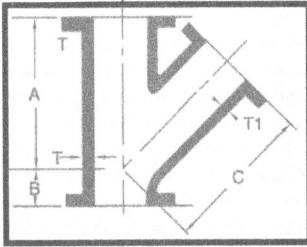

45° REDUCING LATERAL WYES

SIZE	Wt.	A	B	C	T	T₁		SIZE	Wt.	A	B	C	T	T₁
4 x 3	65	12.0	3.0	12.0	0.52	0.48		18 x 16	935	32.0	7.0	32.0	0.75	0.70
6 x 4	105	14.5	3.5	14.5	0.55	0.52		20 x 8	995	35.0	8.0	35.0	0.80	0.60
8 x 4	165	17.5	4.5	17.5	0.60	0.52		20 x 10	1025	35.0	8.0	35.0	0.80	0.68
8 x 6	175	17.5	4.5	17.5	0.60	0.55		20 x 12	1065	35.0	8.0	35.0	0.80	0.75
10 x 4	235	20.5	5.0	20.5	0.68	0.52		20 x 14	1110	35.0	8.0	35.0	0.80	0.66
10 x 6	250	20.5	5.0	20.5	0.68	0.55		20 x 16	1155	35.0	8.0	35.0	0.80	0.70
10 x 8	270	20.5	5.0	20.5	0.68	0.60		20 x 18	1315	35.0	8.0	35.0	0.80	0.75
12 x 4	350	24.5	5.5	24.5	0.75	0.52		24 x 8	1470	40.5	9.0	40.5	0.89	0.60
12 x 6	365	24.5	5.5	24.5	0.75	0.55		24 x 10	1505	40.5	9.0	40.5	0.89	0.68
12 x 8	390	24.5	5.5	24.5	0.75	0.60		24 x 12	1550	40.5	9.0	40.5	0.89	0.75
12 x 10	470	24.5	5.5	24.5	0.75	0.68		24 x 14	1590	40.5	9.0	40.5	0.89	0.66
14 x 6	475	27.0	6.0	27.0	0.66	0.55		24 x 16	1640	40.5	9.0	40.5	0.89	0.70
14 x 8	500	27.0	6.0	27.0	0.66	0.60		24 x 18	1685	40.5	9.0	40.5	0.89	0.75
14 x 10	525	27.0	6.0	27.0	0.66	0.68		24 x 20	1750	40.5	9.0	40.5	0.89	0.80
14 x 12	570	27.0	6.0	27.0	0.66	0.75		30 x 12	2795	49.0	10.0	49.0	1.03	0.75
16 x 6	620	30.0	6.5	30.0	0.70	0.55		30 x 14	2850	49.0	10.0	49.0	1.03	0.66
16 x 8	645	30.0	6.5	30.0	0.70	0.60		30 x 16	2905	49.0	10.0	49.0	1.03	0.70
16 x 10	675	30.0	6.5	30.0	0.70	0.68		30 x 18	2960	49.0	10.0	49.0	1.03	0.75
16 x 12	715	30.0	6.5	30.0	0.70	0.75		30 x 20	3040	49.0	10.0	49.0	1.03	0.80
16 x 14	755	30.0	6.5	30.0	0.70	0.66		30 x 24	3205	49.0	10.0	49.0	1.03	0.89
18 x 8	780	32.0	7.0	32.0	0.75	0.60		36 x 16	4455	54.0	15.3	54.0	1.15	0.70
18 x 10	810	32.0	7.0	32.0	0.75	0.68		36 x 18	4505	54.0	15.3	54.0	1.15	0.75
18 x 12	850	32.0	7.0	32.0	0.75	0.75		36 x 20	4575	54.0	15.3	54.0	1.15	0.80
18 x 14	885	32.0	7.0	32.0	0.75	0.66		36 x 24	4725	54.0	15.3	54.0	1.15	0.89

Dimensions in inches. Weights in pounds.

Figure 7.1 – U.S. Pipe specifications

As we build out our parts, we're going to use a combination of basic AutoCAD 2D and 3D modeling tools to develop our model for eventual incorporation into the custom pressure network catalog.

Creating the 2D base components

Using the following steps, let's go ahead and create the base 2D components:

1. Draw a line horizontally (at an angle of 0 degrees) that is 30 inches long to represent our main centerline.

2. Draw another line at an angle of 45 degrees that is 24 ½ inches long to represent our branch centerline (reducing pipe connection).

3. Place the 24 ½ inch branch centerline on the 30-inch main centerline 5 ½ inches in, with the end result looking similar to that shown in *Figure 7.2*.

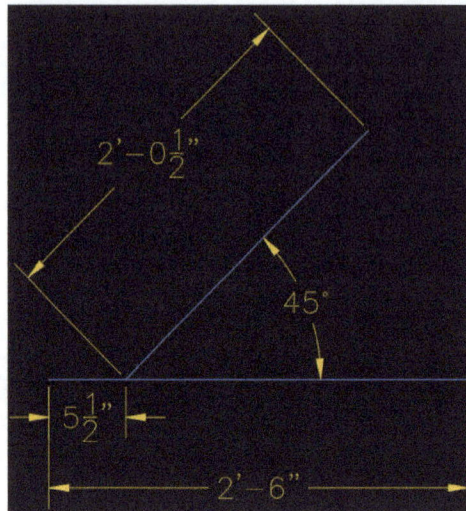

Figure 7.2 – Centerline representation of custom fitting (with dimensions)

With our 2 Centerlines created as the base of our `12x8 Reducing Wye` fitting, we'll want to make a duplicate copy of this linework, where the first version will represent our single line fitting while the second version will represent our 3D modeled part.

When building our pressure network, we will be required to identify not only the full 3D custom part that we build, but also a 2D Block representation of our part. That said, let's go ahead and create a standard AutoCAD Block of our Centerlines, with the basepoint at the intersection of the main and branch lines, giving it a Block name of `12x8 Reducing Wye_2D`, with the end result appearing similar to that shown in *Figure 7.3*.

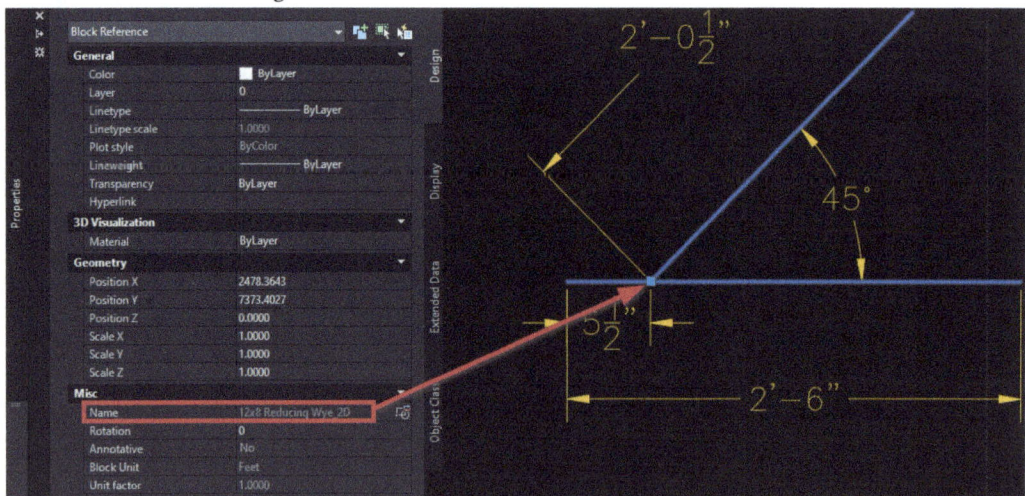

Figure 7.3 – Block depicting centerlines of custom fitting

After our Block has been created, we'll next move over to creating our flanged connections and our 3D modeled part as follows:

1. Using the flange thickness value (found on a separate page from the U.S. Pipe fitting catalog: USP_Flanged_Fittings_C110_Submittal_digital.pdf (uspipe.com)), let's go ahead and create 3 new lines, with 2 of them being 1.25 inches long to represent the flange thickness for our 12-inch main and the last line at a length of 1.12 inches long to represent the flange thickness for our 8-inch branch.

2. Next, we'll initiate the CIRCLE command to create three circles to represent the outer diameter of our flanged pieces, the first two being 19 inches and the last being 13.5 inches to represent the flanged pieces at each end of 12x8 Reducing Wye.

3. We'll also go ahead and create four more circles to represent the inner and outer diameter of our fitting. That said, let's go ahead and create our 12-inch and 8-inch diameter circles to represent the inner diameter of our pipes and then use the OFFSET command to apply the thickness of our pipes according to the table, which are 0.75 and 0.60 respectively, leaving us with a setup similar to that shown in *Figure 7.4*.

PIPE 2D COMPONENTS   FLANGE 2D COMPONENTS

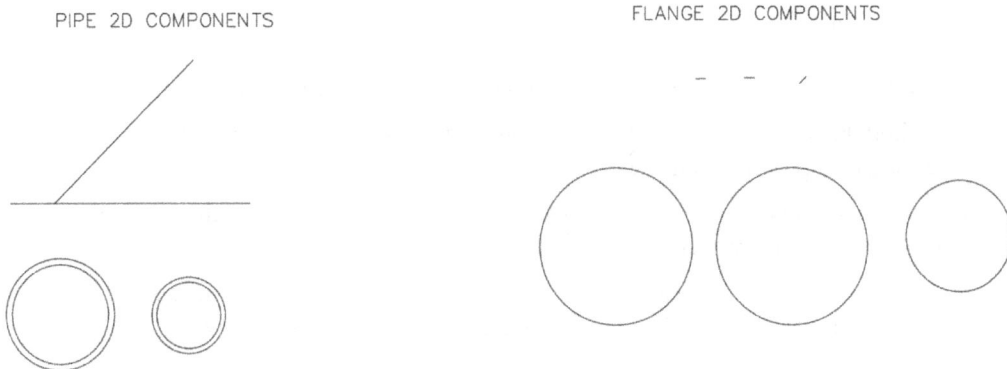

Figure 7.4 – 2D components that will be used to create the 3D model of our custom fitting

With all 2D components created, let's go ahead and start using these to create the 3D model of our custom 12x8 Reducing Wye fitting.

Creating the 3D model

This part of the 3D model creation process can be easier when viewed in an Isometric View. Using ViewCube in model space, let's change our view to the southwest Isometric View, as shown in *Figure 7.5*.

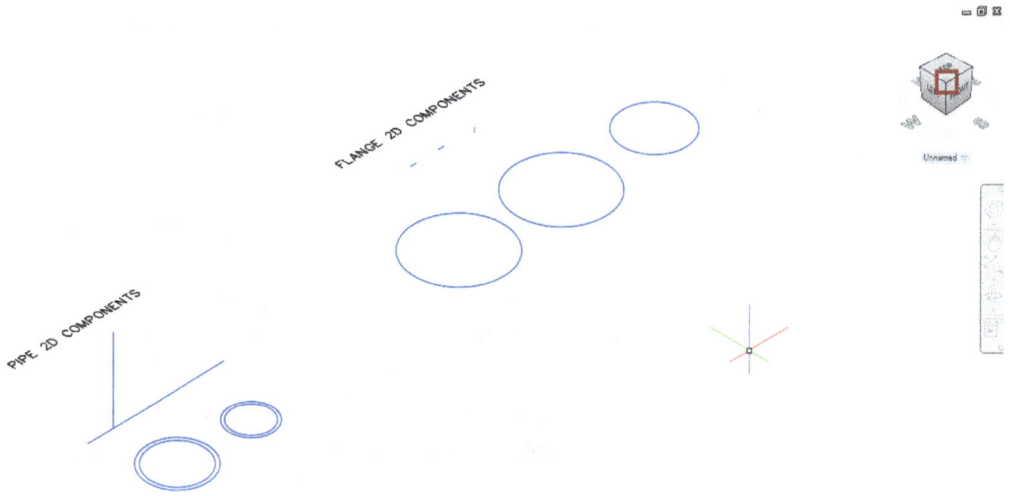

Figure 7.5 – Southwest Isometric View

Now, let's use the following steps to create the 3D model of our 12x8 Reducing Wye fitting:

1. Starting with the pipe 2D components area, go ahead and initiate the AutoCAD SWEEP command.

2. Select both larger diameter circles when prompted at the command line to select objects to sweep and then press the *Enter* key to accept.

3. Select the main centerline when prompted at the command line to select the sweep path and then press the *Enter* key to accept.

4. With our main pipeline modeled, let's go ahead and apply the SWEEP command again on the branch pipeline, with the result appearing similar to that displayed in *Figure 7.6*.

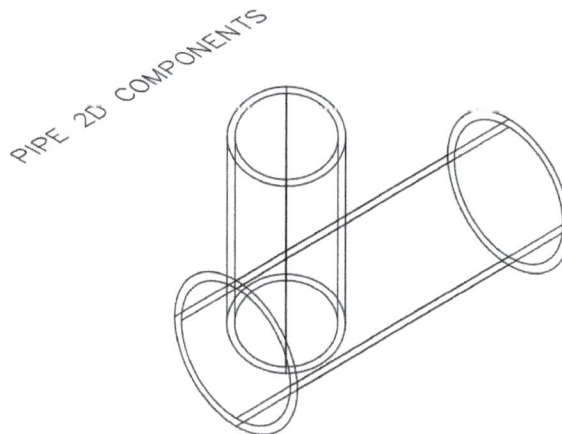

Figure 7.6 – Result of pipe sweeps along centerlines

5.   Continuing, let's apply the SWEEP command to each of the flanged pieces as we just did to our pipe 2D components area, with the result appearing similar to that displayed in *Figure 7.7*.

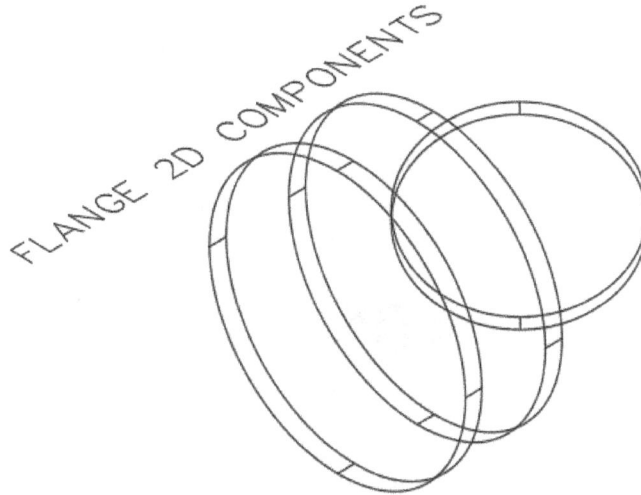

Figure 7.7 – Result of flange sweeps along centerlines

6.   Using the AutoCAD MOVE command, let's now move the 3D flange cylinders over to each of the ends of our 3D pipe cylinders, with the result looking similar to that displayed in *Figure 7.8*.

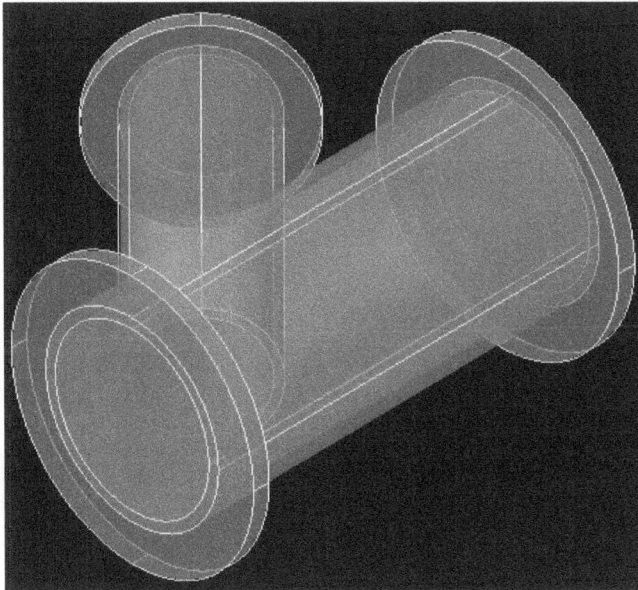

Figure 7.8 – Result after copying flanges to pipes

7. With all individual 3D pieces in place, let's go ahead and initiate the AutoCAD SUBTRACT command to create one complete 3D model representation of our `12x8 Reducing Wye` fitting. Let's follow these steps to walk through this process:

   • When prompted to select objects at the command line after initiating the SUBTRACT command, select the three flanges as well as the outer diameter main and branch pipes, and click *Enter*.

   • When prompted to subtract certain objects at the command line, we'll select the inner diameter pipes only (may need to use the 3D orbit tool below the ViewCube to select the inner diameter pipe for the branch) and press *Enter*, with the final result looking similar to that displayed in *Figure 7.9*.

Figure 7.9 – Final 3D representation of custom 12x8 Reducing Wye fitting

With both 2D (Block) and 3D (Solid) representations created, let's go ahead and align them such that the centerlines of the pipes from the 2D (Block) line up exactly with the 3D model, as displayed in *Figure 7.10*.

Figure 7.10 – 2D (Block) and 3D depictions of our custom part combined

With the 2D and 3D depictions of our custom `12x8 Reducing Wye` fitting created, we'll now need to define it as a part and publish it so that it can be recognized in the Content Catalog Editor.

Defining and publishing parts

With that, let's use the following steps to complete this workflow:

1. Activate the **Insert Ribbon** option.

2. Navigate over to the **Connection Point** panel.

3. Select the **Insert Connection Point** tool (refer to *Figure 7.11*).

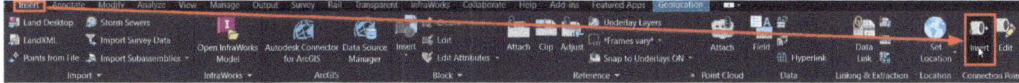

Figure 7.11 – Accessing the Insert Connection Point command

4. Select the 3D model of our part when prompted at the command line to pick an object to add the connection point to.

5. When prompted to pick the insertion point on the object at the command line, select one of the endpoints (either along the main or branch lines) to establish the locations where pipes will eventually connect to our custom 12x8 Reducing Wye fitting.

6. After selecting one of the connection points, we'll next be prompted to pick the first point followed by the second point for the direction of the connection. At this point, we'll want to make sure that we select the opposite side of the part and then reselect the same insertion point we initially selected to establish the direction for connecting pipes to exit our new fitting. We'll do this for all three exit locations (refer to *Figure 7.12*):

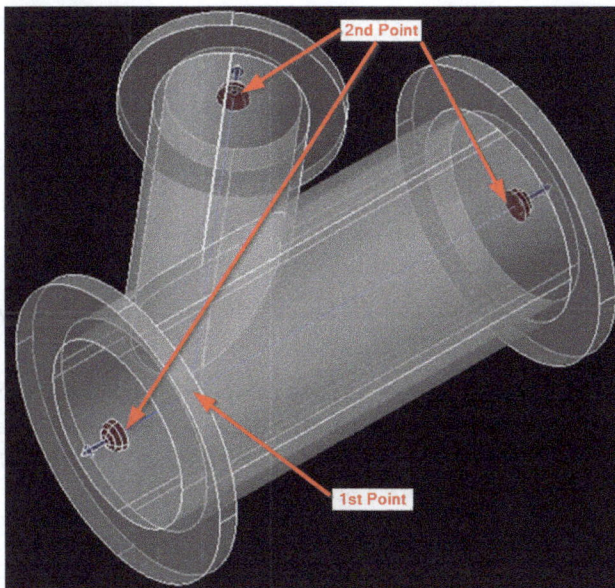

Figure 7.12 – Establishing the connection points and direction for our custom part

Figure 7.13 displays the final result:

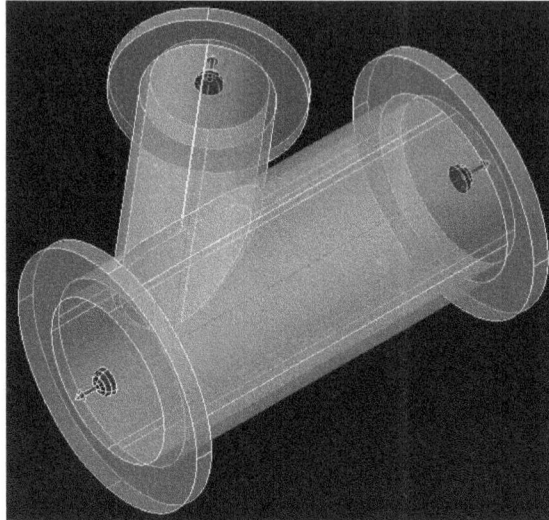

Figure 7.13 – Connection points added to our custom part

7. The next step is to make sure that our part's insertion point is set to 0,0,0. If not done so already, let's go ahead and move our part and set connection points to 0,0,0, where the main and branch lines connect (you can snap to the intersection of the 2D Block representation of our part).

8. Continuing to publishing our part, we'll type PUBLISHPARTCONTENT at the command line to initiate the publishing command.

9. When prompted to select a solid for publishing, go ahead and select our 3D model.

10. When prompted to select a centerline entity that represents the previously selected solid, go ahead and select **12x8 Reducing Wye Block**, which we created earlier on in this section.

11. When prompted to choose an option for the **Specify the Measuring Unit** section, we'll go ahead and select **Inches** for this exercise.

12. When prompted to select an option for the **Specify the Part Type** section, we'll go ahead and select **Wye** for this exercise.

13. We'll next be prompted with a dialog box asking us where we want to save our new part. Let's navigate to C:\ProgramData\Autodesk\C3D 2025\enu\Pressure Pipes Catalog\Imperial\Imperial_Custom_Pressure_Parts\DWG, give it a name of 12x8 Reducing Wye.CONTENT, and click **Save**.

We have now officially created the foundation of our custom 12x8 Reducing Wye fitting using a combination of AutoCAD, Civil 3D and 2D drafting, and 3D modeling techniques. We're now ready to take our design to the next level for eventual incorporation into our design models.

Defining parts in Content Catalog Editor

With our `12x8 Reducing Wye.dwg` and `12x8 Reducing Wye.content` files created, let's open up our Content Catalog Editor. To do so, we'll go to **Start Menu**, expand **Autodesk Civil 3D 2025 – English**, and select **Content Catalog Editor**. Once selected, **Content Catalog Editor** will appear on the screen, as displayed in *Figure 7.14*.

Figure 7.14 – Content Catalog Editor

To get started, let's go up toward the top-left corner of our **Content Catalog Editor** session and click on the new catalog icon (or go to **File | New**). Once selected, our **New Catalog File** dialog box will appear, and this is where we'll want to select **Imperial** as the **Catalog Unit** setting and then select **OK** (as shown in *Figure 7.15*).

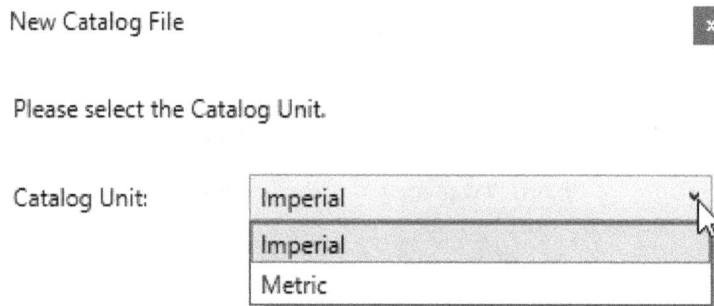

Figure 7.15 – The New Catalog File dialog box

While we're at it, let's go ahead and save our new catalog as well by either clicking on the save icon or going to **File** | **Save As**. When the **Save As** dialog box appears, we'll navigate to our `C:\ProgramData\Autodesk\C3D 2025\enu\Pressure Pipes Catalog\Imperial\` root folder, give it a name of `Imperial_Custom_Pressure_Parts.sqlite`, and click on **Save**.

We can now begin importing the custom parts that we've previously created to build our new custom pressure network catalog. To import our part, we can either click on the import part icon or go to **Edit** | **Import Part** in the top-left corner of **Content Catalog Editor**. Alternatively, you can right-click on the particular part listed along the left side of our catalog that you would like to import a part for.

Once initiated, our **Import Part** dialog box will appear, with the first section, **Catalog File**, selected. Here, we'll want to import the `12x8 Reducing Wye.CONTENT` file by clicking on the ellipses icon, navigating to `C:\ProgramData\Autodesk\C3D 2025\enu\Pressure Pipes Catalog\Imperial\Imperial_Custom_Pressure_Parts\DWG`, and selecting the corresponding file, as shown in *Figure 7.16*.

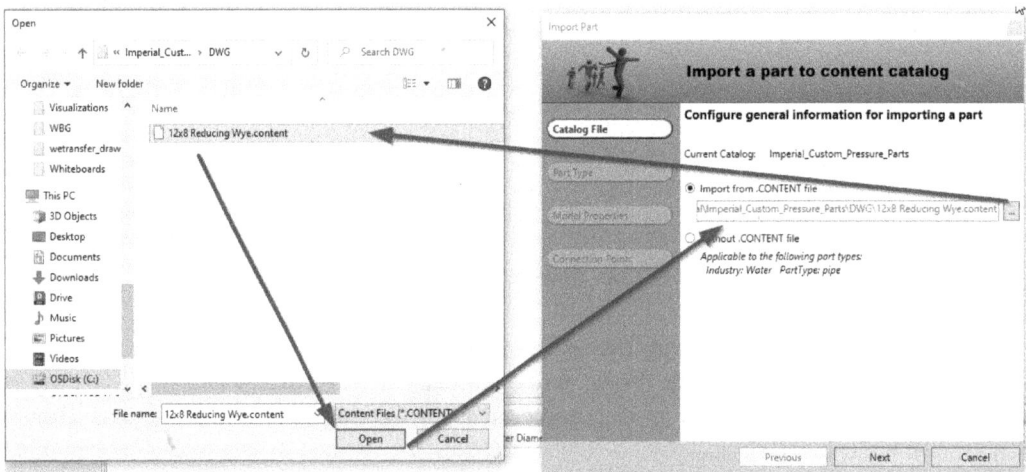

Figure 7.16 – Import a part to content catalog

When the **Import from .CONTENT file** field has been filled out, go ahead and click on the **Next** button in the lower right-hand corner of the **Import Part** dialog box.

In the **Part Type** section of the **Import Part** process, we'll want to fill out the necessary fields as follows (refer to *Figure 7.17*), and click **Next**:

1. **Industry**: **Water** (selection)
2. **Part Type**: **wye** (selection)
3. **Part Family Name**: `Reducing Wye` (manual entry)

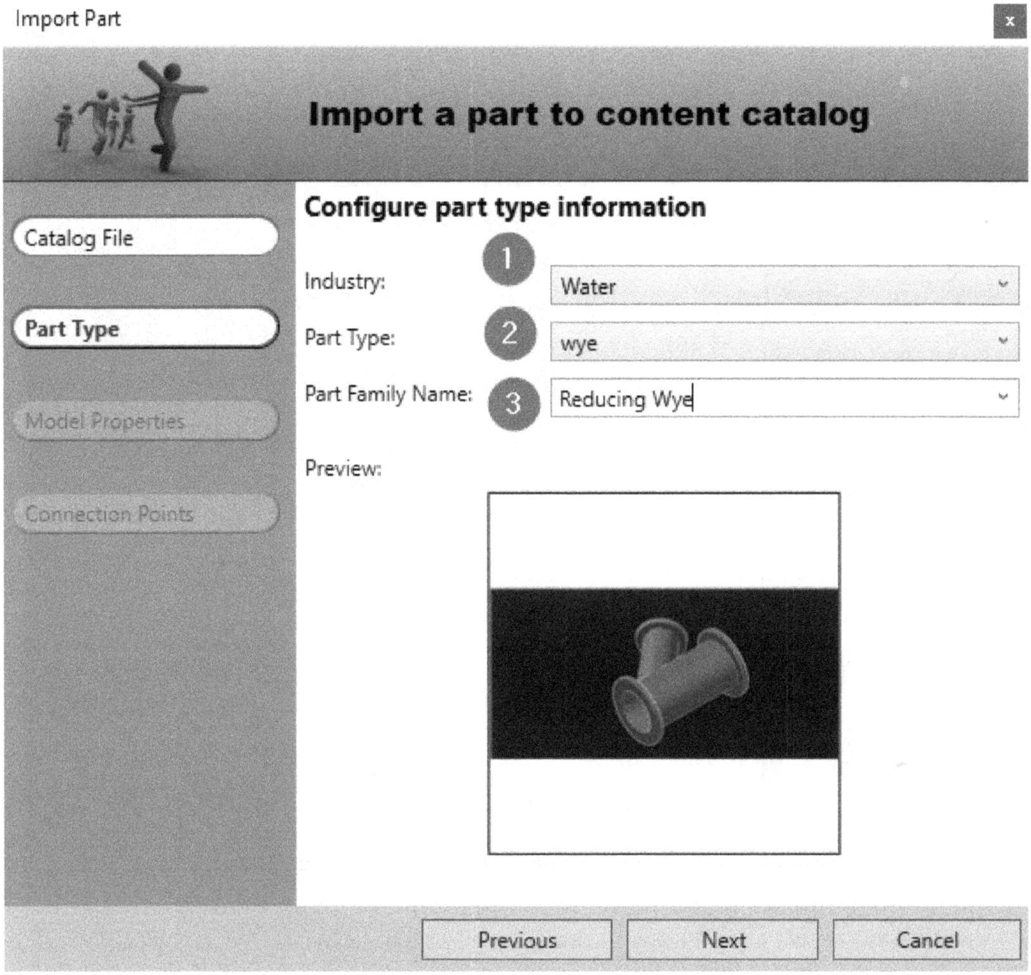

Import Part

Figure 7.17 – Import Part | Part Type

In the **Model Properties** section of the **Import Part** process, we'll want to fill out the necessary fields as follows (refer to *Figure 7.18*), and click **Next**:

1. **Id Type: wye** (selection)
2. **Id Material: ductile iron** (selection)
3. **Description**: `12x8 Reducing Wye` (manual entry)

Figure 7.18 – Import Part | Model Properties

Note

The first four properties are required to be filled out, as indicated by the * symbol at the beginning of each property definition. Although additional fields aren't required, filling these out to the best of our knowledge will only help to improve our LOD and information that can be interrogated and tracked later on down the road (construction, asset management, operations, etc.).

In the **Connection Points** section, we'll want to make sure we fill out the following columns with the following criteria, also shown in *Figure 7.19*, and click **Finish** to create our first official custom part:

1. **Nominal Diameter (in)**: 12, 12, and 8
2. **Joint End Type ID**: flanged, flanged, and flanged
3. **Wall Thickness (in)**: 0.75, 0.75, and 0.6
4. **Outer Diameter (in)**: 13.5, 13.5, and 9.2
5. **Deflection (⁰)**: 5, 5, and 5

Connection Points

Nominal Diameter (in)	Engagement Length (in)	Joint End Type ID	Wall Thickness (in)	Outer Diameter (in)	Deflection (°)
12	0	flanged	0.75	13.5	5
12	0	flanged	0.75	13.5	5
8	0	flanged	0.6	9.2	5

Figure 7.19 – The Connection Point criteria

Now, with our new custom part imported into our new custom **Pressure Network** catalog, let's go ahead and save the current state of the catalog and close out of the **Content Catalog Editor** window. Upon completion, we have successfully utilized the Content Catalog Editor to create unique parts that we can use for special circumstances and projects as well as leverage in the future for other designs. Next, we will look at integrating them into our projects.

Integrating custom parts into our BIM designs

With our Custom Catalog now created to contain our newly created custom part, we'll now jump into our dataset and review the steps we'll need to take to incorporate it into our utility design model. That said, let's open up our Utility Model.dwg file within our Autodesk Civil 3D 2025 Unleashed\Chapter 7\Model folder. Once opened, we'll jump over to our **Toolspace | Settings** tab, expand our Pressure Network category, then click Parts List, right-click on Water, and select the **Edit** option, as shown in *Figure 7.20*.

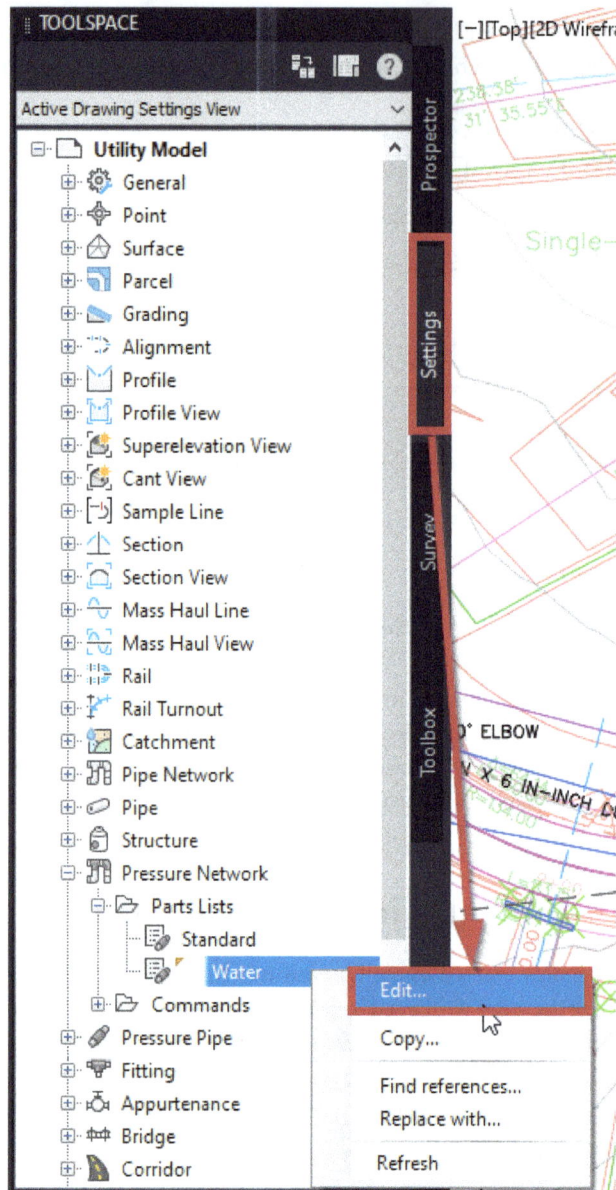

Figure 7.20 – Edit the pressure network parts List

When the **Pressure Network Parts List - Water** dialog box appears, we next want to load our custom catalog. In the **Information** tab, select the **Load new catalog** button and navigate to, select, and open the `Imperial_Custom_Pressure_Parts.sqlite` file, at which point this new catalog will appear in the second line of the **Catalogs** that are being referenced into our current parts list (as shown in *Figure 7.21*).

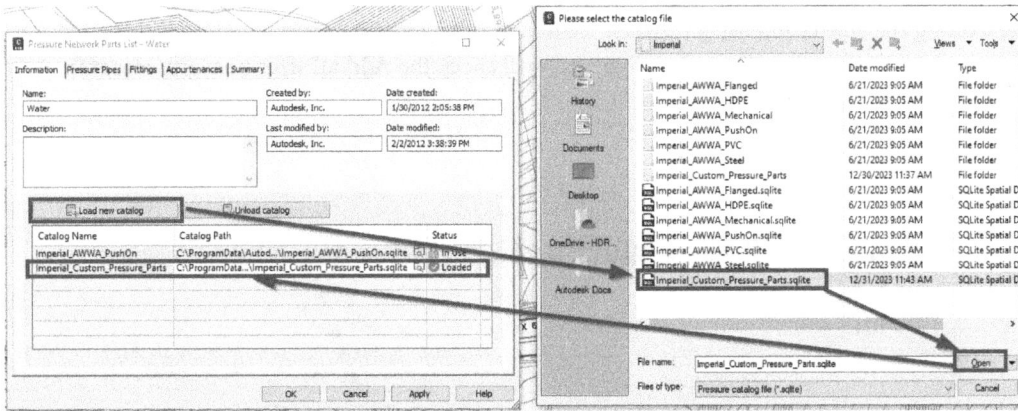

Figure 7.21 – Load new catalog into the existing parts list

> **Note**
>
> This is a relatively new feature made available within Civil 3D where multiple catalogs can be loaded into one comprehensive parts list. Earlier versions of Civil 3D would only allow for one catalog to be loaded per parts list.

With our new custom catalog loaded, let's hop over to our **Fittings** tab within the **Pressure Network Parts List - Water** dialog box, right-click on the main **Water** header, and select **Add Type**. When the next dialog box appears, we'll change our Catalog name to `Imperial_Custom_Pressure_Parts`, Check the `Wye` fitting under `ductile iron`, and select **OK** (as shown in *Figure 7.22*).

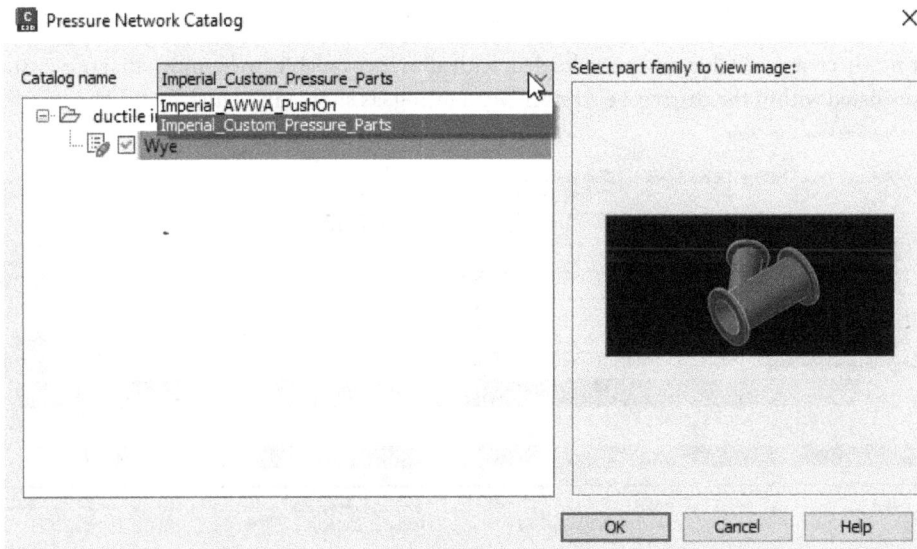

Figure 7.22 – Add custom wye fittings to our parts list

After loading the wye fitting, we'll right-click on **Wye** in our **Fittings** section and select **Add Size**. When the **Add Fitting Sizes** dialog box appears, we'll check the **Add all sizes** box and select **OK**, as shown in *Figure 7.23*.

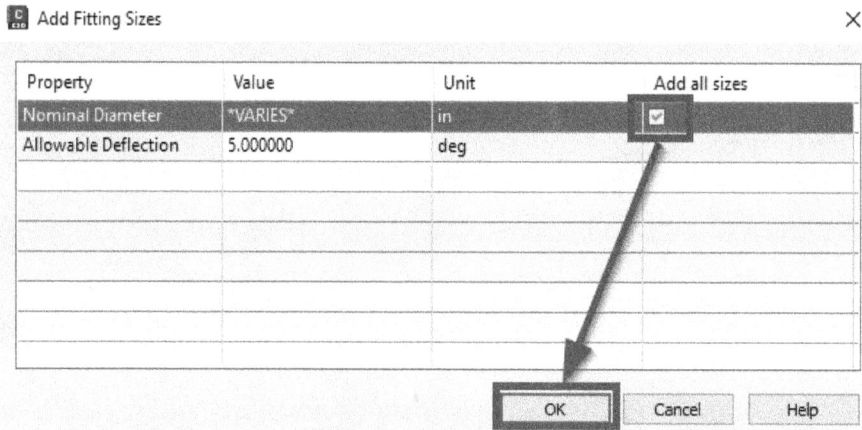

Figure 7.23 – Add all sizes of the parts available in the catalog

> **Note**
>
> Checking the box for **Add all sizes**, whether it be when adding part sizes for gravity or pressure network parts lists, is a quick and easy way to build out a fully comprehensive parts list that includes all possible configurations. This practice, however, can also lead to unnecessary file bloat if only a handful of part sizes are needed for your design. That said, be cautious when using these options.

With our newly created catalog and parts loaded, with all sizes available, we'll notice that our custom part is now listed within the `ductile iron` Wye fitting, as shown in *Figure 7.24*.

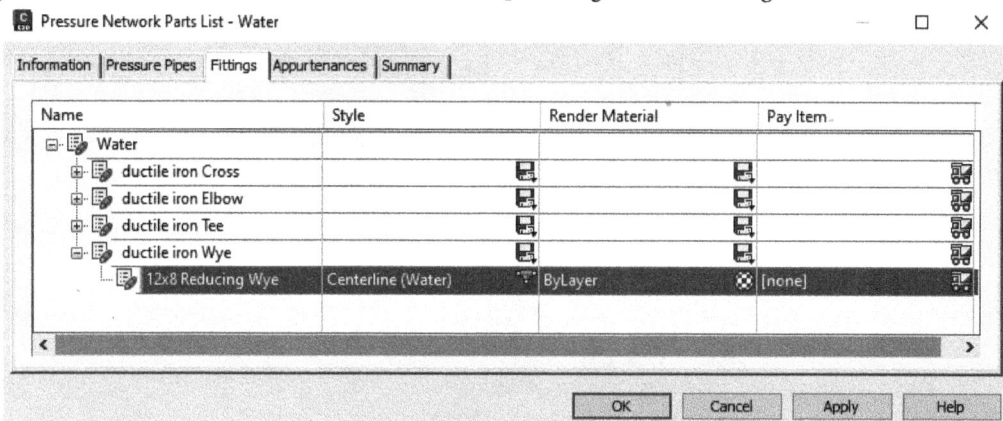

Figure 7.24 – 12x8 Reducing Wye Fitting added to our water pressure network parts list

We are now all set to incorporate our new `12x8 Reducing Wye` fitting into our design as needed (refer to our *Autodesk Civil 3D 2024 from Start to Finish* book for suggested workflows), with an example shown in *Figure 7.25* displaying both Top and Isometric Views of our new custom part connected to our water main.

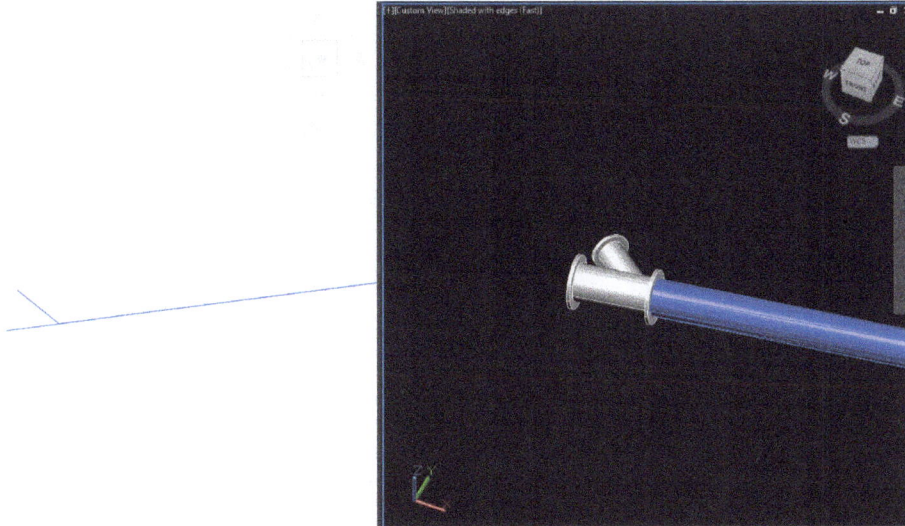

Figure 7.25 – The final result of adding our 12x8 Reducing Wye Fitting to our design model

Now that we can see the 3D representation of our work, that completes our workflows for integrating custom parts into our civil design projects. This can be incredibly useful and impactful to projects, ensuring they stay accurate and on schedule.

Summary

As we worked through this chapter, we learned a very simple way to leverage our AutoCAD and Civil 3D design capabilities to up our game in the pressure network utility design department. Not only did we create a custom 3D part from basic 2D components, but we were also able to understand the workflows for creating new pressure network catalogs using Content Catalog Editor to build out a custom pressure network catalog using these 3D parts. We also learned how to save these parts, best practices for sharing these catalogs across a design team or organization, and how we can incorporate these parts long after a utility design model has been created and designed.

In the next chapter, we'll take a slightly different approach to creating custom pressure and gravity network parts and catalogs by using **Infrastructure Parts Editor**. As we discovered with Content Catalog Editor, the tool is primarily supportive of pressure network catalogs. Infrastructure Parts Editor, on the other hand, is a versatile tool and also allows us to incorporate additional parametric parts that are developed in Autodesk Inventor.

8

Empowering Utility Modeling with Infrastructure Parts Editor

In the previous chapter, we provided an introduction to custom content development using **AutoCAD**, **Civil 3D**, and **Content Catalog Editor** functionality. This method has historically been relied upon by Civil 3D experts as it is the most straightforward method to enhance our approach to custom pressure network design solutions. We also brushed over the concept of making our custom catalogs shareable across an entire project design team or organization, resulting in a seamless integration of our custom content throughout the design modeling process.

As we venture into the upcoming chapter, our focus shifts toward an alternative approach for constructing custom pressure and gravity network parts and catalogs. This time, we will harness the power of **Infrastructure Parts Editor**. Unlike its counterpart, Content Catalog Editor, which primarily supports pressure network catalogs, Infrastructure Parts Editor will prove to be a versatile tool. It not only facilitates the creation of custom parts but also enables the inclusion of parametric parts developed in **Autodesk Inventor**, expanding our capabilities in the realm of utility design. That said, in this chapter, we'll cover the following topics:

- Familiarizing ourselves with Infrastructure Parts Editor
- Creating new parts in Infrastructure Parts Editor
- Integrating custom parts into our BIM designs

Technical requirements

We will be using the same hardware and software requirements as discussed in the *Technical requirements* section of *Chapter 1*.

With that, let's go ahead and launch Civil 3D 2025 and start a new drawing using our Company Template File.dwt file located within Civil 3D 2025 Unleashed\Chapter 8. Once opened, we'll bring our attention up to the home ribbon; move over to the **Create Design** panel, click on the down arrow, and select the **Infrastructure Parts Editor** option, as shown in *Figure 8.1*.

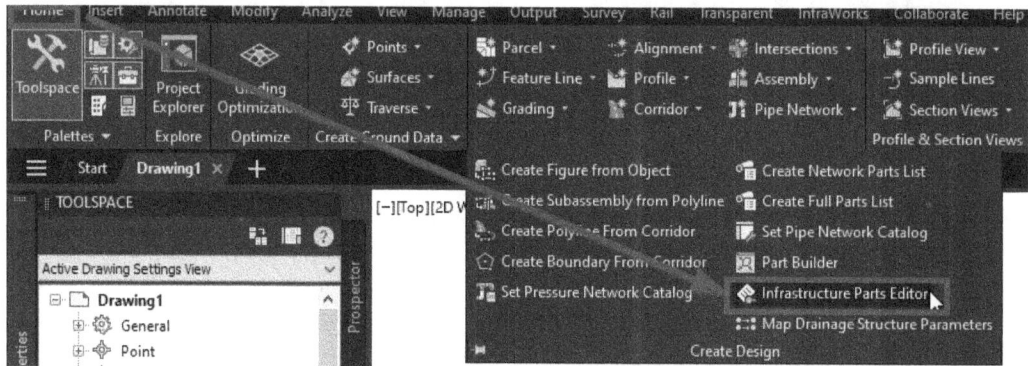

Figure 8.1 – Accessing Infrastructure Parts Editor

Once Infrastructure Parts Editor has been launched, the Infrastructure Parts Editor program will be displayed on the screen. We are now ready to begin familiarizing ourselves with Infrastructure Parts Editor and learn how we can utilize it to our advantage to develop custom utility design solutions.

Familiarizing ourselves with Infrastructure Parts Editor

When the Infrastructure Parts Editor program is launched, we are presented with a few options to begin exploring and developing new catalogs. Along the top, we have **Catalogs**, **Parts**, and **Publish** dropdown menus, which are all greyed out at this point (these will become available for selection and interaction once a catalog has been opened). On the left-hand side of the program (shown in *Figure 8.2*), our first two selections, **Save** and **Save As**, are greyed out and unable to be selected currently, with the remaining options that are available to us listed here:

- **Open my parts Lib**: This allows us to open either an out-of-the-box or custom catalog already developed using Infrastructure Parts Editor.

- **Open by Model**: This allows us to open either an out-of-the-box or custom catalog being utilized on a particular project file.

- **Open by File**: This allows us to open an out-of-the-box or custom catalog by its `.icbt` or `.ACCat` extension. This workflow would typically be used if you receive a custom catalog or are accessing and updating a shared catalog on a hosted server location.

- **Recent**: This allows us to quickly open a recently worked-on catalog.

- **New**: This allows us to start from scratch to create a brand-new custom catalog.

- **Import from Plant3D**: This allows us to import catalogs generated with Plant3D using the `.pspc`, `.pspx`, or `.pcat` extension.

- **Help**: This provides quick access to *Autodesk Help* and the *Overview* documentation on Infrastructure Parts Editor.

Figure 8.2 displays all these options:

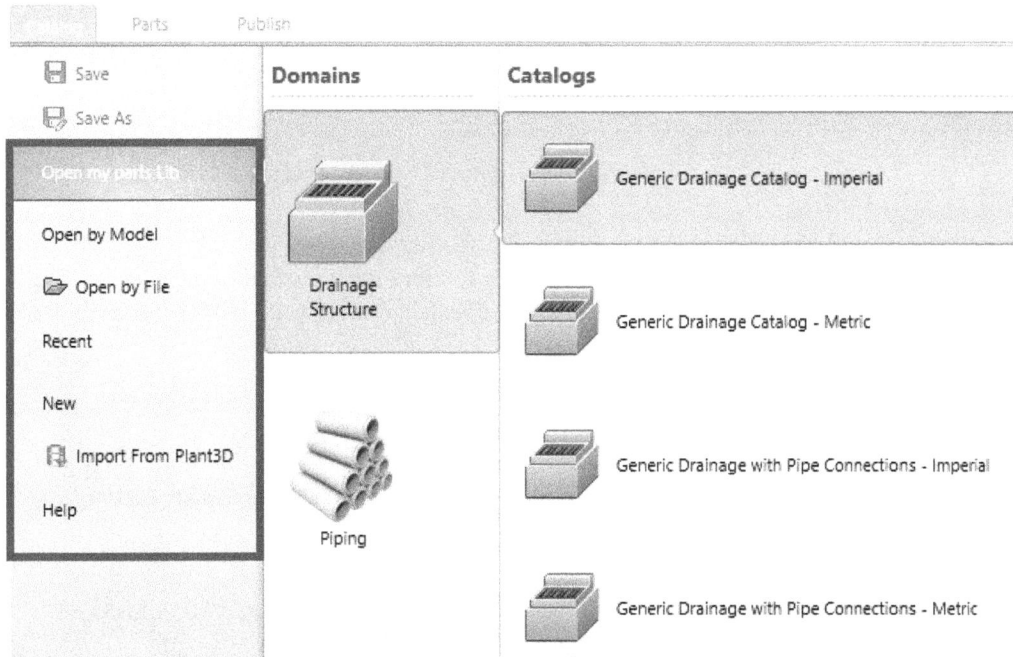

Figure 8.2 – Accessing Infrastructure Parts Editor

To get started, let's go ahead and select the **Open my parts Lib** option. Directly next to this option, we'll see the **Domains** and **Catalogs** columns with different selection options. Depending on the utility type and purpose that we select in the **Domains** column, our list of available **Catalogs** will filter out as well. To prepare us for the rest of the chapter, let's switch our **Domains** selection from **Drainage Structure** to **Piping**. Once selected, we'll then select the first option of **AWWA Pipes Fittings Valves Sample Catalog** in the **Catalogs** column and then click **OK** to open, as shown in *Figure 8.3*.

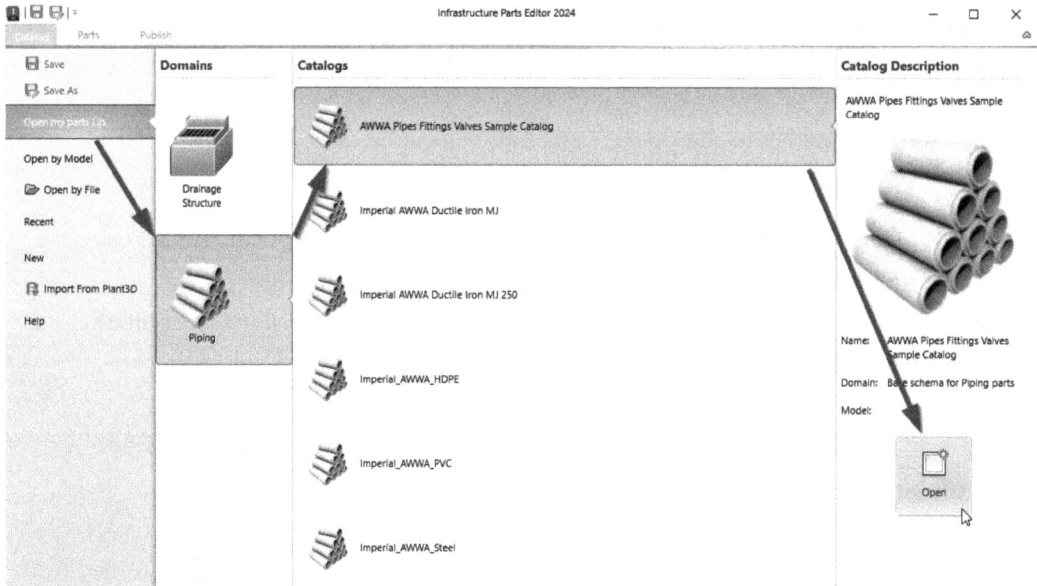

Figure 8.3 – Opening AWWA Pipe Fittings Valves Sample Catalog

After selecting **Open**, you will likely be prompted with another dialog box indicating that there is a draft catalog already in existence and asking if you'd like to overwrite it (as shown in *Figure 8.4*). Feel free to click **Yes** if this prompt appears to continue the opening process.

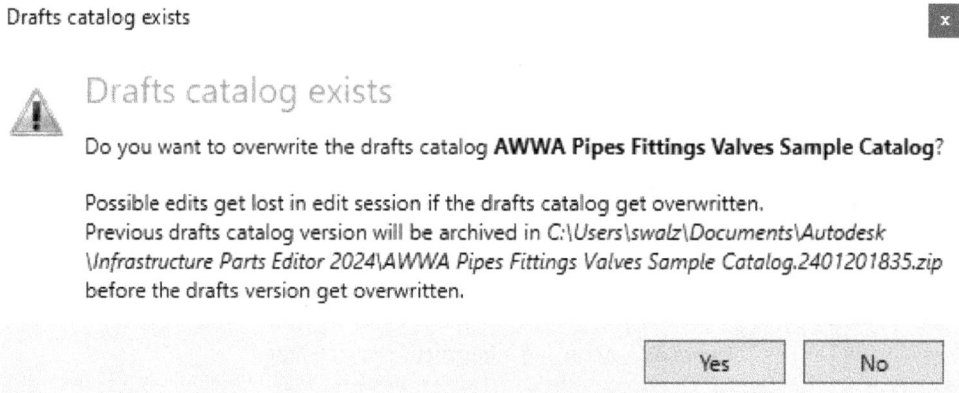

Figure 8.4 – The Draft catalog exists prompt

Once the **AWWA Pipe Fittings Valves Sample Catalog** has been opened, we'll want to immediately take steps to save it as a new catalog to ensure that any changes and updates we make do not affect the standard out-of-the-box catalog. We can always go back and use this as a starting point for any future custom catalogs we want to create. That said, we'll take the following steps to create our new custom catalog:

1. Select the save as icon along the top of our Infrastructure Parts Editor session to access the **Create Piping Catalog** dialog box.

2. In the **Specify the Catalog** file name field, we'll place the following text: `Company Pressure Network Catalog_2025.icbt` (note that `Company` can be replaced with your actual company name or intent of using).

3. In the **Specify the Catalog Description** field, we'll place the following text: `Customized Company Pressure Network Catalog.`

4. In the **Specify the Catalog Units** section, we'll select **Imperial** for this particular workflow.

5. Finally, after all fields have been filled out and selections made, we'll click on the **OK** button to complete our task of creating our first custom pressure network catalog (as indicated along the top bar of our Infrastructure Parts Editor session).

These steps are shown in *Figure 8.5*:

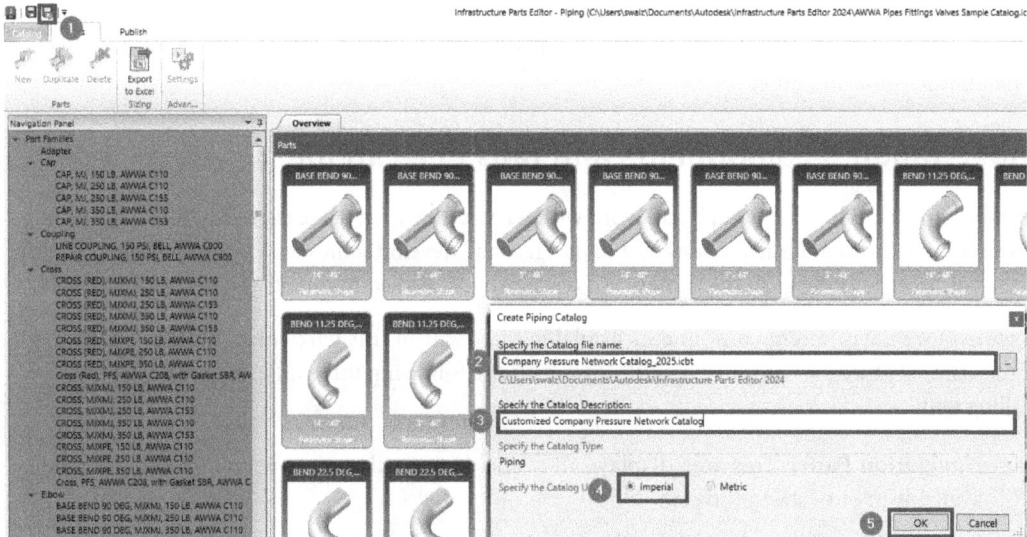

Figure 8.5 – Saving our piping catalog as Company Pressure Network Catalog_2025.icbt

Now that we have our custom catalog created, we can begin familiarizing ourselves with what parts are actually available to us out of the box to build out our custom pressure network catalog.

Creating new parts in Infrastructure Parts Editor

Exploring these out-of-the-box catalogs can be a bit overwhelming as they really make a lot available to us in one location. There are pros and cons to this but it will require us to do a bit of housekeeping and cleanup to make these catalogs usable across a project design team or an organization. Unlike the standard pressure network catalogs that are available to us with all standard Civil 3D 2025 installations, we have the ability to make all the parts we need available to us in one comprehensive catalog, without having to guess which out-of-the-box catalog a particular part might be available in. Another huge benefit of using Infrastructure Parts Editor to develop our custom catalogs is the ability to quickly generate additional sizes of parts that might not be available in one of the out-of-the-box catalogs.

Beyond the standard parts that we have available directly within Infrastructure Parts Editor, we can further develop new parts using Autodesk Inventor and apply various parameters to them, where we can specify a multitude of values to them to create multiple sizes, connections, and so on.

Autodesk Inventor allows designers to create two main types of parts for infrastructure content. First, there are **single-body parts**, which have a part file with one solid body and act as an independent collection of features, such as faces, edges, surfaces, and different shapes. Second, there are **multi-body parts**, which consist of a part file with more than one solid body. Each body can have its own features or share some with others. Importantly, each solid body from a multi-body part can be saved as a separate part file when finishing the design. Creating multi-body parts involves an efficient top-down design workflow, using common modeling commands to add a new body within the existing bodies. This method improves the design process and makes it easier to integrate into assemblies. But this is all for another time and book.

Understanding components and tabs in a catalog

To take a more simplified approach and work with what we have available currently, let's make some sense of what we're looking at and how best to navigate and develop some easy custom solutions. Right off the bat, we're looking at all the contents of our newly created components.

As we explore the different components available to us within this particular catalog, `Company Pressure Network Catalog_2025`, we have the following three different areas that we'll be working with:

1. **Navigation Panel**: This area displays all available parts that have been created and made available in our catalog thus far.
2. **Part Families**: This area displays all categories of parts that we can create/modify within our catalog. As we select different part categories, our **Navigation Panel** will display all parts within that particular category.

3. **Overview**: This area displays all types of parts within the **Part Families** area. As we select each individual part within **Navigation Panel**, our **Overview** area will display various parameters and definitions that we can apply to our parts to build them out.

Refer to *Figure 8.6*, which shows these areas:

Figure 8.6 – Area definitions of our custom catalog within Infrastructure Parts Editor

Going back to the idea of keeping things simple, especially for those of you who are hopping into this program for the first time, let's simplify the view a bit and select the **Pipes** category in **Part Families**. Once selected, we should only be seeing one part family (**Pipe**) and the **Generic Pipe AWWA C151 MJ** part within the part family, as shown in *Figure 8.7*.

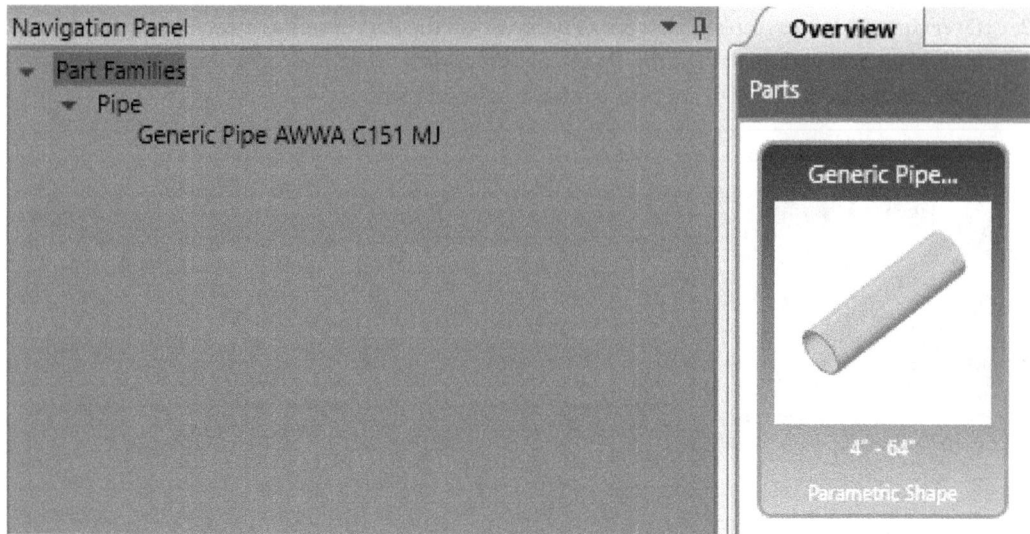

Figure 8.7 – The Part Families | Pipes selection

Let's go ahead and now select the **Generic Pipe AWWA C151 MJ** part listed in **Navigation Panel**. Once selected, you'll notice that our **Overview** area now has several different options for us to modify and analyze our part in the form of tabs along the top of the **Overview** area, which include the following:

- **Part Family Properties**: In this tab, we can specify **Primary End Types** and **Connections** to other parts, as well as **Size Ranges** and **Part Family Naming**.

- **Shapes**: In this tab, we can specify which parametric parts (pre-developed) we are applying as the part we're creating/modifying/updating. This is where we can also import a custom part created from Autodesk Inventor if we were to go down that road.

- **Detailed Properties**: In this tab, we can further specify information related to the particular part, using a fixed set of criteria including, but not limited to, descriptions, materials, **Design Standards**, **Manufacturers**, **Classes**, and **Schedules**.

- **Size Range**: In this tab, we can preview all sizes available within the range we had specified in the **Part Family Properties** tab.

- **Part Editing**: In this tab, we will be able to define multiple parameters and dimensions for the particular part that we're creating/modifying. In this tab, we have the ability to export these parameters and dimensions directly to an Excel file and re-import them back in. This functionality comes in handy for a quick population of values for each part size.

- **Size Validation**: In this tab, we can preview each part size that we have created and are now available to us in a 3D view.

These options are shown in *Figure 8.8*:

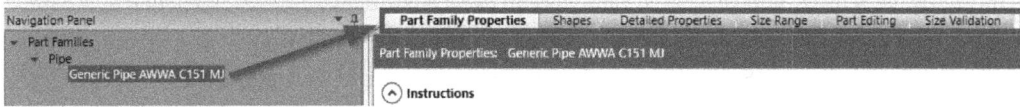

Figure 8.8 – The Pipe part family selection

Exploring the Part Family Properties tab

As we run through this tab, we'll start with the **Part Family Properties** tab. Here, we can keep selections and values as is, or update as needed. Since we're just familiarizing ourselves right now with what options we have available to us, we'll keep selections as is for this particular workflow. After you become more comfortable with the selections, feel free to come back to this section and update the applied **Size Range From** and **Size Range To** values and see how these updates will affect the content displayed in the remaining tabs.

Continuing, we'll keep our preassigned values set in the **Shapes** and **Detailed Properties** tabs. If we were to update the **Size Range To** and **Size Range From** values in the **Part Family Properties** tab, we would see the updated range of pipe diameters listed in the **Size Range** tab.

Moving into the **Part Editing** tab, we'll be presented with a table format that has many parameters and dimensions that will need to be filled in accordingly. Taking a quick look, we can see that the initial size range of 4 " to 64 " diameter pipes has already been filled in. If we had updated the ranges initially, we would need to manually input the values of new sizes based on the manufacturer's specifications as they would be blanked out by default. We'll also notice a handy **Dimension Preview Image** that shows us what each parameter and dimension represents when filling out the table (see *Figure 8.9*).

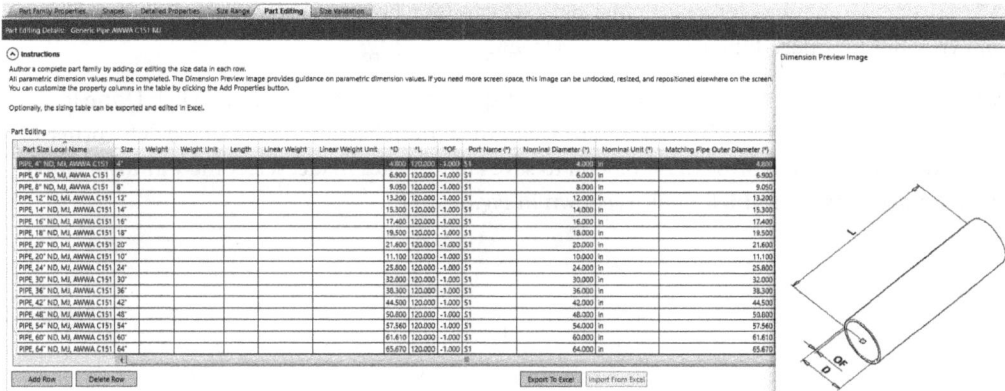

Figure 8.9 – The Part Editing table with Dimension Preview Image

If we were to compare the values already inputted into the table with an industry or manufacturer's specifications, we'll likely find that these values are in close alignment with that particular standard specification for pipes. A simple one to use is the one provided by US Pipe at https://www.uspipe.com/wp-content/uploads/2023/10/FlangedPipe_SubmittalBook_digital.pdf.

Finally, moving over to the **Size Validation** tab, we can visualize the different parts in the 3D view, where we can also rotate our view to get a clear picture of our entire part, as shown in *Figure 8.10*.

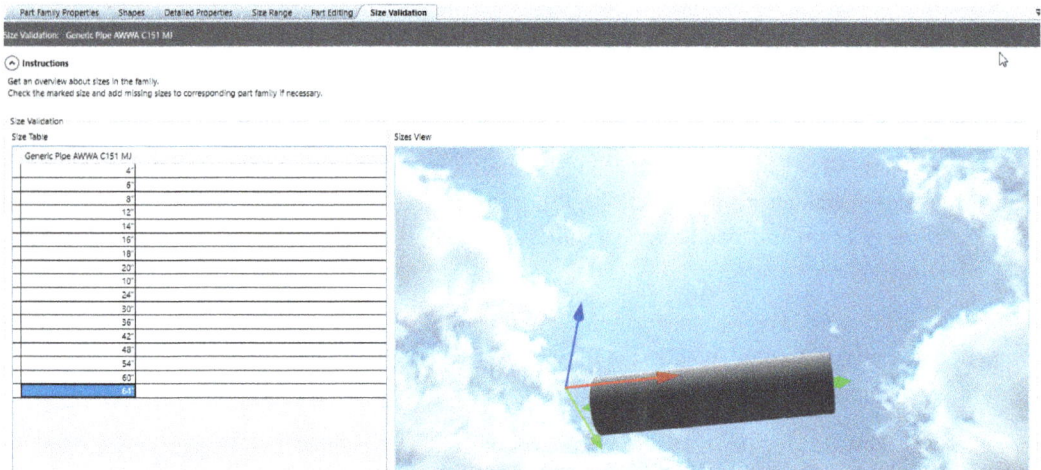

Figure 8.10 – The Size Validation view rotation

It's important to note the views we see here are a reflection of the dimensions we input within our **Part Editing** section. If we hop back into our **Part Editing** tab and change any of the dimensions within the *D, *L, or *OF columns for a particular part, we can then hop back over to our **Size Validation** tab and view our new part dimensions in the 3D view.

Once we're satisfied with our previews and all of the applied values, go ahead and click **Done** in the lower-right corner of the **Size Validation** tab. Then, we'll jump back up to the upper-left corner of Infrastructure Parts Editor, press the save icon to save any changes made, select the **Publish** ribbon, and then finally press the **Publish** tool, as shown in *Figure 8.11*.

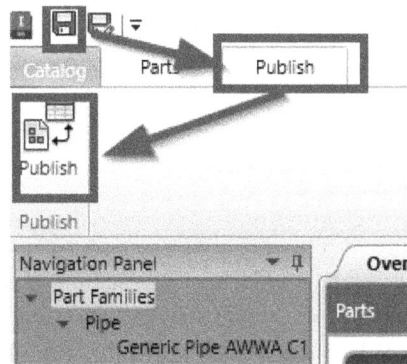

Figure 8.11 – Save and Publish Catalog

Once selected, we'll notice that although we have an enormous list of parts listed in the **Select Assemblies to Publish** dialog box, only the pipes that we just updated are checked. It should be noted that whichever part family we're updating and have selected as current in **Navigation Panel** will automatically be checked in the **Select Assemblies to Publish** dialog box by default. We can, however, check and uncheck any parts listed to publish to a new pressure network catalog.

It's also worth noting that if we were to check all boxes by clicking the **Select All** option within the **Select Assemblies to Publish** dialog box, all 142 parts currently available in this master catalog would be selected for publishing. That's why it's important to do some house cleaning and remove any unnecessary parts within the families to keep everything clean and organized.

For now, we'll just go ahead and check only the following parts (the majority will be listed as 250 LB and C110) to be published and click **Next**:

- **BASE TEE, MJXMJ, 250 LB, AWWA C110, MJF, 250**
- **BEND 11.25 DEG, MJXMJ, 250LB, AWWA C110, MJF, 250**
- **BEND 22.5 DEG, MJXMJ 250 LB, AWWA C110, MJF, 250**
- **BEND 45 DG, MJXMJ 250 LB, AWWA C110, MJF, 250**
- **BEND 90 DEG, MJXMJ 250 LB, AWWA C110, MJF, 250**
- **CAP, MJ, 250 LB, AWWA C110, MJF**
- **CROSS (RED), MJXMJ 250 LB, AWWA C110, MJF, 250**
- **CROSS, MJXMJ 250 LB, AWWA C110, MJF, 250**
- **PLUG MJ, 250 LB AWWA C110, MJP**
- **REDUCER (CONC), MJXMJ 250 LB, AWWA C110, MJF, 250**
- **TEE (RED), MJXMJ 250 LB, AWWA C110, MJF, 250**

- **TEE, MJXMJ 250 LB, AWWA C110, MJF, 250**
- **Generic Gate Valve with UG Nut Actuator Flanged Ports, FL, 0**
- **Generic Gate Valve with Wheel Actuator Flanged Ports, FL, 0**
- **Generic Hydrant with 6" MJ Connection Port, MJF**
- **Generic Pipe AWWA C151, MJ, MJM, 0**

After selecting **Next**, we'll then be asked how and where we want to publish our catalog. Let's go ahead and check the box next to Civil 3D, select **Autodesk Civil 3D 2025 – English** as **Version**, and navigate to `C:\ProgramData\Autodesk\C3D 2024\enu\Pressure Pipes Catalog` to define the location and click **OK** (refer to *Figure 8.12*).

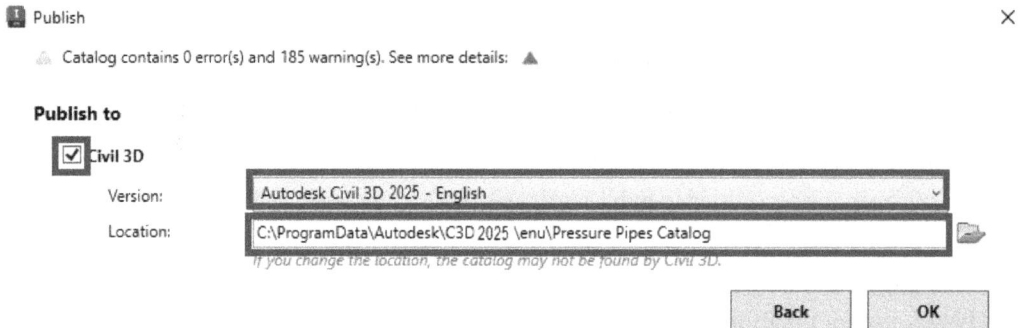

Figure 8.12 – The Publish dialog box

We have now created a new pressure part catalog using Infrastructure Parts Editor. In the next section, we'll run through the steps of connecting to our new catalog, building a parts list, and applying it to our design.

Integrating custom parts into our BIM Designs

With our custom catalog, created from Infrastructure Parts Editor, able to contain a much more comprehensive list of parts and families, we'll now jump into our dataset and review the steps we'll need to take to incorporate this work into our utility design model. That said, let's open up the `Utility Model.dwg` file within our `Autodesk Civil 3D 2025 Unleashed\Chapter 8\Model` folder. Once opened, we'll jump over to our **TOOLSPACE | Settings** tab, expand our `Pressure Network` category, right-click on `Parts Lists`, and select the **New** option, as shown in *Figure 8.13*.

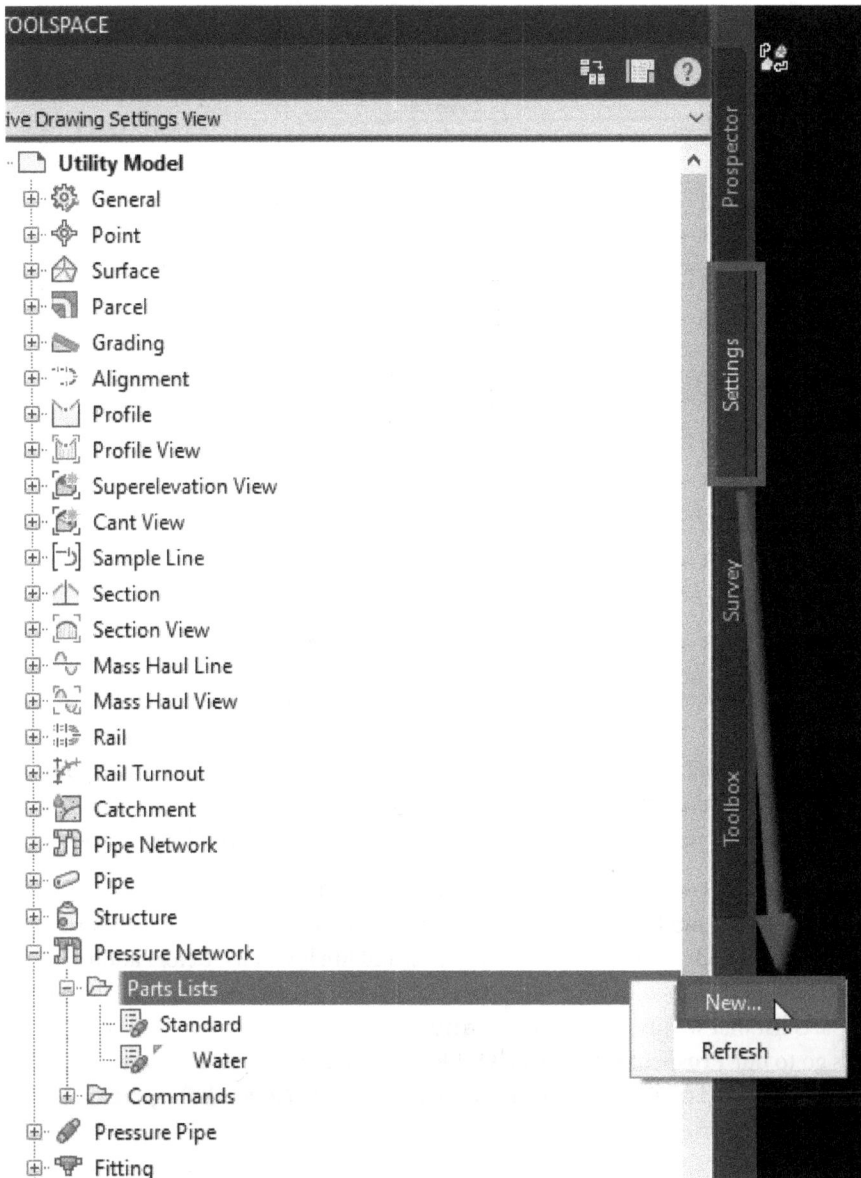

Figure 8.13 – Create a new pressure network parts list

When the **Pressure Network Parts List** dialog box appears, we'll want to select and make the **Information** tab current. We'll then give it a unique name, in our case, we'll call it Company Pressure Network Catalog_2025, remove the Imperial_AWWA_Push On catalog that is currently linked to the parts list by default, and load the new Company Pressure Network Catalog_2025 we just created, as shown in *Figure 8.14*.

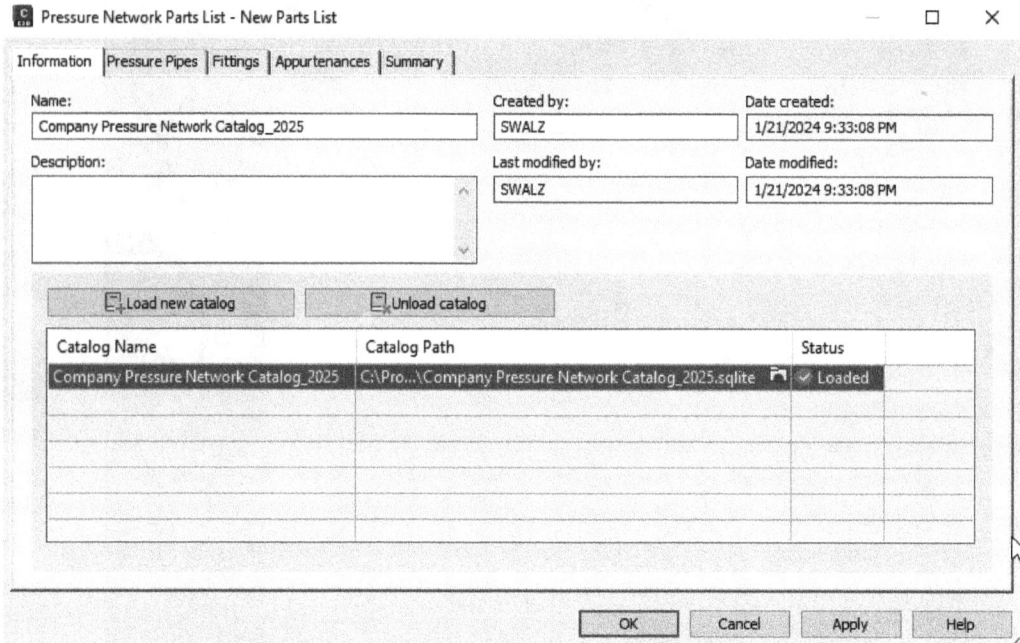

Figure 8.14 – Load a new catalog in the existing parts list

With our new custom catalog loaded, let's hop over to our **Pressure Pipes**, **Fittings**, and **Appurtenances** tabs and add all of the part families and sizes available from `Company Pressure Network Catalog_2025` (refer to *Chapter 7*, for the steps we'll need to take to build out our parts lists).

With our new pressure network parts list created and available to us in our current `Utility Model.dwg` file, let's hop into our model space and update one of our `PPN - Domestic Water Main` pressure networks to include a few 6" x 6" tees, 6" pipes, and fire hydrants. Before adding these parts using a new custom parts list, we'll need to update our pressure network properties to point to the new custom parts list that we just created, `Company Pressure Network Catalog_2025`. That said, let's go to our **Prospector** tab in **TOOLSPACE**, expand `Pressure Networks`, right-click on `PPN - Domestic Water Main`, and select **Pressure Network Properties...**, as shown in *Figure 8.15*.

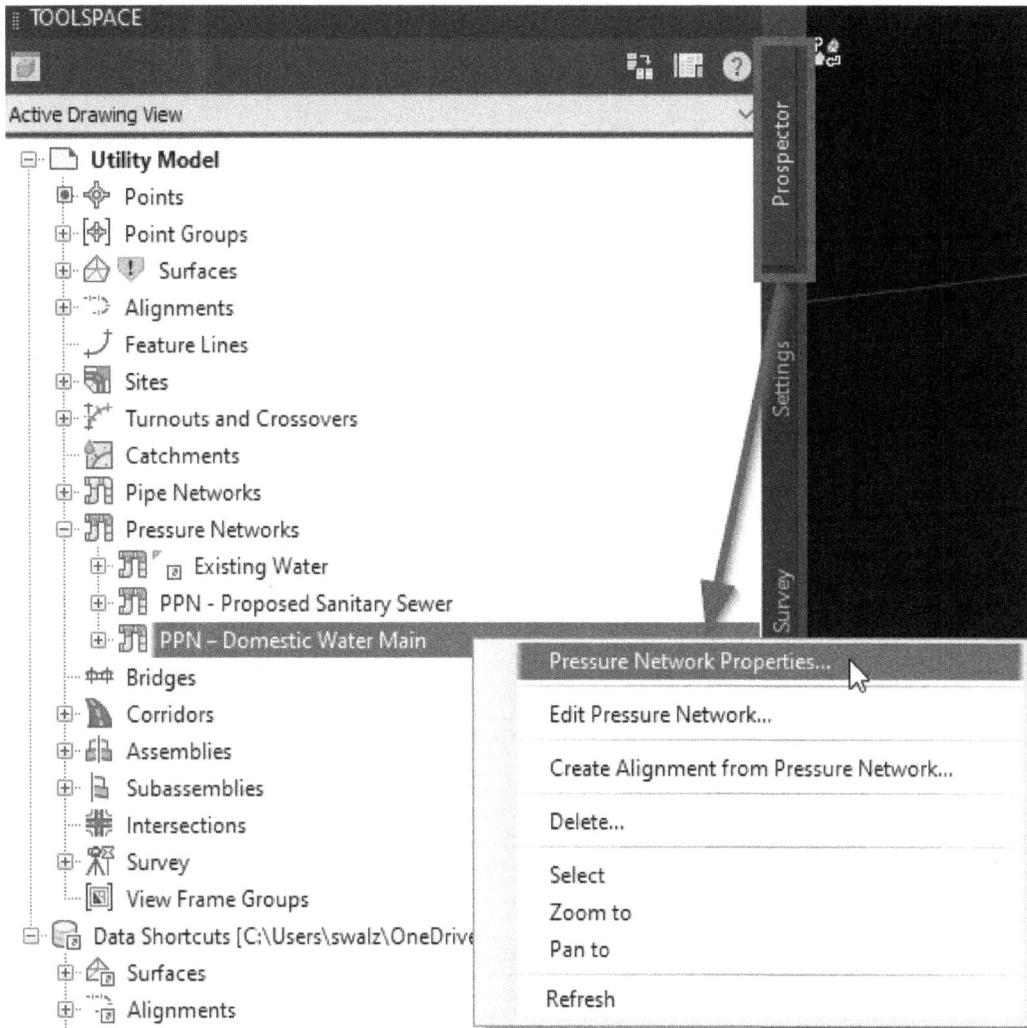

Figure 8.15 – Pressure Network Properties…

When the **Pressure Pipe Network Properties** dialog box appears, go to the **Layout Settings** tab, change the **Pressure Parts List** option from Water to Company Pressure Network Catalog_2025, and select **OK**, as shown in *Figure 8.16*.

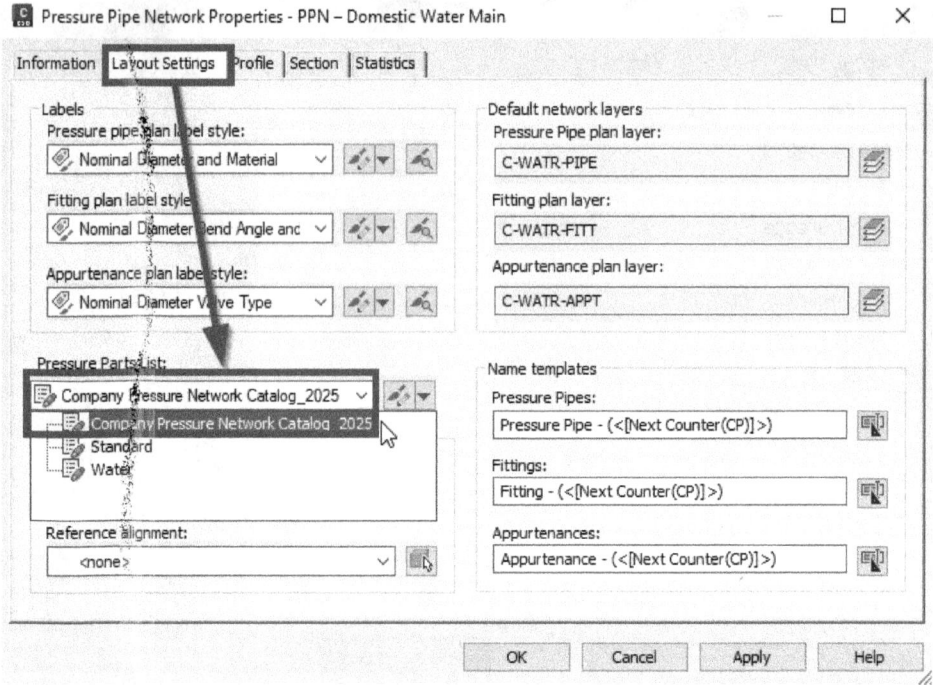

Figure 8.16 – The Pressure Pipe Network Properties dialog box

Once applied, we'll right-click on the `PPN - Domestic Water Main` pressure network again in **Prospector**, but select **Edit Pressure Network…** this time. Once selected, we'll want to bring our attention to the **Pressure Network** contextual ribbon towards the top of our Civil 3D session. We'll immediately want to make the following few adjustments to the default **Pipe**, **Fitting** and **Appurtenance** displayed so that we can quickly lay out our fire hydrant locations:

- `PIPE_6in ND_MJ_AWWA C151` (to connect from tee to the fire hydrant)
- `TEE_6in_250 LB_AWWA C110` (to branch off of the main water line)
- `Generic Hydrant with 6in MJ Connection Port 4.0' Depth` (to insert our fire hydrant structure)

These adjustments are displayed in *Figure 8.17*:

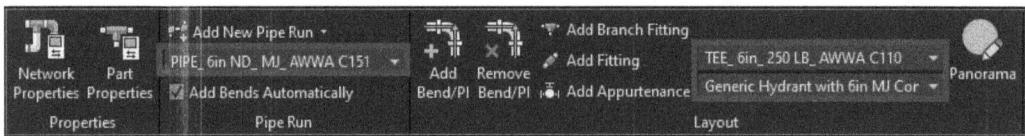

Figure 8.17 – The Pressure Network contextual ribbon selections

Let's go ahead and use the following steps to insert our fire hydrant assembly on our site:

1. Jumping back down to our model, let's navigate to a location where we'd like to insert our fire hydrant assembly.

2. Select the **Add Fitting** icon in the **Layout** tab within our **Pressure Network** contextual ribbon.

3. Jump back down to our model and click on our PPN – Domestic Water Main pressure network in the desired location. Note that to ensure we have a proper connection established, we'll want to hover over the location where we intend to insert our 6" x 6" tee fitting, at which point we should see a connection glyph (similar to that shown in *Figure 8.18*).

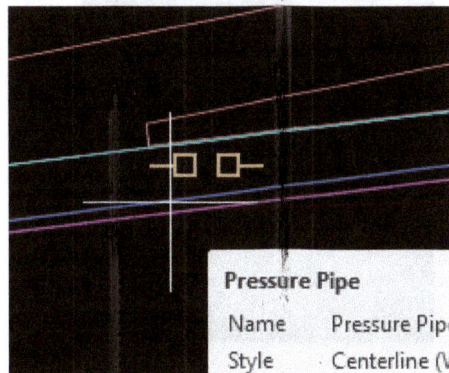

Figure 8.18 – Connection glyph example

4. Depending on the flow of our pressure network, we may need to flip the tee that's inserted using the **Flip** node that displays when the tee is selected (refer to *Figure 8.19*).

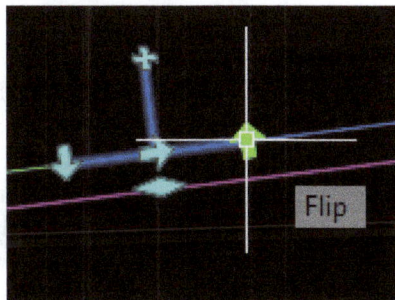

Figure 8.19 – The Flip node

5. Next, we'll simply click on the + node (displayed in *Figure 8.19*) to add a 6" diameter pipe that we'll connect to our fire hydrant.

6. Finally, we'll jump back up to our **Pressure Network** contextual ribbon, select the **Add Appurtenance** icon, and snap to the end of our new 6" pipe that we just added to our model to complete the fire hydrant assembly.

We should now have a fully developed fire hydrant assembly that, when selected and rotated in a 3D view or object viewer, should appear similar to *Figure 8.20*.

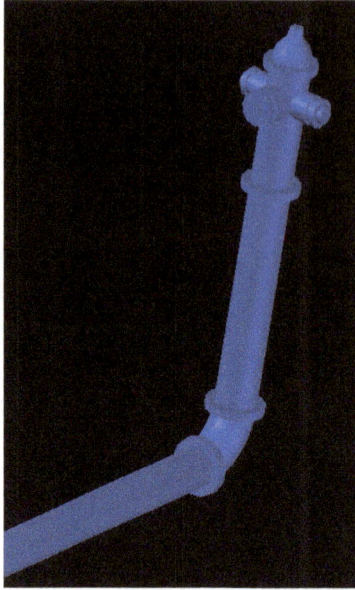

Figure 8.20 – Fire hydrant assembly added to the utility model .dwg file

We have now officially created a new catalog using Infrastructure Parts Editor and incorporated newly created custom families and parts into our utility design model.

Summary

As we worked through this chapter, we were able to get a glimpse into how we can further develop custom utility network catalogs, this time using Infrastructure Parts Editor. This tool is a bit more advanced than Content Catalog Editor and also supports both gravity and pressurized utility networks. We have the ability to import content developed using Autodesk Inventor and apply a multitude of values to parameters embedded in the parts that are imported. This methodology of creating custom parts allows us to simplify our approach a bit more in the sense that we can theoretically develop one part that can accommodate multiple sizes of that given part. If we used Content Catalog Editor, we would need to model each individual size part that we would wish to incorporate into our utility catalogs and designs.

In the next chapter, we'll roll these custom content practices over to the roadway design side and begin exploring **Subassembly Composer**. By using Subassembly Composer, we have the ability to create custom roadway templates that can be imported into our files and applied to corridor models as needed. Proper implementation of custom subassembly solutions will certainly elevate our civil BIM game and set our roadway design teams up for success.

9

Custom Roadway Design with Subassembly Composer

As we worked through the previous couple of chapters, we learned how the Content Catalog Editor and the Infrastructure Parts Editor play a powerful part when it comes to custom utility network designs. With their advanced capabilities, we learned how to harness their potential in developing versatile parts, capable of accommodating various sizes with ease. Instead of making separate models for each size using the Content Catalog Editor, we found a more efficient method using the Infrastructure Parts Editor to develop one part that can fit multiple sizes. This approach simplifies the process and makes our work more effective.

In this new chapter, we transition from the underground utility networks to the open roads, putting on our roadway modeling toolbelt again to further explore advanced tools and workflows that will allow us to design customized roadway solutions. That said, our focus now shifts to Subassembly Composer, a tool that empowers design teams to craft tailor-made solutions for roadway projects.

With Subassembly Composer, the design canvas expands, offering us a platform where creativity meets functionality. It's a realm where we break free from constraints, allowing our design teams to envision and implement solutions that align precisely with project requirements. As we continue our journey, we'll continue to elevate our civil BIM game and set our roadway design teams on a trajectory toward unparalleled success. That said, in this chapter, we'll cover the following topics:

- Familiarizing ourselves with Subassembly Composer
- Creating custom assemblies in Subassembly Composer
- Integrating custom subassemblies into our BIM designs

With that, let's go ahead and launch Subassembly Composer. To do so, we'll go to the Start Menu, expand **Autodesk**, and select **Subassembly Composer 2025**. Once selected, Subassembly Composer will appear onscreen, as displayed in *Figure 9.1*:

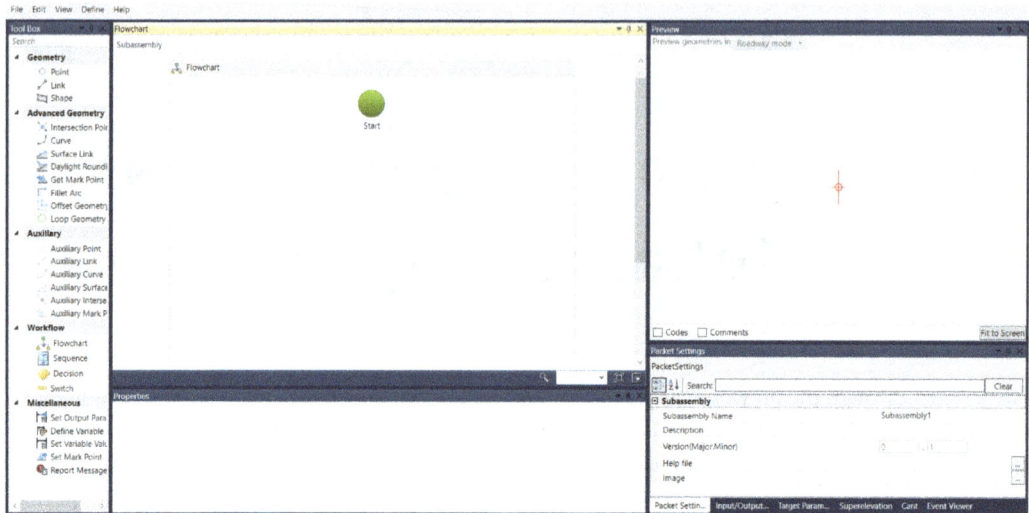

Figure 9.1 – Subassembly Composer

With Subassembly Composer now launched, let's begin familiarizing ourselves with the user interface and options we have available to us to develop custom roadway solutions.

Technical requirements

It's important to note that Autodesk's Civil 3D can often be very taxing on your computer. There is a lot of processing that goes on with modeled design elements, even in the background, that enables the dynamic (connected) capabilities to occur throughout the BIM design lifecycle.

In turn, there are many technical requirements that need to be considered to allow Autodesk's Civil 3D to operate at its full potential. We'll review the minimum requirements that Autodesk recommends, with a few of our suggestions added to increase efficiency and speed throughout the BIM design process:

- **Operating system**: 64-bit Microsoft Windows 10
- **Processor**: 4+ GHz
- **Memory**: 32 GB RAM (we suggest going with either 64 GB or 128 GB)
- **Graphics card**: 8 GB

- **Display resolution**: 1980 x 1080 with true color

- **Disk space**: 20 GB

- **Pointing device**: MS-mouse compliant

Familiarizing ourselves with Subassembly Composer

As we'll quickly find out later in this section, Autodesk Civil 3D Subassembly Composer has five distinct windows, each serving a specific purpose:

- **Tool Box**: This area acts as a repository for the elements essential for constructing the subassembly. It is organized into five sections: **Geometry**, **Advanced Geometry**, **Auxiliary**, **Workflow**, and **Miscellaneous**. To incorporate any of these elements into the subassembly, a simple drag-and-drop action from **Tool Box** to the **Flowchart** panel will suffice (refer to *Figure 9.2*).

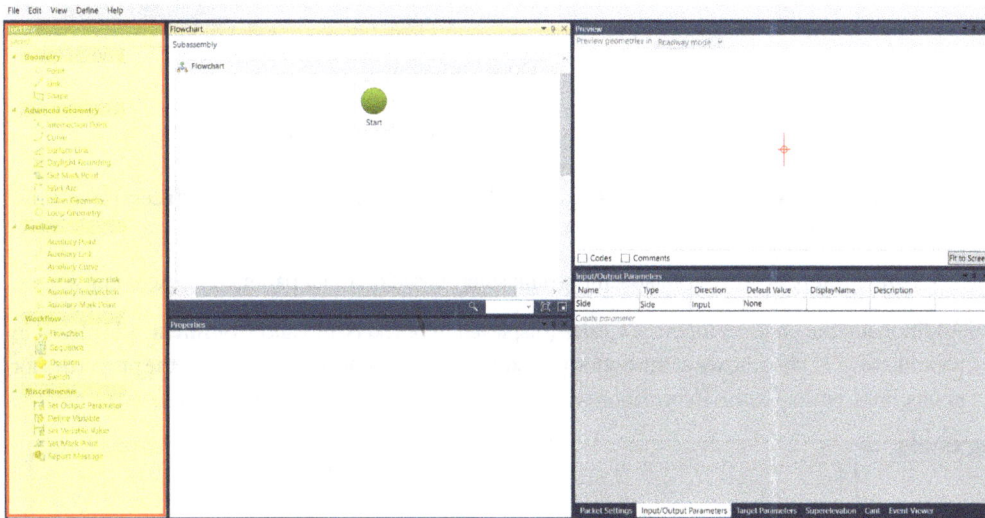

Figure 9.2 – Tool Box area

- **Flowchart**: This panel serves as the space for constructing and arranging the subassembly logic and elements. Whether it's straightforward linear logic or a complex tree of branching decisions, the process always commences at the Start element. In case of issues with the subassembly, a small red circle with an exclamation point will be visible in the upper right-hand corner of the **Flowchart** panel (refer to *Figure 9.3*).

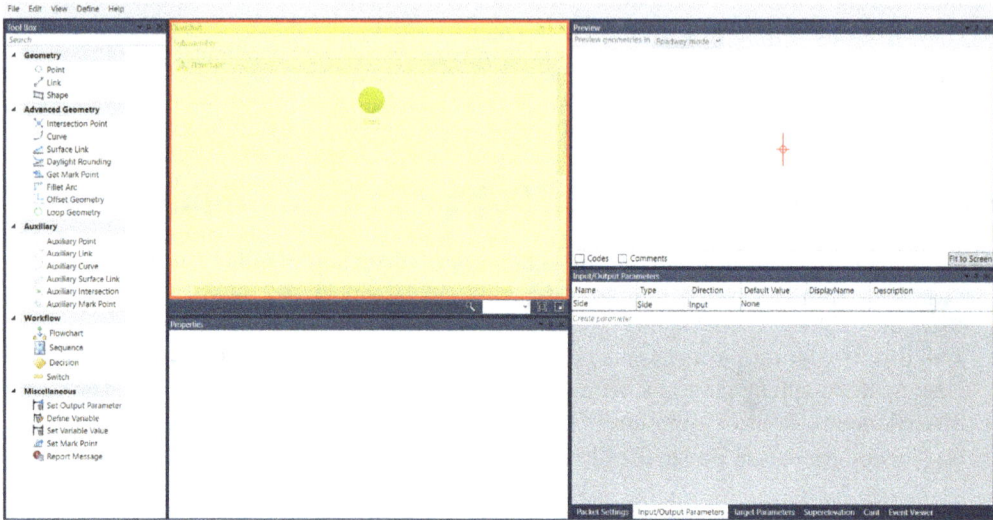

Figure 9.3 – Flowchart area

- **Preview**: Here, you can visualize the current geometry, which offers two modes:

  - **Roadway Model** showcases the subassembly built, considering target surfaces, elevations, and/or offsets.

  - **Layout Mode** displays the subassembly built solely based on input parameters, excluding targets.

  Additionally, the panel features the **Codes** (displayed in brackets [ ]) and **Comments** (displayed in parentheses ( )) checkboxes at the bottom. If Codes or Comments were entered in the properties for points, links, or shapes, this information will be listed next to the relevant geometry (refer to *Figure 9.4*).

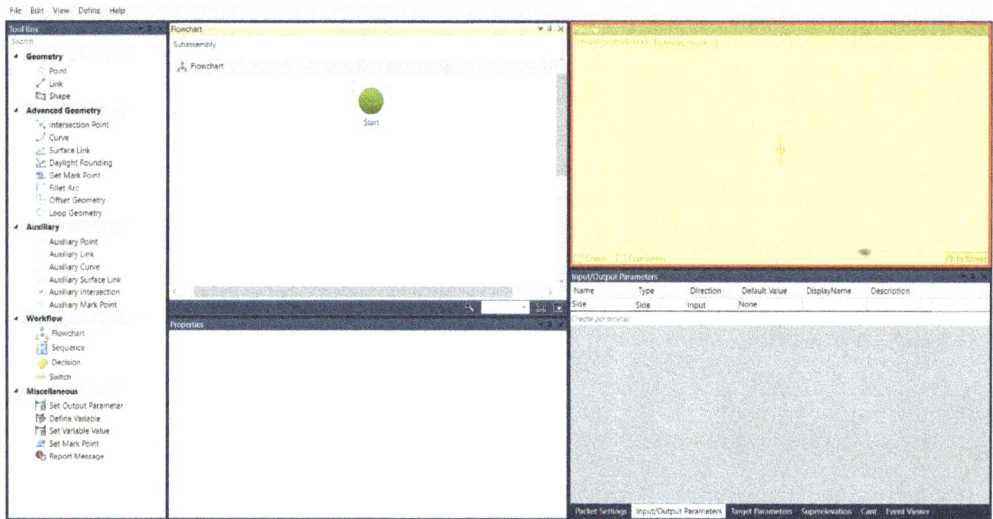

Figure 9.4 – Preview area

- **Properties**: This panel functions as the input location for parameters defining each geometry element (refer to *Figure 9.5*).

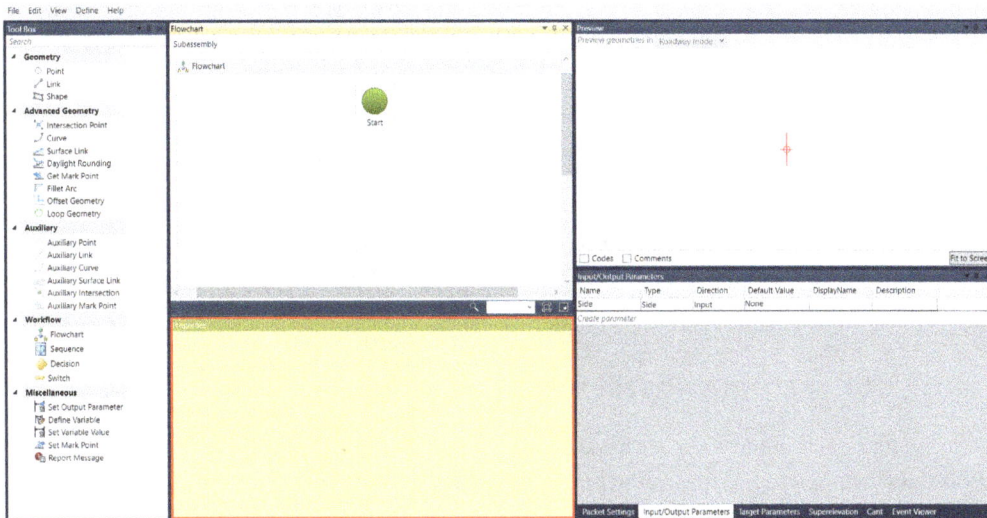

Figure 9.5 – Properties area

- **Packet Settings**: This panel is comprised of six tabs delineating the subassembly: **Packet Settings**, **Input/Output Parameters**, **Target Parameters**, **Superelevation**, **Cant**, and **Event Viewer**. Mastery of these panels and their functionalities will empower you to efficiently navigate and utilize Autodesk Civil 3D Subassembly Composer for your design needs (refer to *Figure 9.6*).

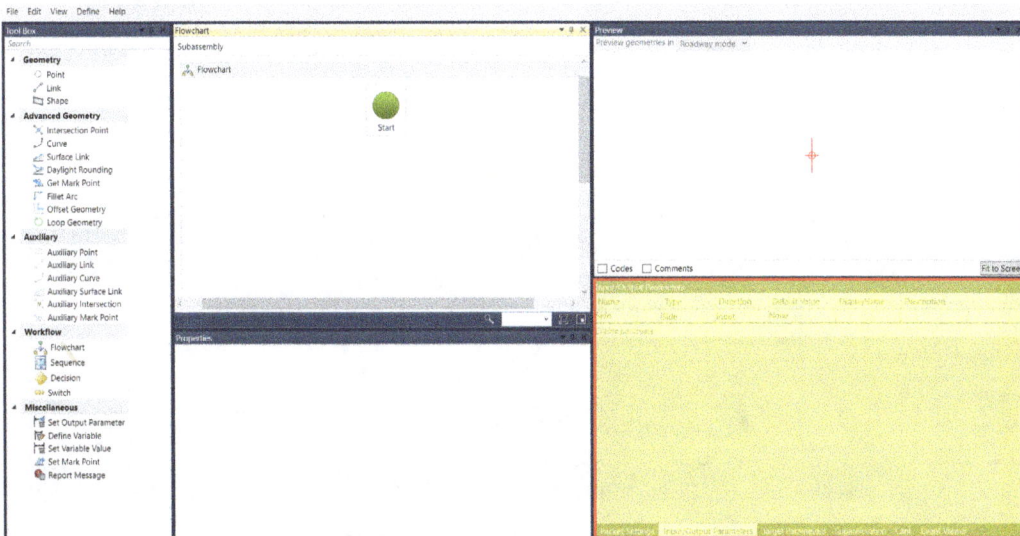

Figure 9.6 – Packet Settings area

> **Important note**
>
> We have the flexibility to reposition any of these windows independently through the use of docking controls. If there's a need to revert to the default layout, it can easily be accomplished by navigating to **View | Restore Default Layout**.

Along the top of Autodesk Subassembly Composer for Civil 3D 2025, we have some basic pulldowns, similar to other supplemental programs we've explored throughout this book, that allow us to save and access custom parts, edit our view, add custom enumerations and codes.

Now that we have a very rudimentary understanding of what we're looking at, let's go ahead and define the name and description of our custom subassembly, and then save the custom subassembly that we're going to create.

Defining our custom subassembly

To define the name and description, we'll want to bring our attention to the **Packet Settings** panel in the lower right-hand corner. With the **Packet Settings** tab selected, we'll fill out the fields as follows:

- **Subassembly Name**: `VaryingHeightRetainingWall`
- **Description**: **Retaining wall that varies in height based on elevation difference at daylight** (the description will appear within Civil 3D for you as you hover your mouse over the imported subassembly in your **Subassembly** tool palette)
- **Version**: Leave it as is for now
- **Help file**: Leave blank for now (we can attach guidance documentation here if available)
- **Image** (refer to *Figure 9.7* for full path): `Sketch_VaryingHeightRetainingWall.png`

These fields are shown in *Figure 9.7*:

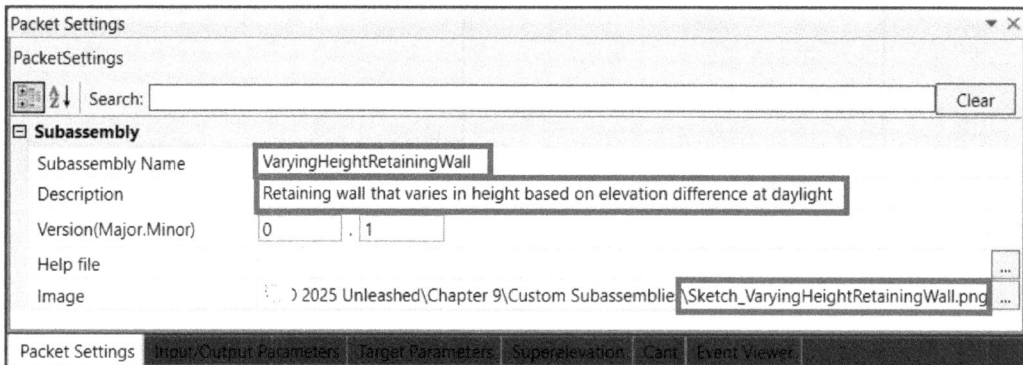

Figure 9.7 – Sketch of VaryingHeightRetainingWall

After filling out **Packet Settings**, we'll now want to save our custom subassembly.

Saving our custom subassembly

To do so, we'll go up to **File | Save As** in the top left-hand corner of Autodesk Civil 3D Subassembly Composer. Once the **Save As** dialog box appears, we'll navigate to the `Civil 3D 2025 Unleashed\ Chapter 9\Custom Subassemblies` folder and give it the name `VaryingHeightRe-tainingWall.pkt`.

As you've probably guessed, we're going to create a custom subassembly that will have a varying height. This height will be determined by the daylight tie-in location to existing grade. Before we get started with any sort of custom subassembly creation, it's best to have a plan and sketch out what we're trying to create so we have a clear direction going into it. As shown in *Figure 9.8*, we have sketched out our subassembly in a section view:

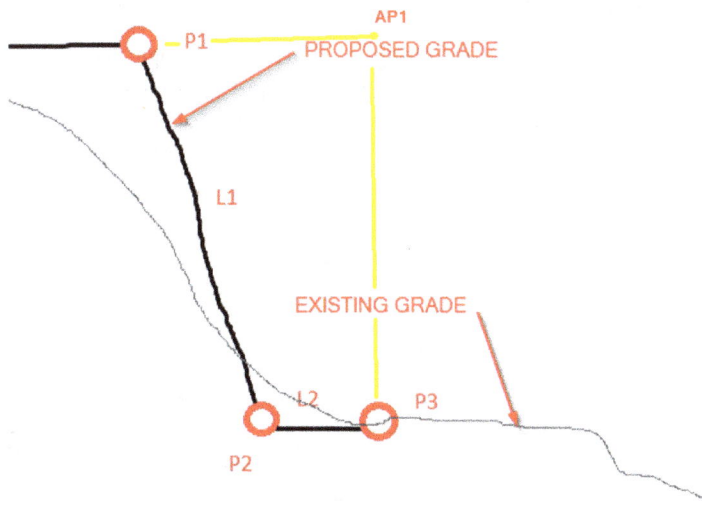

Figure 9.8 – Sketch of VaryingHeightRetainingWall

In the sketch shown in *Figure 9.8*, we have the following types of geometry that will need to be defined in our custom subassembly:

- **P1**: A hinge point that will connect to the edge of the curb in our assemblies and represent the top of our retaining wall

- **P2**: A hinge point that will represent the bottom of our retaining wall

- **P3**: A hinge point that will represent our tie-in to the existing grade

- **L1**: A varying height line that will represent the exposed side of our retaining wall

- **L2**: A varying width line that will represent the proposed grade that ties into the existing grade

- **AP1**: An auxiliary point that will be used to determine the overall width required to tie into the existing grade

With our basic understanding of the layout of this product, and what we want to accomplish by using it, let's get a little more familiar with the various components of Autodesk Subassembly Composer. That said, let's go ahead and jump into the next section to develop our custom subassembly.

Creating new parts in the Infrastructure Parts Editor

Referring to our sketch of **VaryingHeightRetainingWall** (*Figure 9.8*), we'll want to start populating **Input/Output Parameters** within the **Packet Settings** panel. It's best to capture and populate as much of **Input/Output Parameters** as possible at the beginning to limit the back and forth as much as possible while we build our custom subassembly. That's why it is always good to start with a simple sketch that we can use to make sure we have a pretty solid plan in place and a good idea of how we want to create our custom subassembly.

That said, let's jump back down to the **Packet Settings** panel in the lower right-hand corner of Autodesk Subassembly Composer. Once there, we'll select the **Input/Output Parameters** tab so that we can start pre-populating our subassembly parameters. Let's go ahead and add the following input/output parameters:

Input/Output Parameters				
Name	Type	Direction	Default Value	DisplayName
Side	Side	Input	Right	
P1	String	Input	0	CurbHingePoint
P2	String	Input	0	BottomWallHingePoint
P3	String	Input	0	DaylightConnectionPoint
SlopeofWall	Grade	Input	-1000.00%	SlopeofWall
WidthatBottomofWall	Double	Input	5	WidthatBottomofWall
SlopeofGroundatBottomofWall	Grade	Input	-2.00%	SlopeofGroundatBottomofWall
LinkCodes	String	Input	Top	Link Codes
Create parameter				

| Packet Settings | Input/Output Parameters | Target Parameters | Superelevation | Cant | Event Viewer |

Figure 9.9 – Input/Output Parameters

Since we're planning on tying our custom subassembly into a surface as well, we'll want to add a target parameter to our configuration. Let's move over to and activate the **Target Parameters** tab in the **Packet Settings** panel and add the following target parameter:

Target Parameters				
Name	Type	Preview Value	DisplayName	Enabled In Previe
TargetSurface	Surface ∨	0	Existing Grade	☑
Create parameter				

| Packet Settings | Input/Output Parameters | Target Parameters | Superelevation | Cant | Event Viewer |

Figure 9.10 – Target Parameters

Now, we have pretty much everything in place to get started with our custom subassembly creation process. We will now add points and links to our custom subassembly.

Adding point P1

Going back up to **Tool Box**, let's drag and drop the **Geometry | Point** object into **Flowchart** and fill out **Properties** as follows:

- **Point Number**: P1

- **Point Codes**: P1

- **Type**: Delta X and Delta Y

- **From Point**: Origin

- **Delta X**: 0

- **Delta Y**: 0

Figure 9.11 shows these properties:

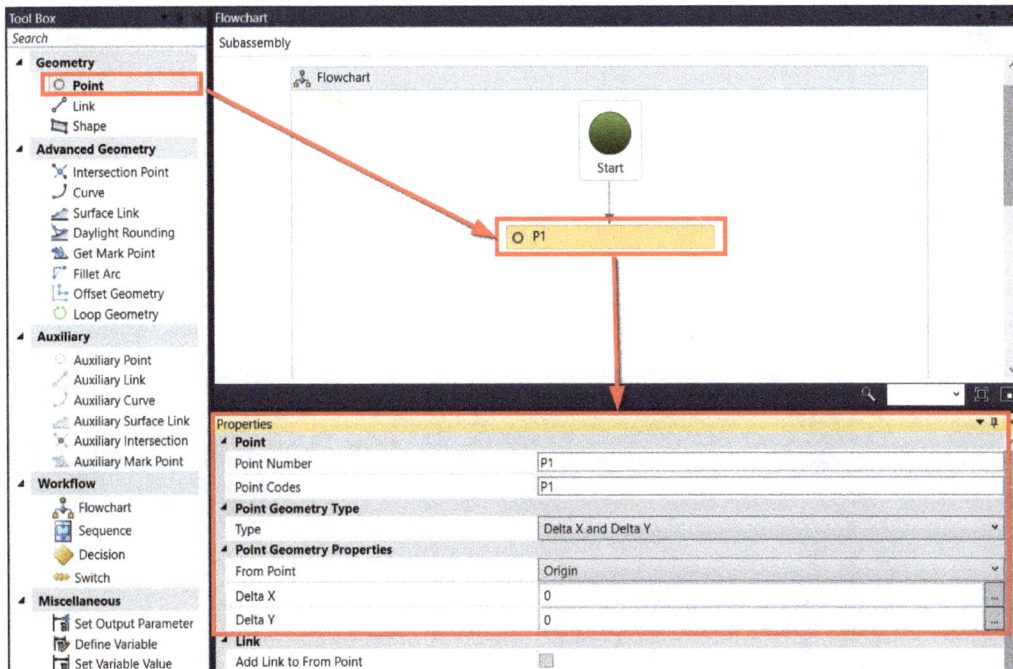

Figure 9.11 – P1 properties

If we direct our attention over to the **Preview** panel, we'll notice that we now have our P1 point displayed in the preview. We'll want to keep an eye on this area as we continue to build out our custom subassembly to make sure it matches our sketch as closely as possible.

Adding point AP1

Let's now go ahead and drag and drop **Auxiliary | Auxiliary Point** from **Tool Box** into **Flowchart** and fill out **Properties** as follows:

- **Point Number**: AP1

- **Type**: Slope and Delta X

- **From Point**: P1

- **Slope**: SlopeofGroundatBottomofWall(this is where we want to match the parameter that we defined in *Figure 9.4*)

- **Delta X**: WidthatBottomofWall (this is where we want to match the parameter that we defined in *Figure 9.4*)

Figure 9.12 shows these properties:

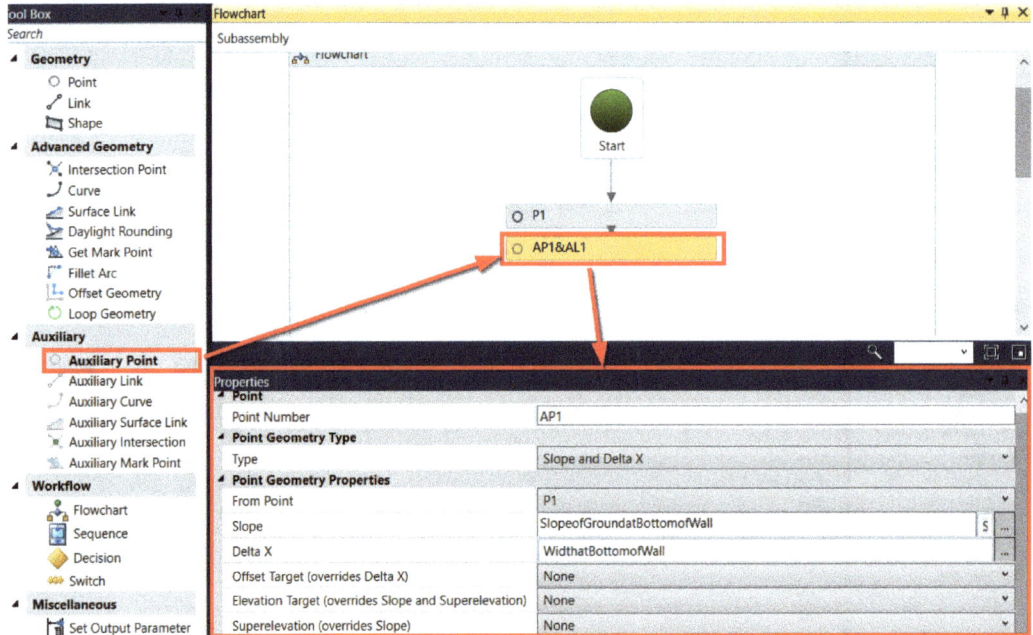

Figure 9.12 – AP1 properties

Adding point P3

Next, we'll go back up to **Tool Box**, drag and drop another **Geometry | Point** object into **Flowchart**, and fill out **Properties** as follows:

- **Point Number**: P3

- **Point Codes**: P3

- **Type**: Slope to Surface

- **From Point**: AP1

- **Slope**: SlopeofWall

- **Surface Target**: TargetSurface

Figure 9.13 shows these properties:

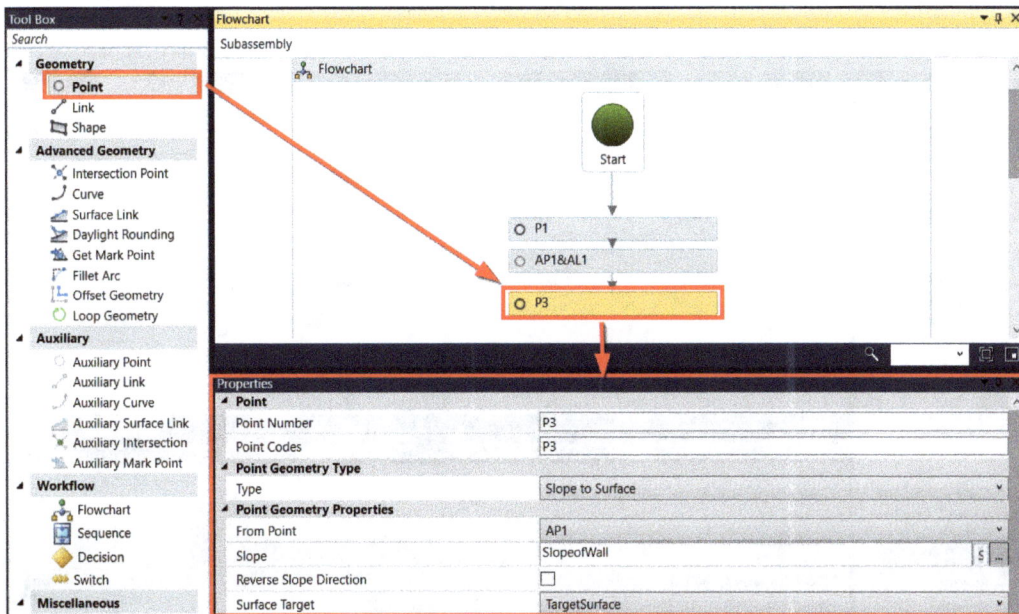

Figure 9.13 – P3 properties

Adding point P2

Finally, we'll go ahead and add our last point, P2, by dragging and dropping another **Geometry |
Point** object from **Tool Box** into **Flowchart** and filling out **Properties** as follows:

- **Point Number:** P2

- **Point Codes:** P2

- **Type:** Slope and Delta X

- **From Point:** P3

- **Slope:** -SlopeofGroundatBottomofWall (we add a - sign at the beginning of the
parameter to indicate the reverse since we want to work backward from geometry point P3)

- **Delta X:** -WidthatBottomofWall (we add a - sign at the beginning of the parameter to
indicate the reverse since we want to work backward from geometry point P3)

Figure 9.14 shows these properties:

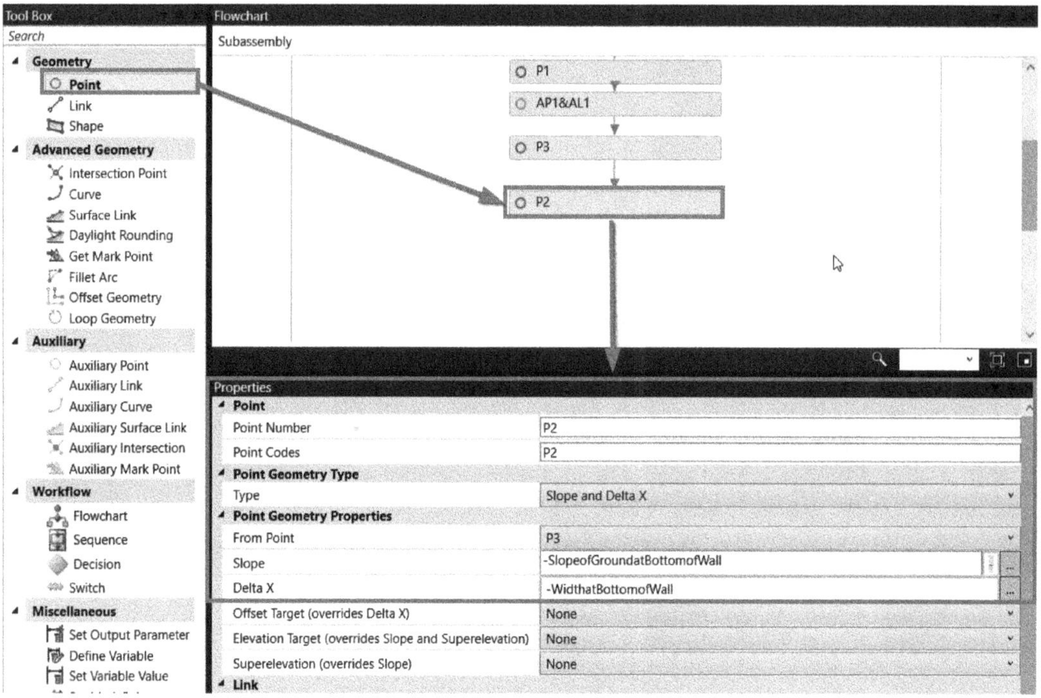

Figure 9.14 – P2 properties

> **Important note**
>
> Whenever we add points to our custom subassembly, the default setting is to add a link between those points. Because we are defining our beginning and end points first, we need to make sure that the **Add Link to From Point** property for all points (P1, P2, and P3) is unchecked in the last section of **Properties**.

Adding link L1

With all points defined, let's go ahead and add a link from the **Tool Box** panel under **Geometry** | **Link** and fill out **Properties** as follows:

- **Link Number**: L1
- **Link Codes**: LinkCodes
- **Start Point**: P1
- **End Point**: P2

Figure 9.15 shows these properties:

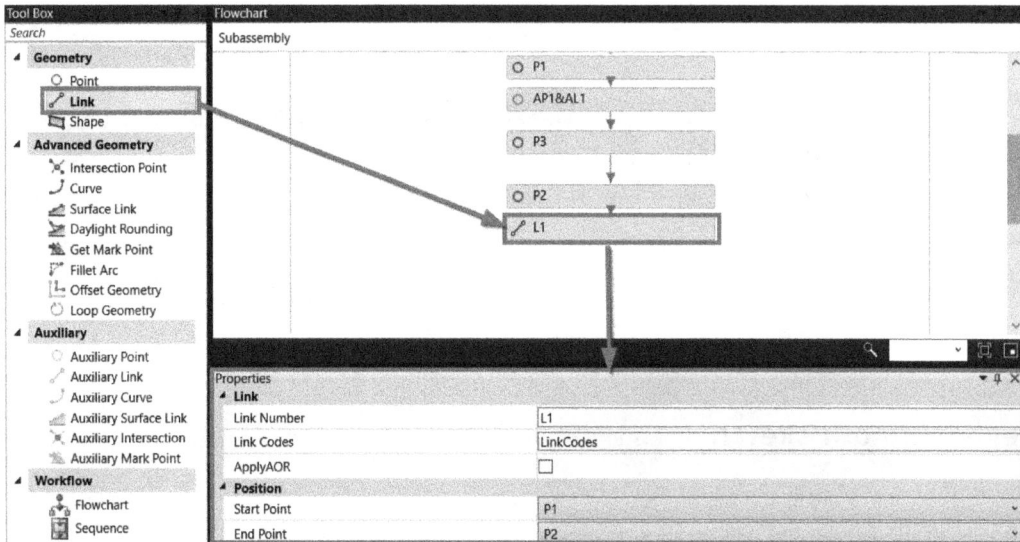

Figure 9.15 – L1 properties

Adding link L2

To finish up our custom subassembly, go ahead and add another link from the **Tool Box** panel under **Geometry** | **Link** and fill out **Properties** as follows:

- **Link Number**: L2
- **Link Codes**: LinkCodes
- **Start Point**: P2
- **End Point**: P3

Figure 9.16 shows these properties:

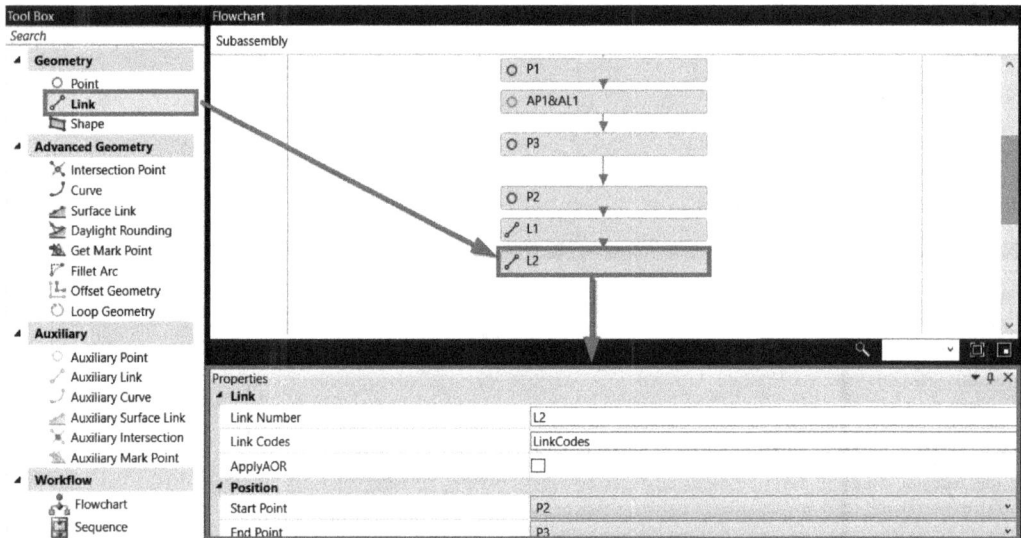

Figure 9.16– L2 properties

Previewing our sketch

If we shift our focus back over to the **Preview** panel, we should have a custom subassembly created that looks pretty similar to *Figure 9.17*:

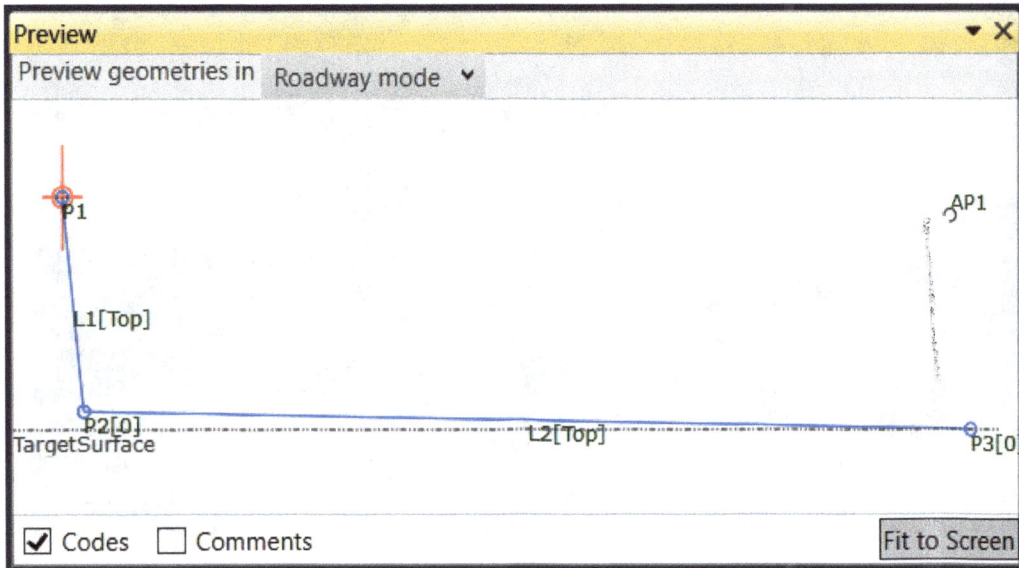

Figure 9.17 – Preview panel with codes displayed

Important note

Another great self-check to make sure the subassembly works, where all points and links are dynamically linked in our custom situation, is to left-click the mouse on **Target Surface** in the **Preview** panel and drag it up and down to see how the height varies. In doing so, you may notice that if the target surface is above geometry point P1, the subassembly configuration will not work. That said, to account for scenarios where P1 is in a cut situation, as compared to our target surface, we'll need to create a new custom subassembly that can be applied in cut situations as required.

With that, let's hop back into our grading model and learn how we can integrate our newly created custom subassembly into our roadway design.

Integrating custom parts into our BIM designs

Let's go ahead and launch Civil 3D 2025 and open up the `Grading Model_Start.dwg` file located within `Civil 3D 2025 Unleashed\Chapter 9\Model`. Once opened, let's navigate over to the **ASM - Subdivision Main Road – Access** assembly. Because we have several subassemblies in our file, we tend to rely on **Toolspace | Prospector** to navigate to our assemblies so that we can quickly identify which one we need to update. To do so, we'll expand our Assemblies, right-click with the mouse on the **ASM – Subdivision Main Road – Access** assembly, and select **Zoom To**.

Once there, we'll want to select and delete our proposed **BasicSideSlopeCutDitch** subassemblies, which are located after our **BasicCurbAndGutter** subassemblies, as shown in *Figure 9.18*.

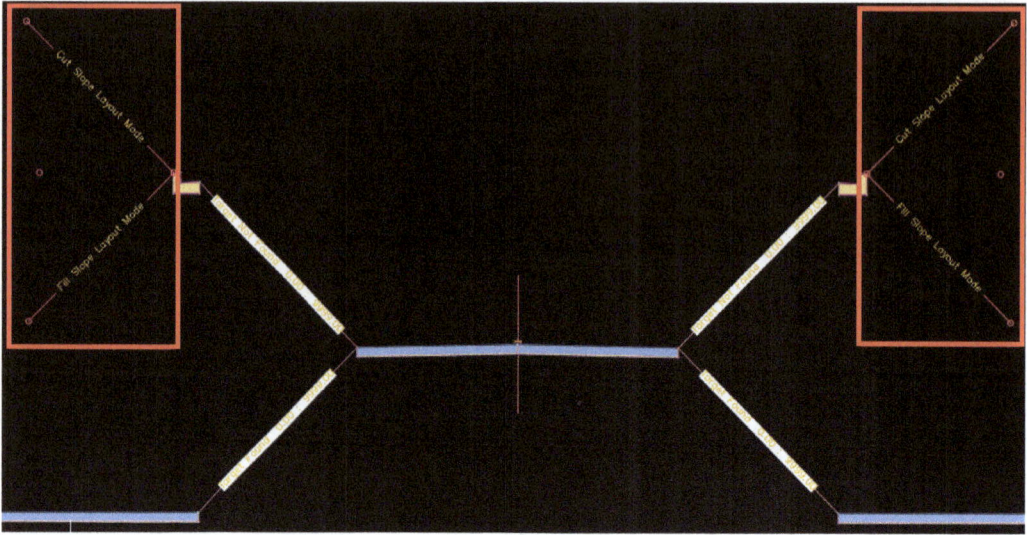

Figure 9.18 – Preview panel with codes displayed

Next, we'll select our **ASM – Subdivision Main Road – Access** assembly baseline to activate our assembly's contextual ribbon toward the top of our Autodesk Civil 3D 2025 session. Once activated, we'll navigate over to the **Launch Pad** panel and select **Tool Palette** to pull up **Subassembly** tool palette.

When the **Subassembly** tool palette appears, select the **Retaining Walls** category to make that current. Then right-click with the mouse on the tab itself and select the **Import Subassemblies…** option. After selecting this option, the **Import Subassemblies** dialog box will appear, where we want to navigate to where we originally saved our newly created custom `VaryingHeightRetainingWall.pkt` subassembly and import it into our **Retaining Walls** tool palette as shown in *Figure 9.19*:

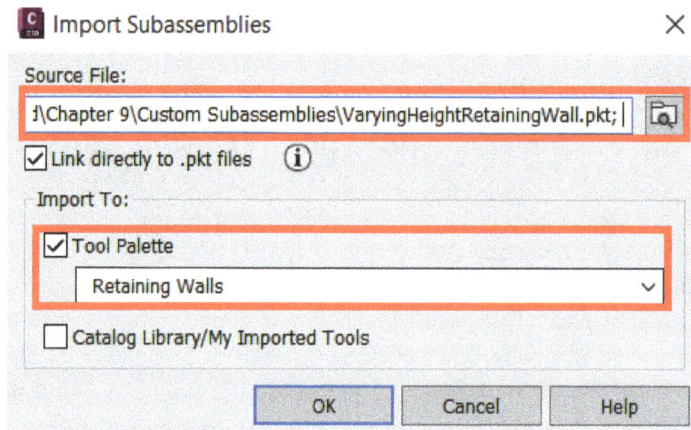

Figure 9.19 – Import VaryingHeightRetainingWall.pkt subassembly

> **Important Note**
>
> If we check the box directly under our **Source File** path – **Link directly to .pkt files**, our newly imported subassembly will now be dynamic in the sense that any updates made to the `.pkt` file moving forward will be synced with our model. Leaving this box unchecked will simply import the subassembly into our model as a static version.

After clicking the **OK** button in the **Import Subassemblies** dialog box, we'll notice that our custom **VaryingHeightRetainingWall** subassembly is now accessible in the **Retaining Walls** tool palette. Additionally, we also have our sketch image displayed for the preview next to it, along with a description of our subassembly if we hover our mouse over it.

Let's now go ahead and insert our **VaryingHeightRetainingWall** subassembly into our **ASM – Subdivision Main Road – Access** assembly as shown in *Figure 9.20*.

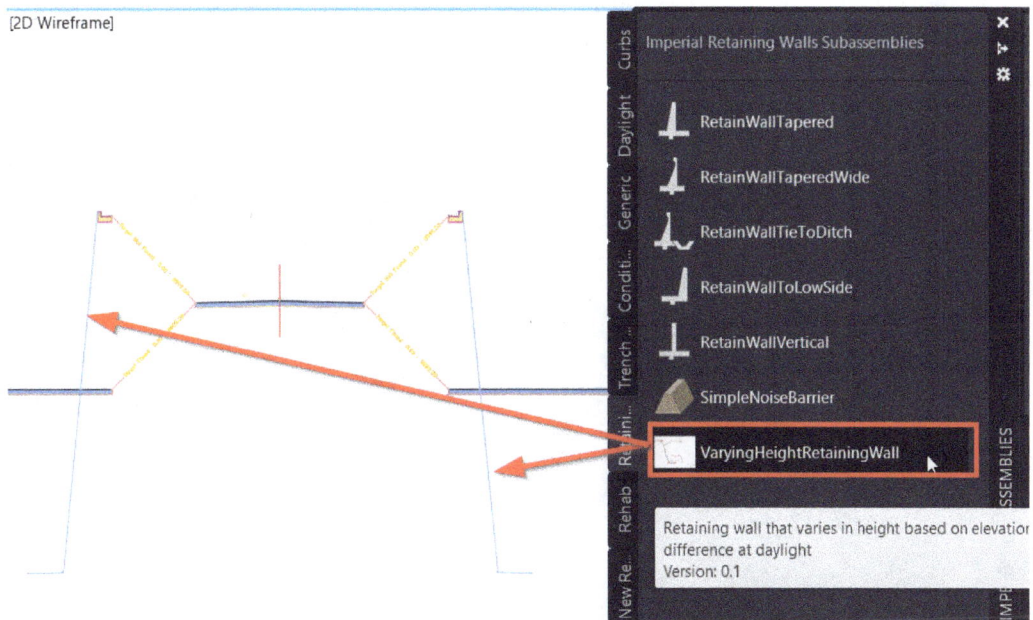

Figure 9.20 – Attach the VaryingHeightRetainingWall.pkt subassembly
to the ASM – Subdivision Main Road – Access assembly

With our **ASM – Subdivision Main Road – Access** assembly updated to include our custom **VaryingHeightRetainingWall** subassembly, we can now jump back over to our **COR - Subdivision Main Road – Access** corridor model and update it accordingly. As we update it, we'll want to remember to apply the surface we wish to target. In our case, we'll target the **SRF - Existing Grade – FromSurveyPoints** surface model in the **Surface** tab within the **Target Mapping** dialog box, as shown in *Figure 9.21*.

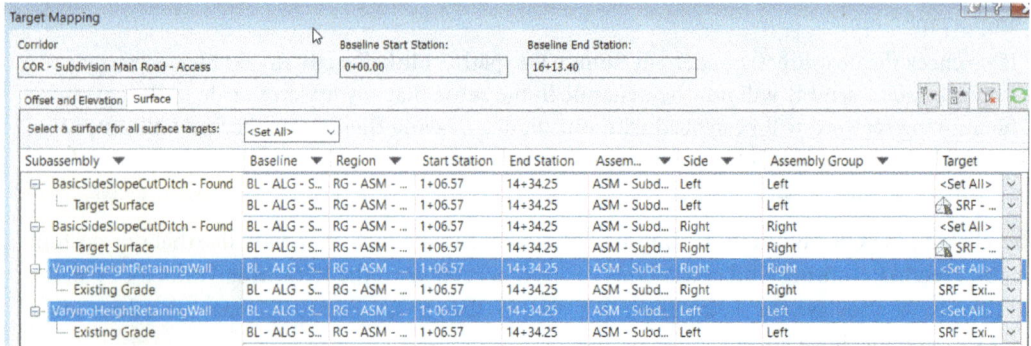

Figure 9.21 – Apply surface targets to the VaryingHeightRetainingWall subassembly

After another rebuild of our corridors and surfaces is applied with these new settings and targets, we should have an update to our **COR - Subdivision Main Road – Access** corridor model, looking similar to that displayed in *Figure 9.22*:

Figure 9.22 – Final display of our COR - Subdivision Main Road – Access corridor model

As you can see in *Figure 9.22*, we have a very smooth retaining wall with varying heights applied in all fill situations along our **COR - Subdivision Main Road – Access** corridor model. As mentioned earlier in this chapter, the custom **VaryingHeightRetainingWall** subassembly was created specifically to account for fill design situations. When in cut design situations, we'll want to create another **VaryingHeightRetainingWall** subassembly specific to those conditions as well. That said, to make sure you're completely comfortable with this tool and the workflow of developing custom roadway solutions, feel free to go back and tweak the workflows a little to develop a **VaryingHeightRetainingWall** subassembly specific to cut design situations.

Summary

As we transitioned from custom underground utility networks to custom roadway designs, our focus shifted to discovering the power of Autodesk Subassembly Composer. As we witnessed throughout this chapter, Subassembly Composer provides a platform where creativity converges with functionality.

Throughout this chapter, we explored and familiarized ourselves with Autodesk Subassembly Composer, all while developing a custom subassembly from a simple sketch we created (just as most design concepts/ideas start with). We also learned how we can take that simple sketch and apply a custom engineering solution to our overall design to further enhance our understanding of the overall capabilities that Autodesk Civil 3D 2025 offers us, all while elevating our knowledge and skills to up our civil BIM game.

In the next chapter, we'll begin exploring ways we can bolster our models and design objects to inform project stakeholders at all levels of pertinent information related to our civil BIM designs. Information related to our models and objects can greatly improve collaboration, the construction process, asset management, and maintenance after the design and even the construction is completed. All of this can be achieved through the application of property sets, provided we apply them correctly and are able to surface the right information for the stakeholder's purpose.

Part 3:
Managing Information Models and Automating Workflows

To take our new custom design solutions to another level, we'll begin exploring what it means to build pertinent information into our models, better analyze our civil BIM designs using the ultimate civil Model Manager Tool, and increase efficiency through automation. This is where we really begin morphing into that model manager role in our projects, taking our careers to the next level.

This part contains the following chapters:

- *Chapter 10, Information Modeling with Property Sets*
- *Chapter 11, Introduction to Project Explorer*
- *Chapter 12, Automating Routine Workflows with Dynamo and Scripting*

10

Information Modeling with Property Sets

Starting with this chapter, we will learn how we can take our new custom design solutions to another level. We'll begin exploring what it means to build pertinent information into our models, better analyze our civil BIM designs, and increase efficiency through automation. Here is where the rubber meets the road and we really begin morphing into that BIM model manager role on our projects, taking our careers to the next level.

In this particular chapter, we'll dive into the transformative power of information modeling with property sets. We'll discover how we can enhance the effectiveness of our models and objects, empowering project stakeholders with the tools to access and utilize pertinent information seamlessly. As designers and engineers, our goal extends beyond the mere creation of designs; we strive to enrich them with actionable insights that foster collaboration, streamline construction processes, and ensure optimal asset management and maintenance long after the design phase concludes.

Throughout this chapter, we will equip ourselves with essential skills that lie at the heart of information modeling with property sets. Firstly, we will unravel the intricacies of Style Manager within Civil 3D, mastering the art of accessing and navigating this vital tool. From there, we will journey into the realm of crafting custom property sets, learning how to harness a diverse array of definitions to tailor our data to the specific needs of our projects. Finally, we will explore the practical application of property sets within our civil BIM designs, understanding how to seamlessly integrate them to elevate collaboration and optimize project outcomes. By honing these skills, you will emerge prepared to leverage the full potential of property sets, transforming your civil BIM designs into dynamic and informative assets that drive success in every phase of project development.

That said, in this chapter, we'll cover the following topics:

- Introduction to property sets and Style Manager
- Creating property sets with an array of property definitions
- Applying property sets to label our modeled objects

Technical requirements

We will be using the same hardware and software requirements as discussed in the *Technical requirements* section of *Chapter 1*.

With that, let's go ahead and open up our `Utility Model_Start.dwg` file located within `Civil 3D 2025 Unleashed\Chapter 10\Model`. Once opened, you'll notice that the file has been zoomed in to our site, as displayed in *Figure 10.1*.

Figure 10.1 – Utility Model_Start.dwg display

With our `Utility Model_Start.dwg` file now opened up, let's begin familiarizing ourselves with how we can apply custom property definitions to our modeled objects within Autodesk Civil 3D 2025.

Introduction to property sets and Style Manager

Although there is a lot of added value and benefit to utilizing property sets in our design files, it must be done so with intention. We certainly don't want to have unnecessary information and work being done that could lead to budget overruns or even file bloat. It's recommended to make sure that whatever custom property sets are defined in our files are in close alignment with downstream uses either in the design phase for labeling and reporting data or future construction, asset, and/or operations needs and requirements.

Thinking through and mapping out what you would like to achieve with property sets will allow you to stay focused on your end goal. Property sets can add many benefits to your projects, including the following:

- Additional tagging and placement for metadata

- Labeling and tabling

- Quick reference/linking to pertinent details (cut sheets, specs, etc.)

- Reporting

- Supported via **Industry Foundation Classes (IFC)** exchanges

- Preparation for construction management

- Preparation for asset management

Although some property definitions will require manual input and configuration, there are far fewer instances of this required today than there were several years ago. Autodesk Civil 3D 2025 and additional Autodesk-based products continue to evolve, while more improvements/enhancements are being made available to streamline the automation of property set creation and assignments to support our (Civil) needs.

Before jumping into it, we should recognize that there are nine different types of property definitions that can be applied to objects within Autodesk Civil 3D 2025, although only three of these types are truly supported for Civil objects being developed. These three types of property definitions are as follows:

- **Manual**: Where the population of values assigned will require manual input

- **Automatic**: Where the population of values assigned is automatically defined

- **Formula**: Where the population of values assigned is automatically defined based on formulaic equations associated with each property definition

Now, to configure property sets in our current drawing, there are a couple of ways to access the main dialog box where these can be configured: type either PROPERTYSETDEFINE or STYLEMANAGER at the command line to pull up the property set configuration dialog box. It is important to note that typing STYLEMANAGER allows for additional ways to configure some level of automation and predefined lists.

Let's dive into this workflow using the following steps:

1. Once you have typed either one of these into your command line at the bottom of your session, the **Style Manager** dialog box will appear, as shown in *Figure 10.2*:

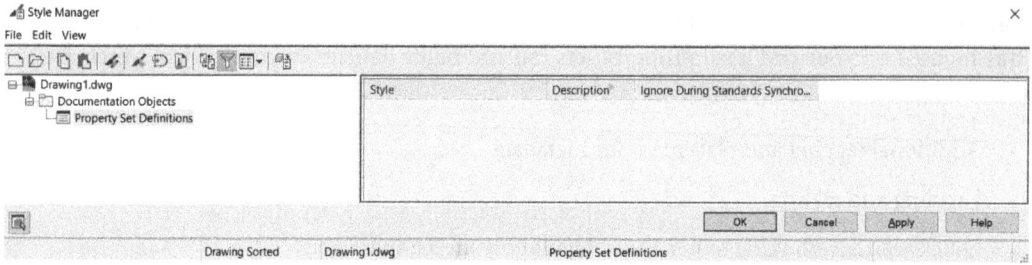

Figure 10.2 – Style Manager dialog box

2. After the **Style Manager** dialog box appears, we'll right-click on the **Property Set Definitions** option listed and select **New**, as shown in *Figure 10.3*.

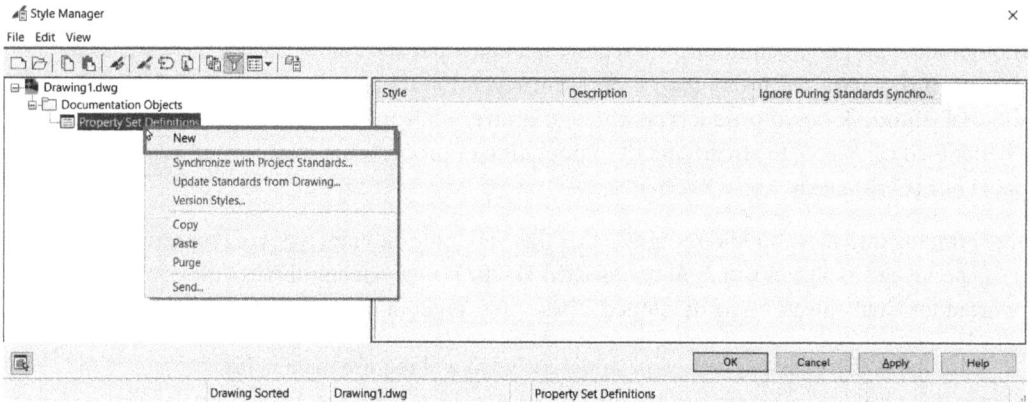

Figure 10.3 – Create new property set definition

3. Once created, a new property set will be created with the **General** tab activated. Here, we'll want to give it a new name; in this case, we'll call it `Storm Drainage` (as shown in *Figure 10.4*).

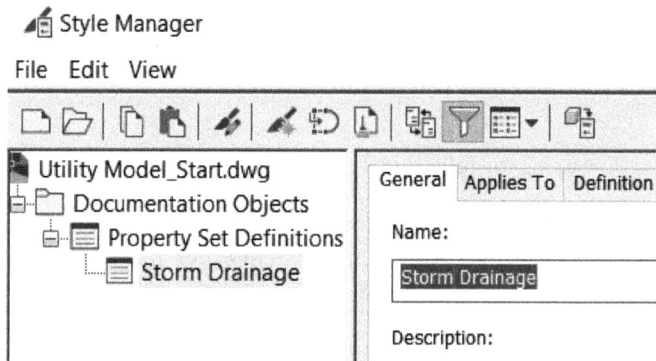

Figure 10.4 – Naming our new property set

4. Next, we'll switch over to the **Applies To** tab. Here, we have the option to select any objects we want to associate our property sets to. In our case, we only want to focus on adding additional information to our **Storm Drainage** pipes. That said, let's go ahead and check the box next to **Pipe**, as shown in *Figure 10.5*.

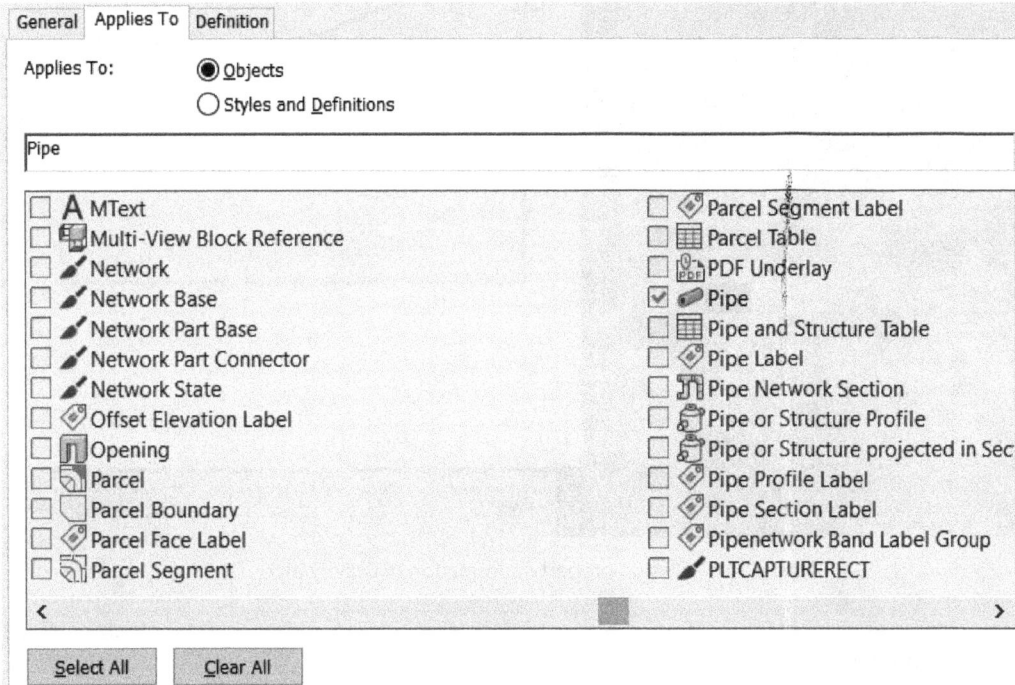

| General | Applies To | Definition |

Applies To: ◉ Objects
               ○ Styles and Definitions

Pipe

☐ A MText   ☐ ⬦ Parcel Segment Label
☐ Multi-View Block Reference   ☐ Parcel Table
☐ Network   ☐ PDF Underlay
☐ Network Base   ☑ Pipe
☐ Network Part Base   ☐ Pipe and Structure Table
☐ Network Part Connector   ☐ Pipe Label
☐ Network State   ☐ Pipe Network Section
☐ Offset Elevation Label   ☐ Pipe or Structure Profile
☐ Opening   ☐ Pipe or Structure projected in Sec
☐ Parcel   ☐ Pipe Profile Label
☐ Parcel Boundary   ☐ Pipe Section Label
☐ Parcel Face Label   ☐ Pipenetwork Band Label Group
☐ Parcel Segment   ☐ PLTCAPTURERECT

[Select All]   [Clear All]

Figure 10.5 – Selecting objects to associate property sets to

5. At this point, we have officially created a new property set that can be applied to our pipe objects (**Gravity**, not **Pressure**, as there is a selection specific to pressure pipes). If we now close out of the **Style Manager** dialog box, select one of the Gravity pipes in our file, and pull up our **Properties** dialog box (accessible by right-clicking on **Properties** or hitting the *F1* key), we'll immediately see some basic information related to pipes available in our **Design** tab within the **Properties** dialog box, similar to that displayed in *Figure 10.6*.

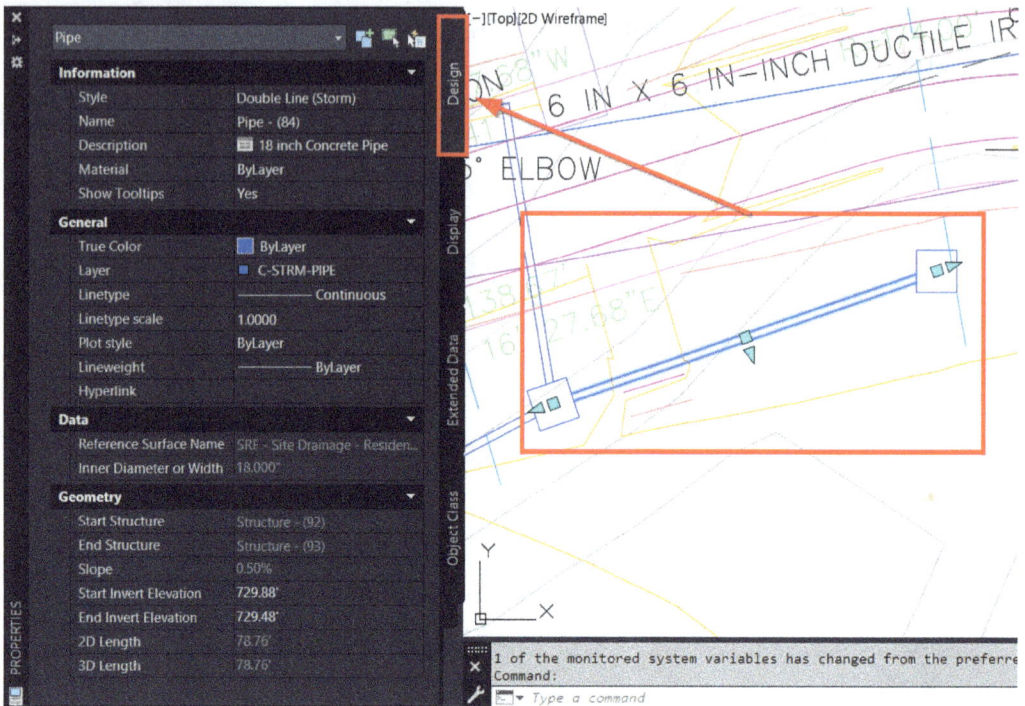

Figure 10.6 – Design property information of Gravity pipe

Important note

As you can see, right off the bat, we're able to see that the properties dialog box contains some basic, but valuable, information related to the pipe design itself. More information related to that particular pipe can be accessed by selecting the pipe, right-clicking, selecting your Pipe Properties selection, and going to the **Part Properties** tab.

6. To assign the newly created property set (**Storm Drainage**) to our particular pipe, we'll switch over to the **Extended Data** tab in the **Properties** dialog box and click on the icon in the lower left-hand corner (when hovered over the icon, **Add Property Sets** will display), as shown in *Figure 10.7*.

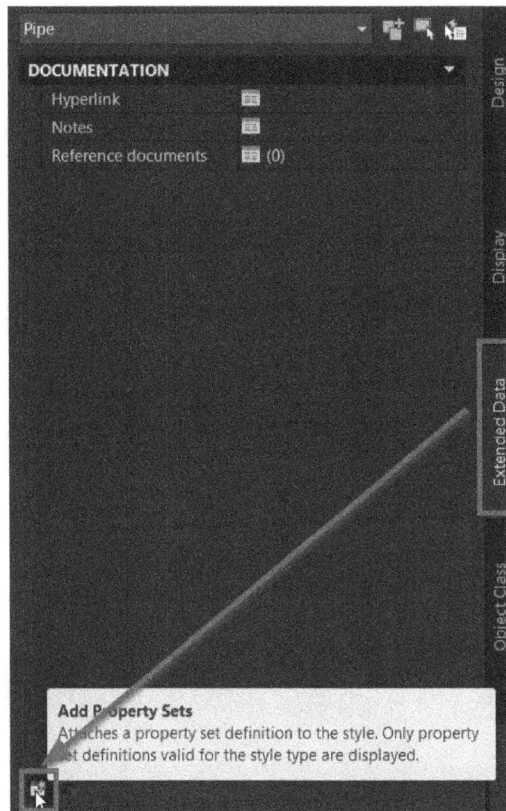

Figure 10.7 – Add Property Sets option in the Properties | Extended Data tab

7.   After clicking the icon, an **Add Property Sets** dialog box will appear, at which point we'll go ahead and check the box next to **Storm Drainage** and click on **OK** (as shown in *Figure 10.8*).

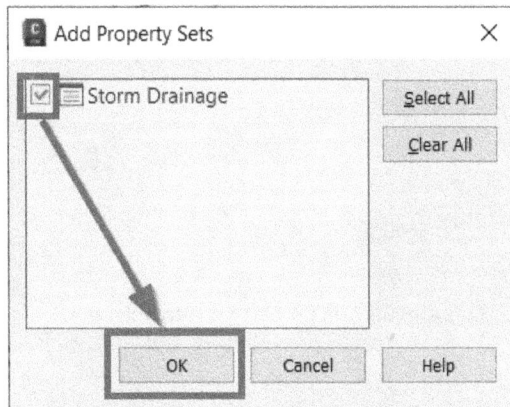

Figure 10.8 – Add property sets to objects

Congratulations! Although it may not appear as such immediately, we have now assigned a property set to our **Pipe** object in our current file! The reason we are not able to see this assignment within the **Properties** dialog box yet is due to the fact that we have not added any property definitions to our property set currently. Rest assured, as we add property definitions to our property sets, we will be able to view this information in the **Extended Data** tab of our **Properties**, so long as the pipe that we have assigned the **Storm Drainage** property set to is selected. Now, we can explore our options for creating property definitions within the **Storm Drainage** property set itself.

Creating property sets with an array of property definitions

Now, to begin adding property definitions to be displayed within our property sets, we'll follow these steps:

1.   We'll begin by typing PROPERTYSETDEFINE (or STYLEMANAGER) at the command line again to pull up our **Style Manager** dialog box.

2.   Once the **Style Manager** dialog box appears, we'll go ahead and expand the **Documentation Objects | Property Set Definitions** sections in our tree on the left side.

3.   Then, we'll select the **Storm Drainage** property set we just created and add a **Manual**, **Automatic**, and **Formula** property definition, as follows:

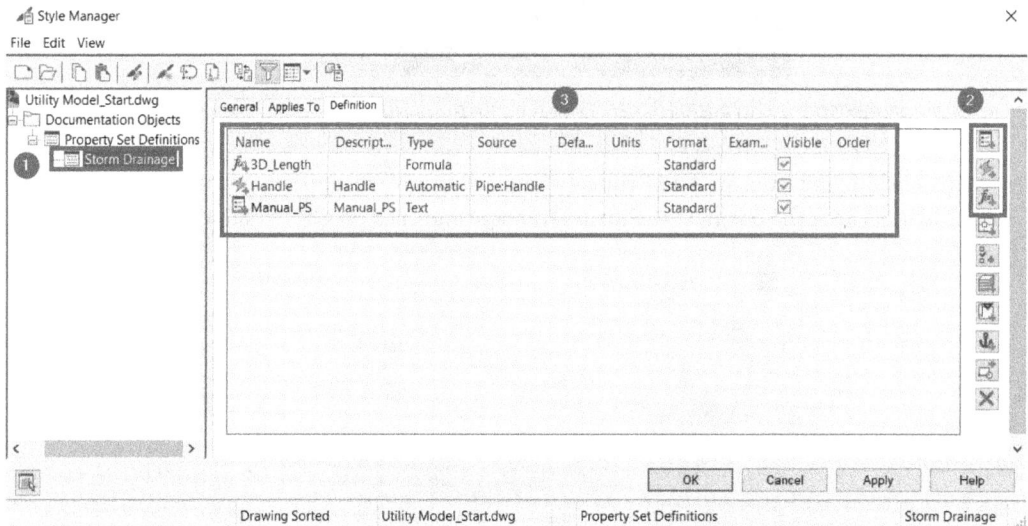

Figure 10.9 – Add property sets to objects

4. Starting with the **Manual** property definition, we'll click on the **Add Manual Property Definition** icon along the right side, give it a name of `Manual_PS` for now, and then click **OK**. Your **Manual_PS** property definition will then display in your **Storm Drainage** property set display (refer to *Figure 10.10*):

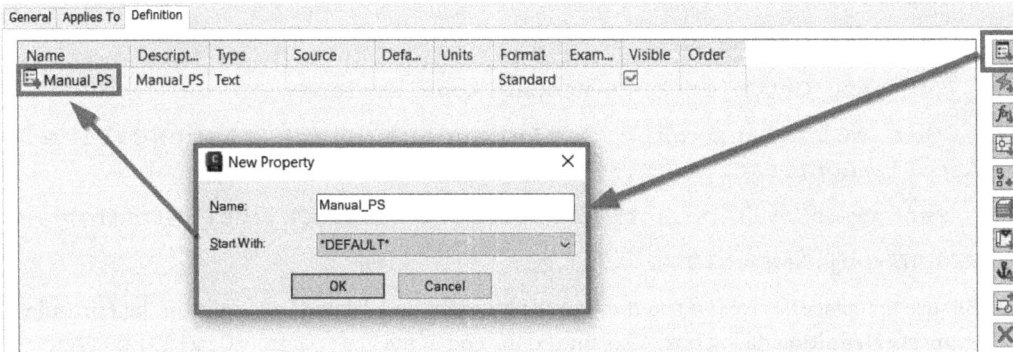

Figure 10.10 – Create Manual property set

5. Next, we'll click on the **Add Automatic Property Definition** icon, check the box for **Handle** listed in the **Pipe** section, and click on **OK**. The **Handle** automatic property definition will be added to our **Storm Drainage** property set display, as shown in *Figure 10.11*:

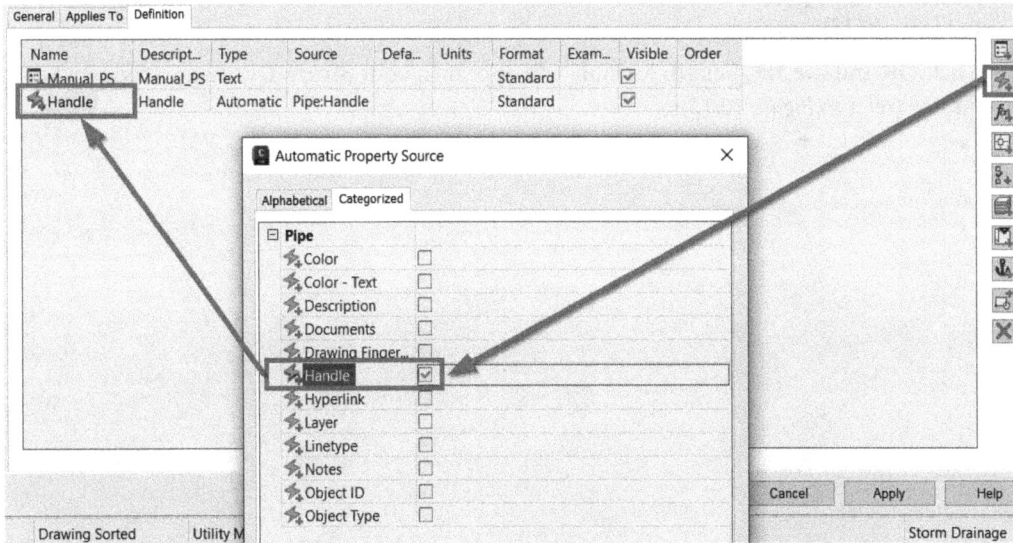

Figure 10.11 – Create automatic property set

6. Finally, let's go ahead and click on the **Add Formula Property Set** icon.

7. When the **Formula Property Definition** dialog box appears, give the property definition a name of 3D_Length and then plug the following formula into the **Formula** section:

```
RESULT="--"

On Error Resume Next

Set oApp=GetObject(, "AutoCAD.Application")

 Set oCivilApp=oApp.GetInterfaceObject("AeccXUiPipe.
AeccPipeApplication.13.7")

 Set obj=oCivilApp.ActiveDocument.HandleToObject("[Handle]")

RESULT=obj.Length3D
```

   Be sure to replace [Handle] in the formula by selecting a [Handle] value in the **Formula Property Definition** dialog box. Also, toward the end of the line   Set oCivilApp=oApp. GetInterfaceObject("AeccXUiPipe.AeccPipeApplication.13.7"), it's important to note that 13.7 represents the Civil 3D product version. If you are ever in doubt about the product version you are currently using, type ABOUT at the command line to view the current Civil 3D version you're running.

8. Then, double-click the **Storm Drainage Handle Automatic** property definition we just created in the **Insert Property Definition** section in the lower left-hand corner of the dialog box (refer to *Figure 10.12*).

9. Click **OK** and the **3D_Length** formula will appear in your **Storm Drainage Property Set** display (refer to *Figure 10.12*):

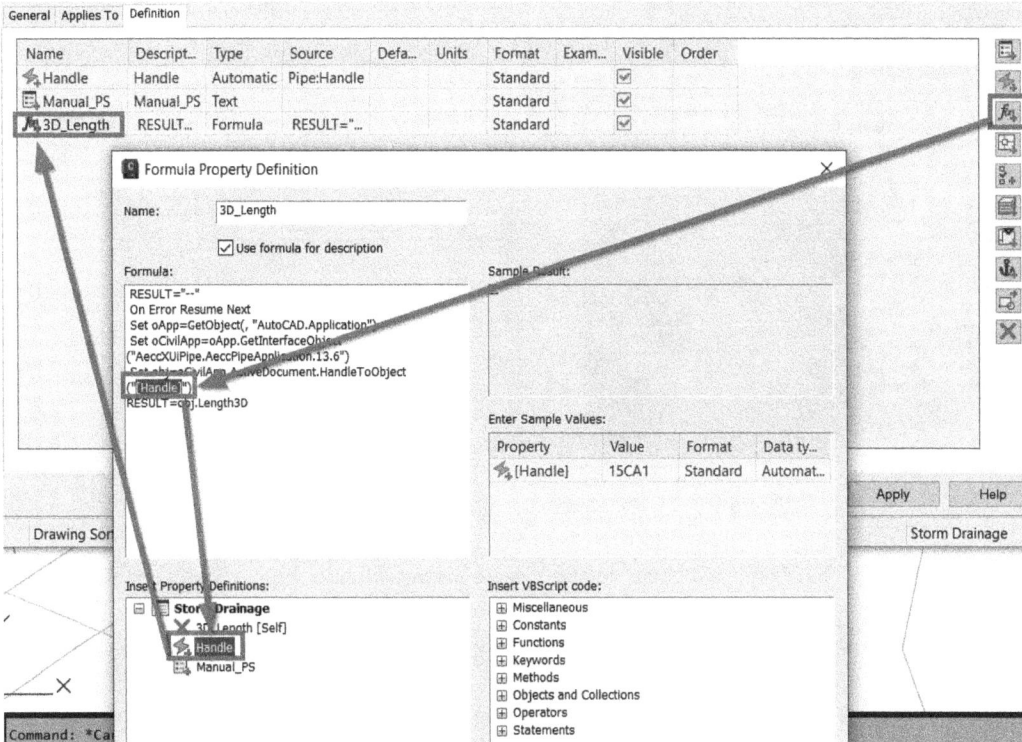

Figure 10.12 – Create a formula property set

10. Now that we've applied several types of property set definitions to our **Storm Drainage** property set, we'll select **Apply** and **OK** in the main **Style Manager** dialog box to close out.

11. Next, we'll hop back down into our model and select the pipe that we previously added to the **Storm Drainage Property Set**.

12. Then, go into the **Properties** dialog box and select the **Extended Data** tab, and all property definitions will be listed and associated with your **Pipe** object, where the values displayed in the **Automatic** and **Formula Property Definition** fields will be automatically populated, as shown in *Figure 10.13*:

Figure 10.13 – Updated property set extended data

With this exercise, we have only just begun to learn about the potential for property sets and property definitions. In the next section, we will learn more about property set definitions and look at the difference between manual and automatic opportunities within Civil 3D.

Learning more about property set definitions

Automatic and **Formula** present opportunities to automate the mapping of metadata and automatically fill in values associated with these types of definitions. Manual property set definitions offer opportunities to manually as well as automatically fill in values, depending on the values you wish to assign.

Within the **Manual** property set definition, we have the following options available to us:

- **Auto Increment – Character**: Automatically sets an alpha value based on the order in which the root property set is assigned to objects

- **Auto Increment – Integer**: Automatically sets a number value based on the order in which the root property set is assigned to objects

- **Integer**: Manually enter a whole number (no decimal placement capabilities)

- **List**: Ability to assign a preset list of values (accessible through **STYLEMANAGER | Multi-Purpose Objects | List Definitions**)

- **Real**: Manually enter numbers; allows for decimal placement

- **Text**: Manually enter values

- **True/False**: Boolean

Within the automatic property set definition, we have the ability to automatically fill in the values of the definitions based on the identified source(s) selected. The list of values can change based on the types of objects we are associating/assigning to. In our example shown in *Figure 10.14*, we can apply several values automatically based on the **Pipe** object:

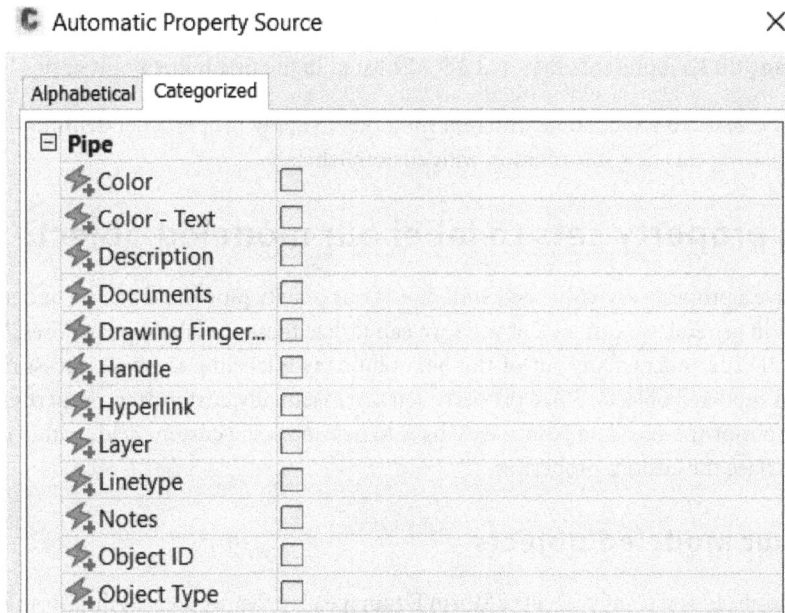

Figure 10.14 – Automatic Property Source

These automatic property sets are necessary in the event we'd like to use the **Formula** property set, as they enable us to tie a formula back to specific objects that we will extract information from and automatically assign as the value.

When it comes to **Formula** property set definitions, we have the ability to apply code and leverage AutoCAD and Civil 3D APIs to automatically assign values associated with each object. In the following example, we are presented with example code with a little explanation as to what the highlighted text represents:

```
RESULT="--"
On Error Resume Next
Set oApp=GetObject(, "AutoCAD.Application")
Set oCivilApp=oApp.GetInterfaceObject("AeccXUiPipe.
AeccPipeApplication.13.6")
Set obj=oCivilApp.ActiveDocument.HandleToObject("[Handle]")
RESULT=obj.Length3D
```

The code applied equates to the following aspects for our understanding:

- `AeccXUiPipe.AeccPipeApplication` represents the Civil 3D API library
- `13.7` represents the version of Civil 3D that is currently in use
- `[Handle]` represent the automatic property set we just created
- `obj.Length3D` represents the Civil 3D API listed in the first bullet point of this list

Now that we have learned about some different methods to apply property set definitions, we will move on to displaying this new information with custom labels.

Applying property sets to label our modeled objects

Now that we have a property set associated with one of our gravity pipes, and a basic understanding of property sets in general, we can look at ways we can utilize these for labeling purposes. We know Autodesk Civil 3D 2025 offers many out-of-the-box solutions for labeling as we can access many data points from our modeled objects. Since property sets are essentially customized properties and not included in our out-of-the-box data points, we'll have to develop some custom fields within our **Label** properties to surface the custom properties.

Labeling our modeled objects

With that, let's go ahead and simply label the **Storm Drainage** pipe that we applied the **Storm Drainage** property set to. To do so, we'll follow these steps:

1. We'll select the **Storm Drainage** pipe in model space, right-click, and select the **Add Label** option. This workflow will quickly label our **Storm Drainage** pipe with the default Civil 3D **Pipe Label Style** set in our file; in our case, it will be **Length Description and Slope Label Style**.

2. To quickly configure a new pipe label style, let's go ahead and select the newly inserted pipe label, right-click, and select the **Edit Label Style** option.

3. Once the **Pipe Label Style** dialog box appears, go ahead and select the **Create New** option, as shown in *Figure 10.15*.

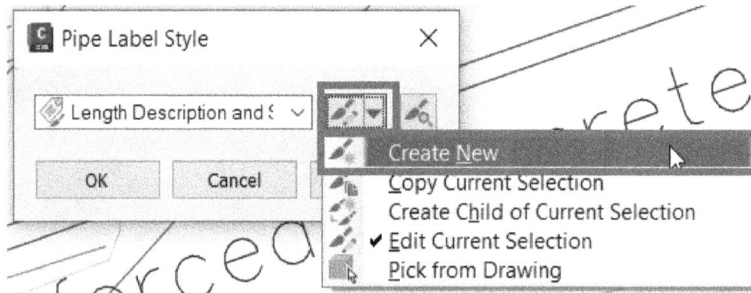

Figure 10.15 – Create new pipe label style

4. When the **Label Style Composer** dialog box appears, we'll fill in the fields in the **Information** tab as follows:

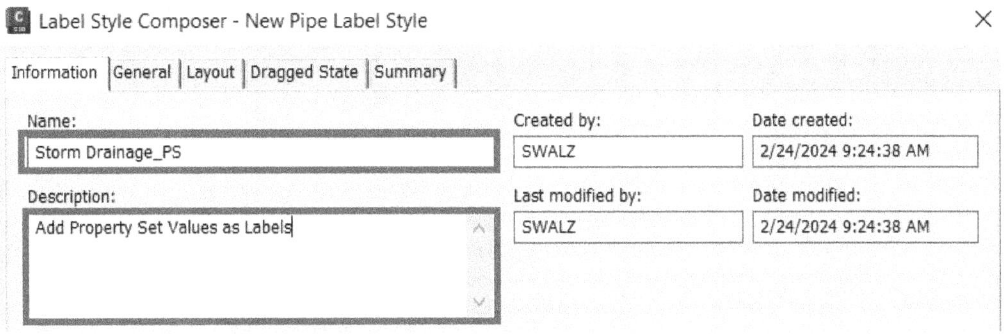

Figure 10.16 – Label Style Composer | Information tab

5. Use the following steps to update information we'd like to include in our pipe labels:

   I.   Switch over to the **Layout** tab in **Label Style Composer**.

   II.   Click on the ellipsis icon in the **Text | Contents** section to pull up the **Text Composer Editor** dialog box.

   III. Switch over to the **Property Sets** tab in the **Text Component Editor** dialog box.

   IV.   Select the property set definitions and properties you'd like to populate the label with and then click on the arrow to add them to the **Text Component Editor** area. In our case, we're going to add the **Manual_PS** and **3D_Length** properties to our label.

> **Note**
> It's recommended that you select and remove any existing content within the **Text Component Editor** area prior to adding new property set definitions.

   V.   Click the **OK** button in the **Text Component Editor** dialog box.

   VI.   Click the **OK** button in the **Label Style Composer** dialog box.

Figure 10.17 displays the preceding steps:

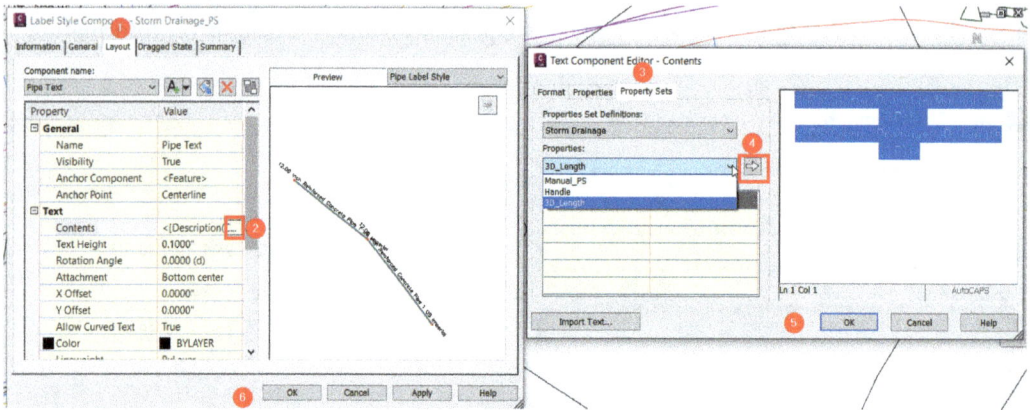

Figure 10.17 – Creating a custom label style using property sets

Our newly created Storm Drainage PS Pipe Label Style should now be applied to our pipe with only the **3D_Length** value displayed in the label itself, as shown in *Figure 10.18*.

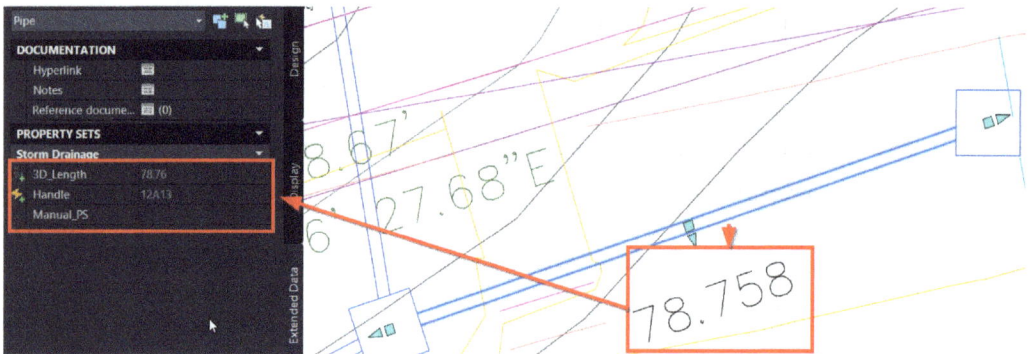

Figure 10.18 – Storm Drainage PS Pipe Label Style applied

Although we added the **Manual_PS** property to our **Pipe Label Style** as well, we actually don't have a value assigned to our **Manual_PS** property. If we were to add one now manually, the pipe label would automatically update to include the newly populated value for the **Manual_PS** property.

Using lists to apply property sets to labels

Alternatively, if we'd like to apply some additional levels of automation to these labels, we can go back to the **Style Manager** dialog box and update it accordingly. In this situation, we'd like to rely on prepopulated lists to fill in the values of our **Manual_PS** manual property set definition. To create a list, though, we'll need to type STYLEMANAGER instead of PROPERTYSETDEFINE at the command line to activate the **Style Manager** dialog box, and then follow these steps:

> **Note**
>
> By typing STYLEMANAGER, we gain access to some additional capabilities within the **Style Manager dialog** box.

1. Once the **Style Manager** dialog box appears, we'll want to expand the **Multi-Purpose Objects** section along with the **List Definitions** section.

2. Right-click on **List Definitions** and select the **New** option.

3. Once we have created our new list definition, let's give it the name Construction Sequencing, as shown in *Figure 10.19*.

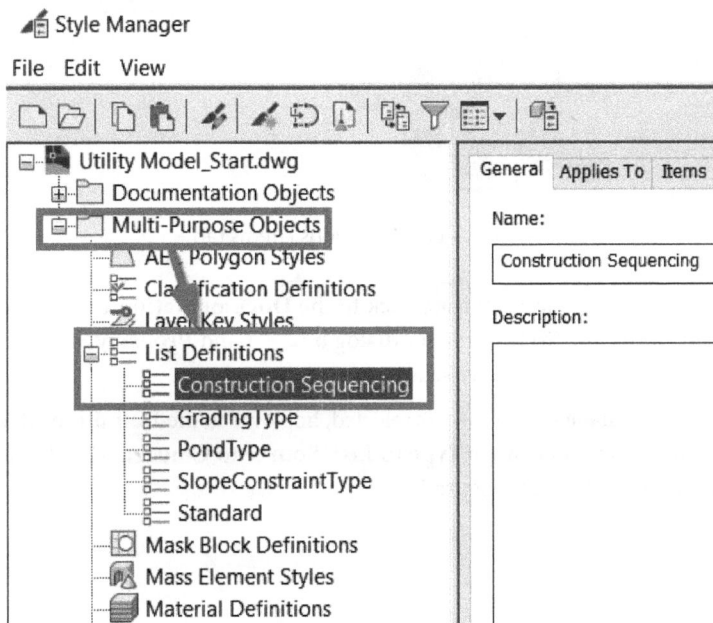

Figure 10.19 – New Construction Sequencing list

4. With the new **Construction Sequencing** list selected, we'll hop over to the **Applies To** tab and check the box next to **Manual Property Definition** (refer to *Figure 10.20*). By checking this box, we will then be able to use this list for our **Manual_PS** property definition.

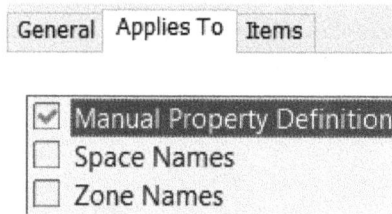

Figure 10.20 – List Definitions | Applies To tab

5. Continuing on, we'll switch over to the **Items** tab and populate our list with a sequence of phased construction that we may see on our projects. We can do so by clicking the **Add** button and typing in each phase of construction that we'd like to account for, as shown in *Figure 10.21*).

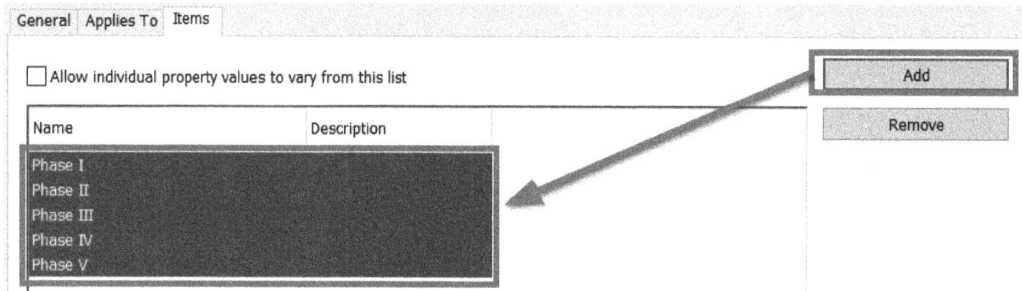

Figure 10.21 – List Definitions | Items tab

6. With our new list created, let's navigate back to the **Documentation Objects** section in our tree on the left side of the **Style Manager** dialog box, expand **Property Set Definitions**, and select our **Storm Drainage** property set.

7. Once the **Storm Drainage** property set is selected, hop over to the **Definition** tab of our **Storm Drainage** property set and change **Type** to **List**, **Source** to **Construction Sequencing**, and **Default** to **Phase I**, as shown in *Figure 10.22*.

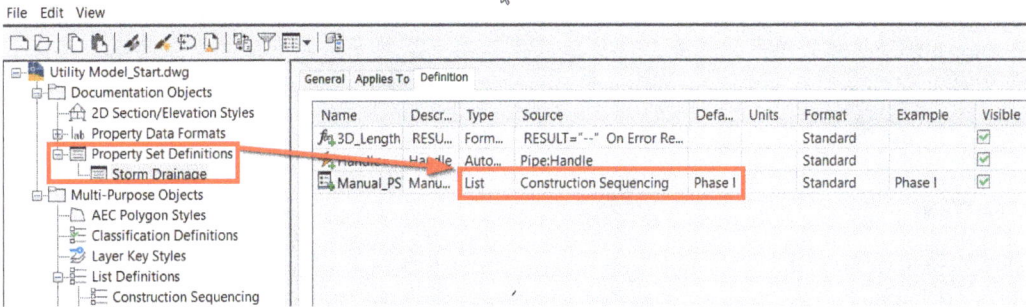

Figure 10.22 – Change type of Manual property set to List

8. Now, let's select **Apply**, close out of the **Style Manager** dialog box, and navigate back to our **Storm Drainage** pipe object.

We should now see our label updated to **Phase I**, which was set as the default value in our **Storm Drainage** property set definition. If we were to select the **Pipe** object, go over to the **Properties** dialog box, and activate the **Extended Data** tab, we would see that we have our newly created construction sequencing list with all fields available to be applied to our **Manual_PS** property, as shown in *Figure 10.23*.

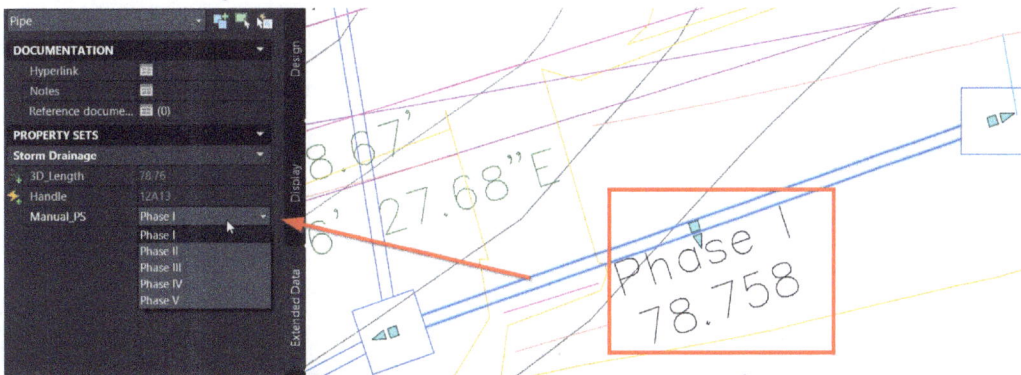

Figure 10.23 – Manual_PS property set label with list values applied

We've now applied another level of automation to our property set value populations by using lists.

Keep in mind that this is just a starting point and is meant to give you an idea of some of the possibilities of how we can apply and leverage custom properties in our designs. As mentioned earlier, it's important to line up the custom properties that we're adding to our design files with the intended use of these files downstream. Maybe we would like to add custom properties that link to the manufacturer's specifications for contractors to know the installation requirements. Or maybe we would like to add information around decommissioning, servicing, and so on for operator awareness and to allow them to quickly filter and/or isolate specific components in their systems that the files are being integrated into. Whatever the requirements are, we now have a good idea and approach to incorporate into our design files moving forward.

Summary

In this chapter, we embarked on a journey into the realm of information modeling by understanding and realizing the value that property sets offer us in the design space. With Civil 3D and its network of connected add-ins and new tools, we can generate custom and exceptional data that when contextualized with other aspects of our models and projects becomes useful information. When we can correlate that to other pieces of information and connect it to other aspects of our projects, our efficiency and confidence climb exponentially.

Discovering their transformative power goes a long way in enhancing the effectiveness of our civil BIM designs. As designers and engineers, our goal extends beyond the mere creation and development of models. As clients become more tech-savvy and realize the benefits of BIM as well, we are met with higher expectations. As things become more digital and the value of data is being discovered, the proper configuration of property sets gives us an advantage in streamlining data exchanges as required for downstream uses and platform integrations.

Looking ahead, in the next chapter, we'll dive into a relatively new tool called Project Explorer. As we explore this new tool, we'll continue developing our BIM model management skills as we understand its benefits in improving model adjustments, design changes, reviews, and reporting capabilities. This tool is highly utilized in model management and by senior designers as it provides a very high-level view of all objects and data within our files, which we can leverage to make quicker and more informed decisions throughout the design process.

11

Introduction to Project Explorer

In the last chapter, we looked at information modeling and why Property Sets are important in design. Knowing how Property Sets can change things helps us make better civil BIM designs. As designers and engineers, we do more than just create models. Clients are getting smarter about technology and see the benefits of BIM, so we have to offer more. In today's digital world, knowing how to set up Property Sets gives us an advantage by making it easier to share data for different uses and systems.

In this chapter, we'll focus on managing models and using a tool called Project Explorer. Project Explorer helps us adjust models, make design changes, review work, and create reports faster. Learning about this tool helps us become better at managing BIM models, which improves our design process. Many experienced designers use Project Explorer because it gives a clear view of everything in our files, helping us make decisions quickly and wisely as we design.

That said, in this chapter, we'll cover the following topics:

- An introduction to Project Explorer and its user interface
- Reviewing and editing Civil 3D objects inside Project Explorer
- Generating reports and tables from Civil 3D object data

Technical requirements

Here are the technical requirements for this chapter:

- **Software**: Autodesk Civil 3D 2024
- **Operating system**: 64-bit Microsoft Windows 10
- **Processor**: 4+ GHz
- **Memory**: 32 GB of RAM (we suggest going with either 64 GB or 128 GB)

- **Graphics card**: 8 GB

- **Display resolution**: 1,980 x 1,080 with True Color

- **Disk space**: 20 GB

- **Pointing device**: Microsoft compliant mouse

With that, let's go ahead and open up our `Utility Model_Start.dwg` file, located within our `Civil 3D 2025 Unleashed\Chapter 11\Model` location. Once the file is opened, you'll notice that it has zoomed into our site, as displayed in *Figure 11.1*.

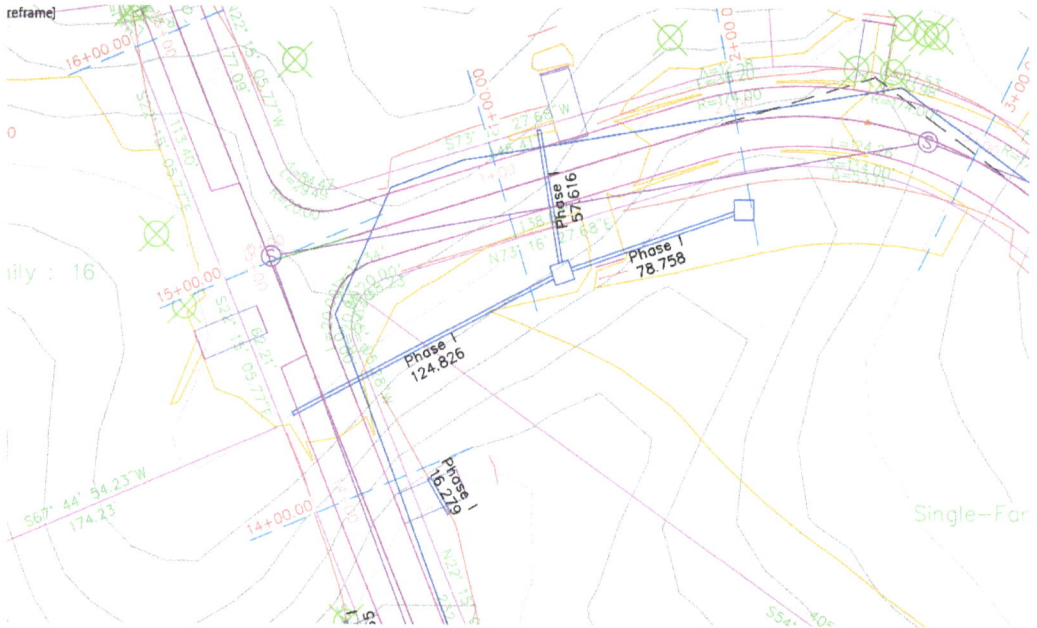

Figure 11.1 – The Utility Model_Start.dwg display

With our `Utility Model_Start.dwg` file now opened up, let's learn how to access and begin familiarizing ourselves with Project Explorer to understand how we can best utilize its capabilities within Autodesk Civil 3D 2025.

An introduction to Project Explorer and its user interface

Project Explorer in Autodesk Civil 3D 2025 is a key tool for managing Civil 3D models efficiently, hence why it is heavily utilized by senior designers and those managing design outcomes. At a very high level, it promotes ease of access to perform various tasks such as reviewing, reporting, and editing models all in one place.

Project Explorer truly helps designers understand their designs better and share geometric information with project teams easily. It does this through a simple tabbed interface that organizes modeled objects within our current files. In Project Explorer, we can also filter, adjust, and review content with ease.

Another great feature is its ability to create object reports in different file formats such as Excel, PDFs, and AutoCAD tables. These reports can be customized to fit specific project needs. Any changes made to the design in Civil 3D are seamlessly updated in tables and exported documents.

As we navigate and familiarize ourselves with Project Explorer, we can also see that it will provide helpful warnings to identify potential design issues, making it easier for us to pinpoint and fix problems early on.

To access Project Explorer in Civil 3D 2025, we'll want to bring our attention up to the top of the screen for our current session of Civil 3D. Click on the **Home** ribbon, and then select **Project Explorer** in the **Explore** panel, as shown in *Figure 11.2*.

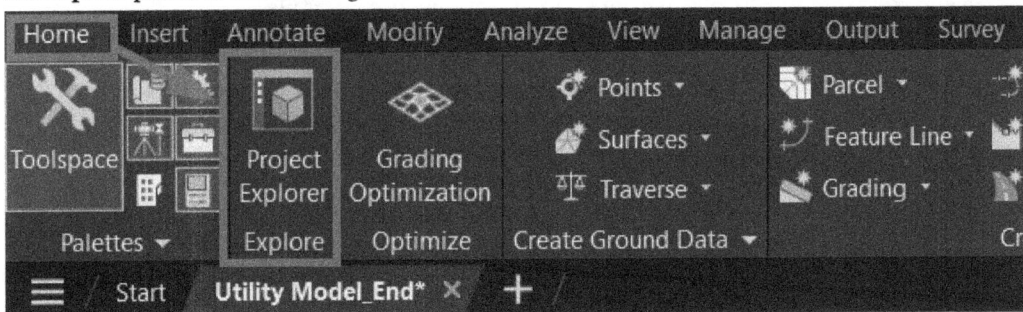

Figure 11.2 – Activating the Project Explorer tool

Once activated, the **Project Explorer** dialog box will appear, where we have access to all types of objects and associated metadata. What's unique about this tool is that it is *modeless*. Modeless, in this case, means that we can still navigate and interact with our current file within model space while also interacting and analyzing objects and data within Project Explorer. To maximize benefits and screen space, it's recommended to utilize two monitors when using Project Explorer so that we can have Project Explorer set up on one screen, while our Civil 3D 2025 session is displayed on the other screen.

With Project Explorer launched, let's get a quick idea of what types of objects and data we have available to us in this modeless dialog box. Starting from top left to right, we have the following (also shown in *Figure 11.3*):

1. **Alignments**: Provides quick access and analysis to all alignments and profiles within our current file.

2. **Assemblies**: Provides quick access and analysis to all assemblies, subassemblies, and the associated corridor regions that they are applied to within our current file.

3. **Subassemblies**: Provides quick access and analysis to all subassemblies and the associated corridor regions that they are applied to within our current file.

4. **Corridors**: Provides quick access and analysis to all corridors, baselines, and associated feature lines within our current file.

5. **Point Groups**: Provides quick access and analysis to all point groups and COGO points within our current file.

6. **Surfaces**: Provides quick access and analysis to all surfaces within our current file.

7. **Feature Lines**: Provides quick access and analysis to all feature lines within our current file.

8. **Parcels**: Provides quick access and analysis to all parcels within our current file.

9. **Catchments**: Provides quick access and analysis to all catchments within our current file.

10. **Pipe Networks**: Provides quick access and analysis to all pipe networks and parts and pipe runs that make up our gravity pipe network designs within our current file.

11. **Pressure Networks**: Provides quick access and analysis to all pressure networks and parts and pipe runs that make up our pressure pipe network designs within our current file.

12. **Sample Line Groups**: Provides quick access and analysis to all sample line groups and sample lines within our current file.

13. **AutoCAD Blocks**: Provides quick access and analysis to all AutoCAD blocks used in our current file.

14. **Property Set Definitions**: Provides quick access and analysis to all Property Sets along with objects that the Property Sets have been applied to in our current file. Note that Property Set data can be viewed in any of the previous object tabs listed at a micro level, whereas the Property Set **Definitions** tab provides a macro-level view of all Property Sets within the current file.

15. **Object Sets**: Groupings of objects we define within Project Explorer for reporting purposes.

Note that the number assigned to each component in the preceding list is the same as the number assigned to the component in *Figure 11.3*.

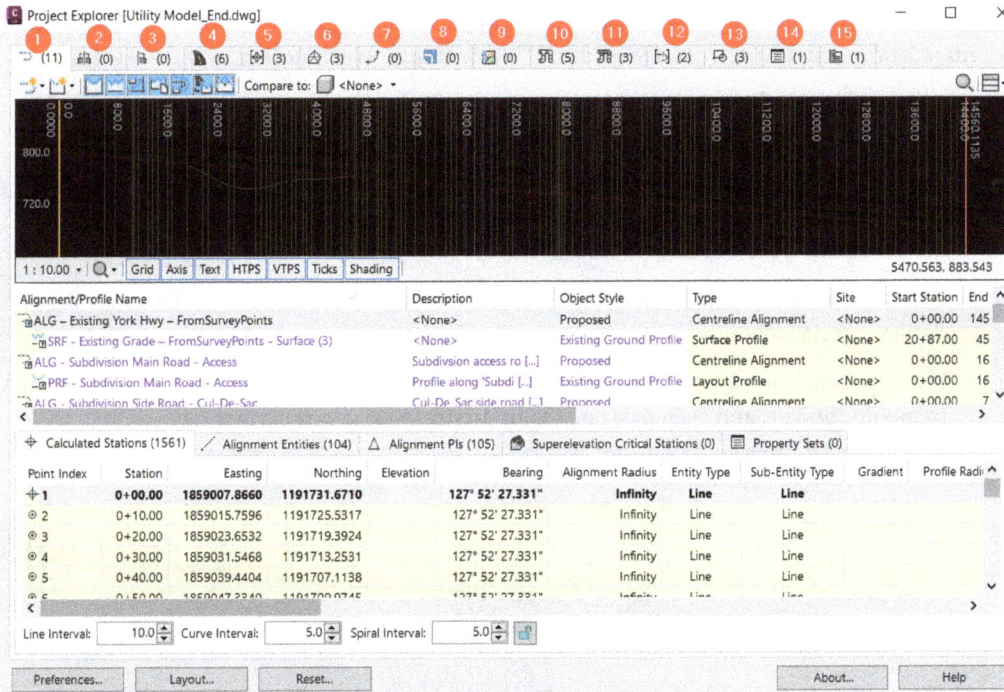

Figure 11.3 – The Project Explorer Object tabs

Let's now hop over to the **Corridor** tab to get an idea of what we're looking at within each tab within Project Explorer. These views will mostly be consistent regardless of the object we analyze; just how the subcomponents are displayed along with the columns and data will change.

From the top down, we have the following key areas within Project Explorer:

- **The profile/section view pane:** This view provides a profile or section view, depending on the object, to give us some additional context of the configuration of the objects we analyze. We also have some quick updating and analysis tools along the top of this pane for ease of use (refer to *Figure 11.4*).

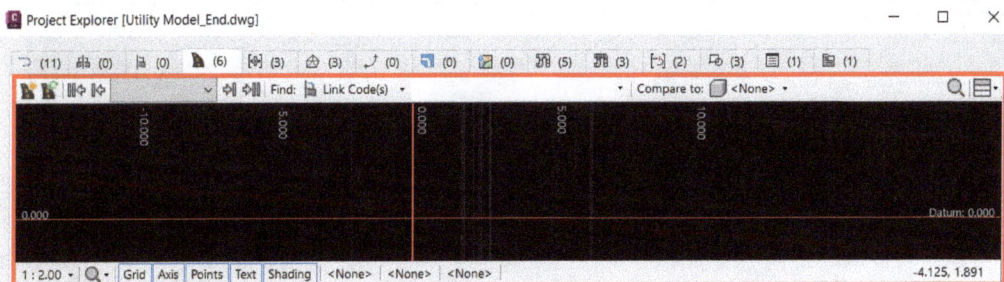

Figure 11.4 – Project Explorer's profile/section view pane

- **The object list pane**: This view provides access to individual objects along with their associated metadata and styles (refer to *Figure 11.5*). A few things worth noting in this view include the following:

  - Within each of the columns, note that there is either a white or yellow background behind the data. If the background behind the data is white, we can update the properties associated with the object directly within Project Explorer. If the background behind the data is yellow, the information is static and not editable.

  - If the text of a listed object is red, this gives us an indication that there may be a warning or design criteria violation.

  - If we right-click on any of the objects listed in the object list pane, we have additional options to modify, update, and even navigate to our model space in our current file.

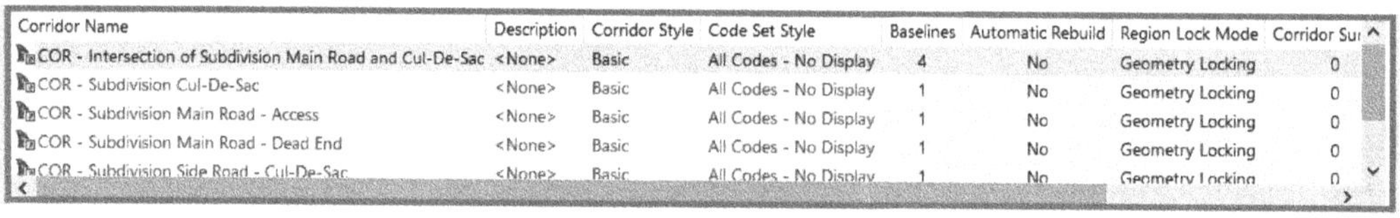

Figure 11.5 – Project Explorer's object list pane

- **The sub-object list pane**: This view provides access to individual objects that are used to define the object selected in the object list pane (refer to *Figure 11.6*). Similar to the object list pane, there are a few things worth noting in this view:

  - Within each of the columns, note that there is either a white or yellow background behind the data. If the background behind the data is white, we can update those properties associated with the object directly within Project Explorer.

  - If the text of a listed object is red, this gives us an indication that there may be a warning or design criteria violation.

  - If we right-click on any of these objects listed in the object list pane, we have additional options to modify, update, and even navigate to our model space in our current file.

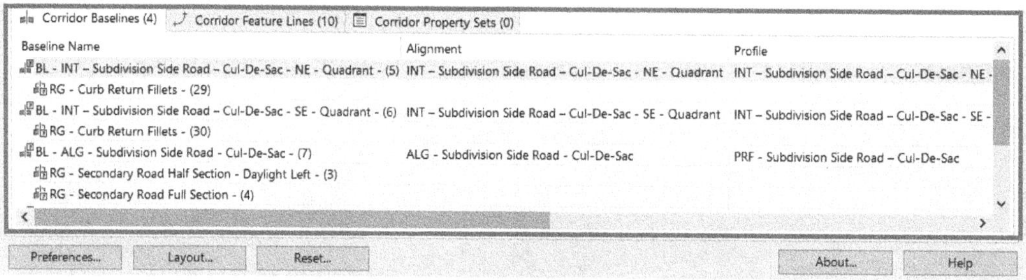

Figure 11.6 – Project Explorer's sub-object list pane

Now that we have a general idea of the types of content we have access to directly within Project Explorer, let's begin exploring the review and editing capabilities in more detail in the next section.

Reviewing and editing Civil 3D objects inside Project Explorer

To provide a good example of some of the capabilities we have available to us within Project Explorer, let's hop over to the **Pipe Networks** tab. Here, we can quickly see all of the *gravity pipe networks* we have in our current file, either designed and modeled directly within our file or those that are data referenced into our current file (refer to *Figure 11.7*).

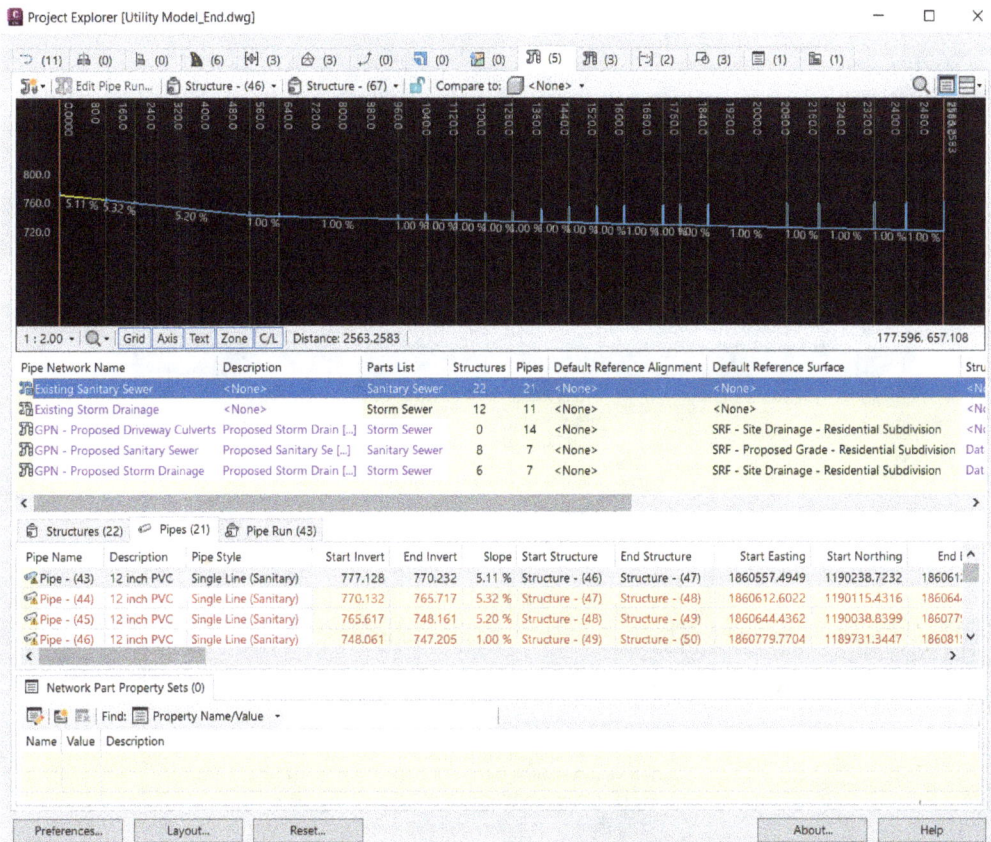

Figure 11.7 – The Pipe Networks Tab within Project Explorer

Looking at the profile/section pane, we can quickly see the full profile of the entire gravity pipe network that's currently selected in our object list pane, giving us a quick overview of how the network has been modeled. In the object list pane, we can quickly view a list of all gravity pipe networks in our current file. This view also provides, at a high level, various statistics and properties associated with each of the gravity pipe networks as a whole. Note that as we select different gravity pipe networks in this list, both the profile/section pane and the sub-object list pane will update.

If we right-click on any of our gravity pipe networks listed in the object list pane, we have the following options available to us (refer to *Figure 11.8*):

- **Object Set**: Provides the ability to add selected objects to an object set for reporting purposes.

- **Export**: Provides the ability to quickly generate reports and new files of the selected objects.

- **Set**: Provides the ability to set a new, or make current, settings and styles associated with the selected objects.

- **Navigation**: Provides the ability to quickly locate the selected objects within model space.

- **Copy to clipboard**: Provides the ability to quickly copy all statistics and properties displayed in Project Explorer associated with the selected object.

- **Delete…**: Provides the ability to delete the selected object within our current file.

- **Properties**: Provides the ability to quickly access the object's **Property** dialog box associated with the selected object.

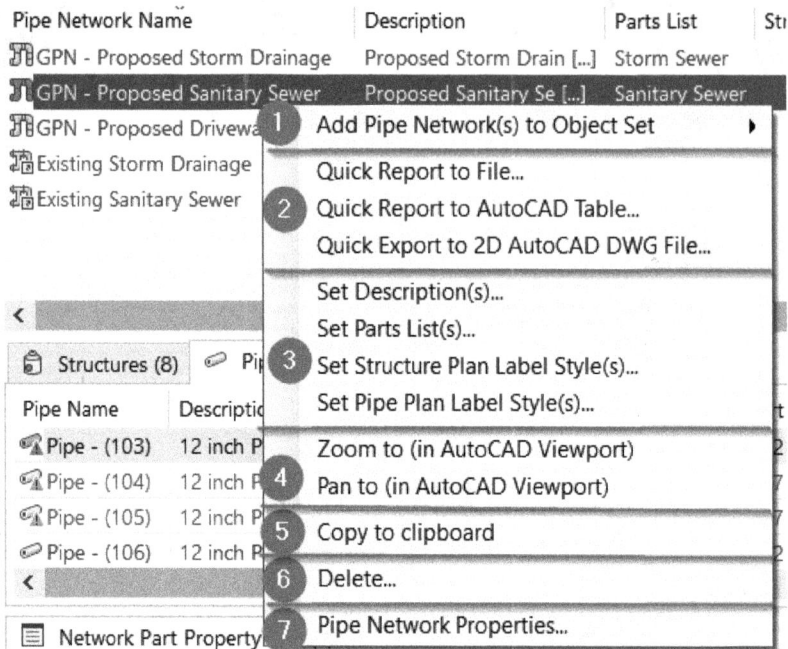

Figure 11.8 – The right-click options of objects in the object list pane

As mentioned earlier, within each of the columns, we can see that there is either a white or yellow background behind the data. If the background behind the data is white, we can update those properties associated with the object directly within Project Explorer. Go ahead and simply double-click on any of these fields to update the network properties as a whole.

Now, let's go ahead and select our **GPN - Proposed Sanitary Sewer** gravity pipe network within the object list pane and shift our focus down to the sub-object list pane. Once selected, we now have visibility of all of the objects that make up our **GPN - Proposed Sanitary Sewer** gravity pipe network. As shown in *Figure 11.9*, note that all of the objects that make up our **GPN - Proposed Sanitary Sewer** gravity pipe network are categorized in additional tabs. These tabs will change based on the properties of the **Object** tab we currently are in and the objects we have selected.

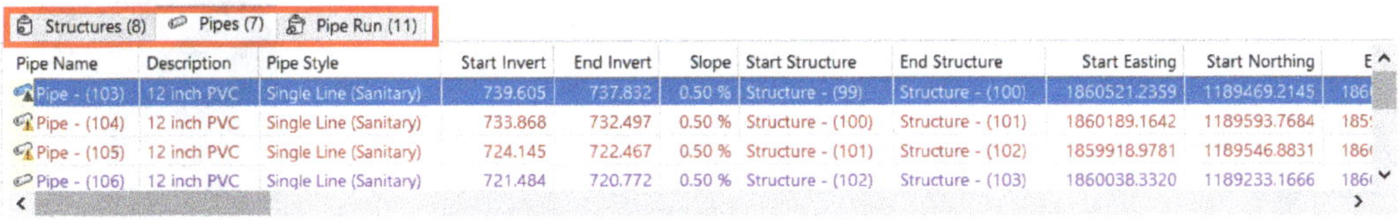

Figure 11.9 – The sub-object pane tabs

In this view, we can quickly see additional properties associated with each of the parts that define our object, selected in the object list pane. As we select individual objects within the sub-object list pane, we can see these objects selected in our profile/section view pane as well.

Similar to the object list pane, the data associated with each part will have either a white or yellow background, indicating whether or not we can edit that particular property. Also, note that some parts are displayed with red text. As we hover our mouse over objects with red text, we can quickly see any warnings or design violations associated with a particular object, as shown in *Figure 11.10*.

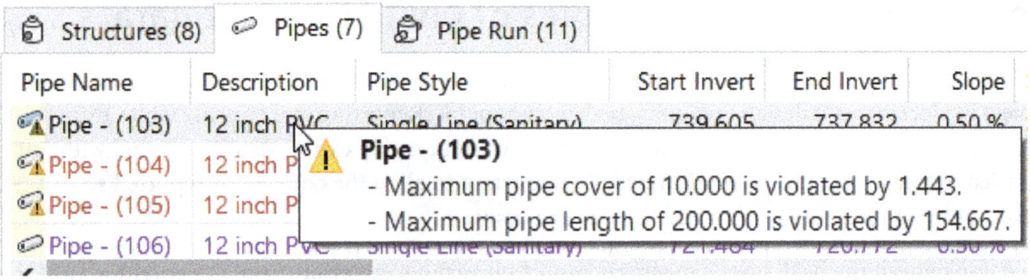

Figure 11.10 – The sub-object pane tabs

Now that we have the ability to quickly focus on design issues using Project Explorer, we can now see the added benefit from a senior designer or BIM model manager's perspective. As we hone in on design warnings and violations, we can either update our design and objects through property modifications within Project Explorer directly or right-click on our object, select **Pipe Properties**, and update the rules associated with that pipe object, as shown in *Figure 11.11*.

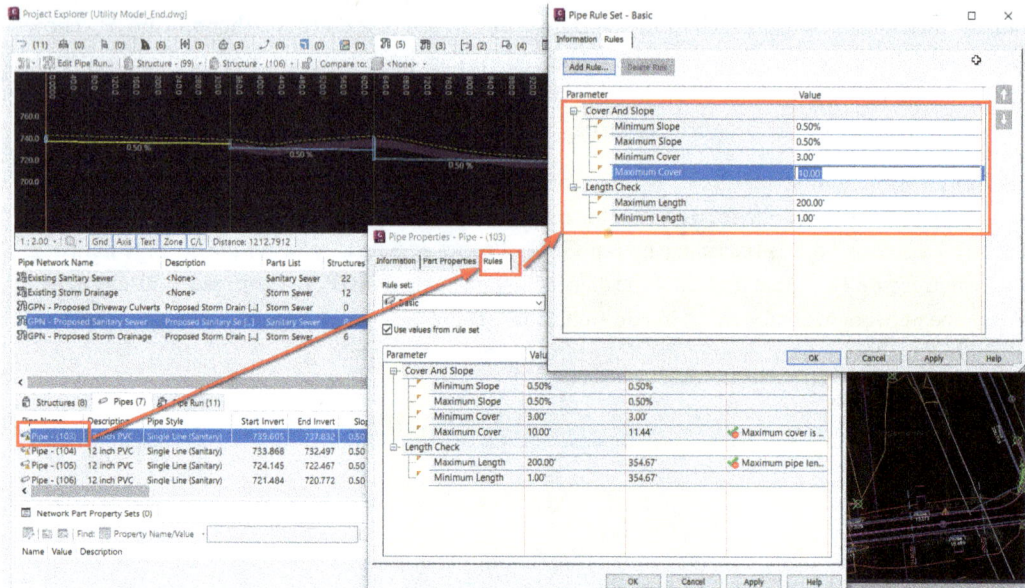

Figure 11.11 – Updating the rules of a pipe from Project Explorer

As you can see in *Figure 11.11*, the more we interact with Project Explorer, the more dialog boxes can potentially appear as we manage and further refine our models to conform to design standards. This workflow of design refinement also highlights the preference to utilize multiple monitors, as mentioned earlier in the chapter. Moving forward, let's explore how we can export our modeled objects to reports and tables.

Generating reports and tables of Civil 3D object data

As I'm sure you've realized at this point, there are many benefits and applications when utilizing Project Explorer from a design and quick review standpoint. Reporting and creating tables from this data is a huge plus as well. Using Project Explorer, we can quickly export data associated with our modeled objects in various ways. However, it's important to align the data that is reported so that it makes sense for the audience that will analyze and assess it.

If we think about the basics, we can simply report on various materials from the corridor and/or utility models, generating quantification reports of AutoCAD blocks used to depict the erosion control measures being applied, or even understanding how much dirt needs to be brought on-site or hauled off-site throughout construction. Once this data is exported, the sky's the limit with what we can do with it, including developing simple cost estimation reports that associate values with the quantities being reported. These types of concepts give all stakeholders greater insight as to how much a project will cost to be constructed.

However, it's true what they say – *garbage in, garbage out*. Our quantification, cost estimates, and so on are only as good as the accuracy of the model objects we generate to build our design. That said, it's important to note that Project Explorer is a great tool that has great power and advantages, but we still need to perform our own due diligence and inspections of the files to ensure that the data being reported is reliable and accurate.

To officially begin our reporting and tabling journey, let's go ahead and activate our **Pipe Networks** tab, if not done so already; jump down to the object list pane and select our **GPN - Proposed Storm Drainage** gravity pipe network this time. Once there, right-click on our **GPN - Proposed Storm Drainage** gravity pipe network, select **Add Pipe Network(s) to Object Set**, and then select the **New Object Set…** option, as shown in *Figure 11.12*.

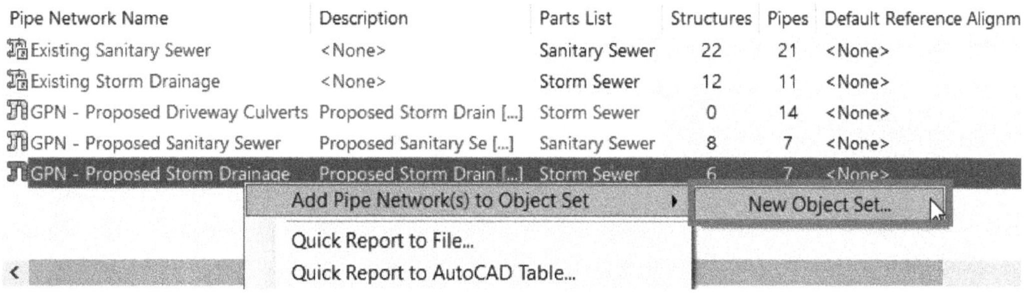

Pipe Network Name	Description	Parts List	Structures	Pipes	Default Reference Alignm
Existing Sanitary Sewer	<None>	Sanitary Sewer	22	21	<None>
Existing Storm Drainage	<None>	Storm Sewer	12	11	<None>
GPN - Proposed Driveway Culverts	Proposed Storm Drain […]	Storm Sewer	0	14	<None>
GPN - Proposed Sanitary Sewer	Proposed Sanitary Se […]	Sanitary Sewer	8	7	<None>
GPN - Proposed Storm Drainage	Proposed Storm Drain […]	Storm Sewer	6	7	<None>

Add Pipe Network(s) to Object Set	▶	New Object Set…
Quick Report to File…		
Quick Report to AutoCAD Table…		

Figure 11.12 – Creating a new object set in Project Explorer

After selecting the **New Object Set…** option, we'll be presented with a **Project Explorer | Select Sub-Object Type** dialog box, at which point we'll follow the following steps to create our new object set (the steps are also shown in *Figure 11.13*):

1. In the **Project Explorer | Select Sub-Object Type** dialog box, we'll change **Sub-Object Type** to **Pipe Run**.

2. In the **Project Explorer | Select Sub-Object Type** dialog box, we'll select **Filter.…**

3. In the **Project Explorer | Apply Pipe Run Filter** dialog box, we'll select **Structure - (81)** as the **Start Structure**.

4. In the **Project Explorer | Apply Pipe Run Filter** dialog box, we'll select **Structure - (85)** as the end structure.

5. In the **Project Explorer | Apply Pipe Run Filter** dialog box, we'll select **OK**.

6. In the **Project Explorer | Select Sub-Object Type** dialog box, we'll select **Next**.

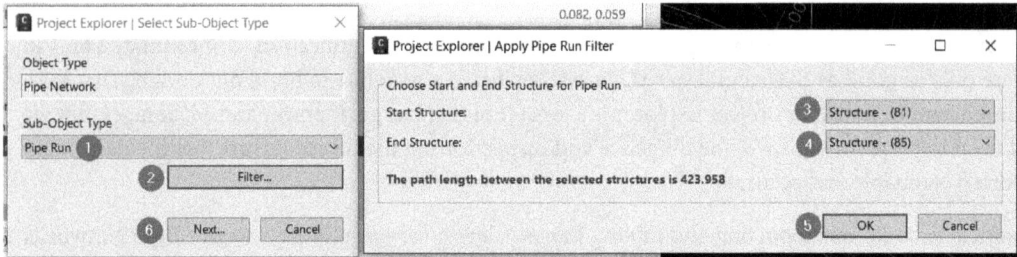

Figure 11.13 – The steps to create a new object set

After making these selections, we'll next be presented with the **Project Explorer | Create Object Set** dialog box, where we'll have a whole slew of options to further customize our output. For our exercise, we're going to simply place a dynamic table directly within our file by filling out the fields, making the following selections (also displayed in *Figure 11.14*), and pressing the **OK** button:

- **Object Set Name**: OS - GPN - Proposed Storm Drainage
- **Object Set Description**: From Structure (81) to (85)
- **Object Set Action**: Export to AutoCAD Table(s) in MODEL Space
- **Object Set Action Type**: Dynamic
- **Layout Style**: Use Layout of Project Explorer Window
- **Title Cell Text Template**: {Object_Name}
- **Table Style**: Use Default Table Style for Object Sets

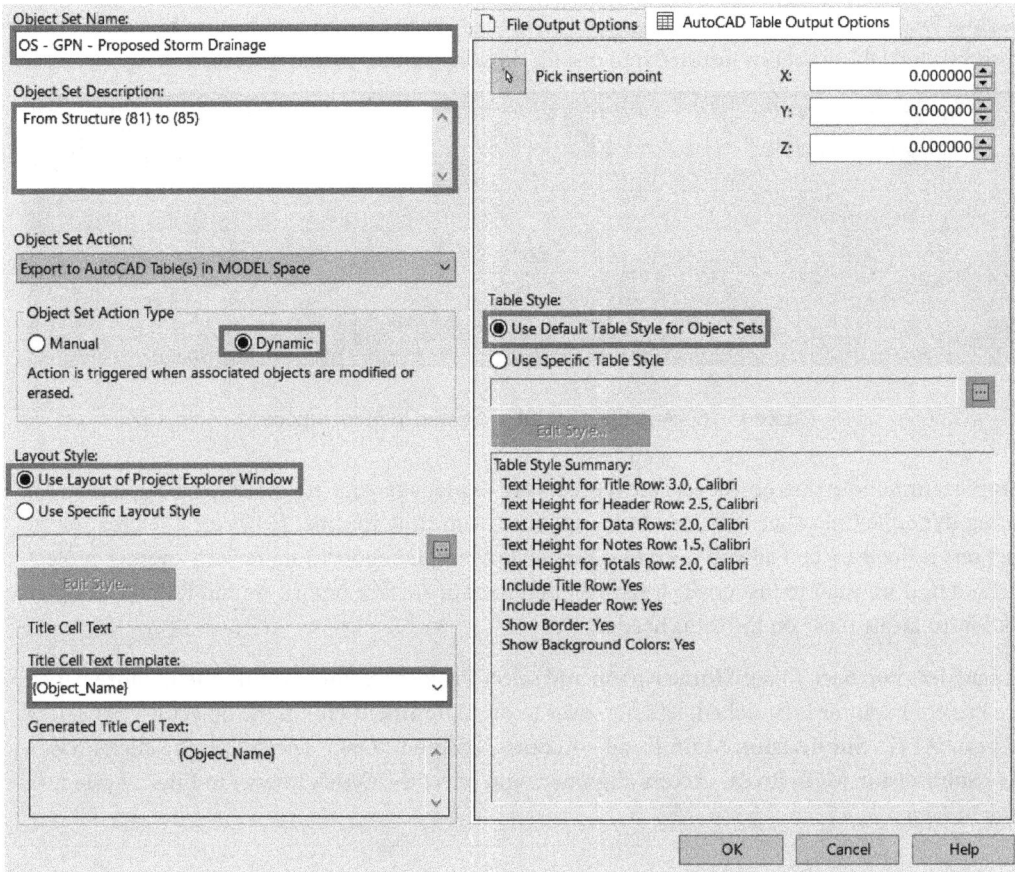

Figure 11.14 – Project Explorer | Create Object Set

After filling everything out as detailed and clicking on the **OK** button in the bottom right of the **Project Explorer | Create Object Set** dialog box, we'll be presented with one final dialog box that lets us know that we've successfully created an object set, as shown in *Figure 11.15*.

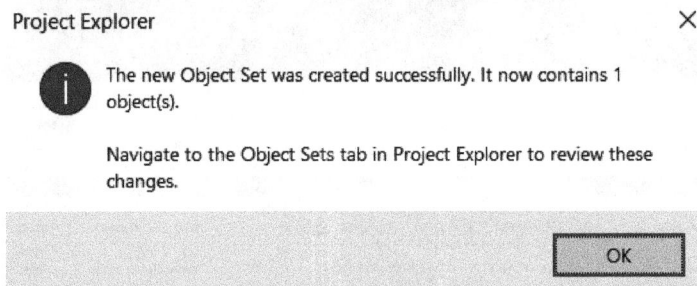

Figure 11.15 – The object set successfully created notification

If we close Project Explorer now and navigate down to the 0 , 0 , 0 location within our file, we'll see that a newly created table has been inserted into our file, looking similar to that displayed in *Figure 11.16*.

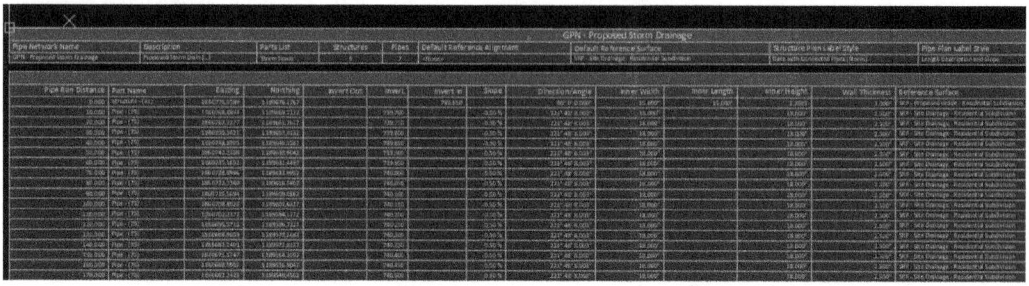

Figure 11.16 – An object set table created in model space

It's important to note that object sets are a great way to manage your reported data if you plan on creating dynamic links that will need to be updated from time to time. However, the creation of object sets is not critical if all you need to do is create a simple report to share with another project stakeholder. If we want to just create a simple alignment or surface report, we can follow a simpler workflow to create these on the fly as needed.

That said, let's hop back to our **Home** ribbon and select **Project Explorer** within the **Explore** panel. Once Project Explorer is launched, let's hop over to our **Alignment** tab. In the object list pane, let's select our **ALG - Subdivision Main Road - Access** alignment. Once selected, right-click on our **ALG - Subdivision Main Road - Access** alignment and select the **Quick Report to File…** option, as shown in *Figure 11.17*.

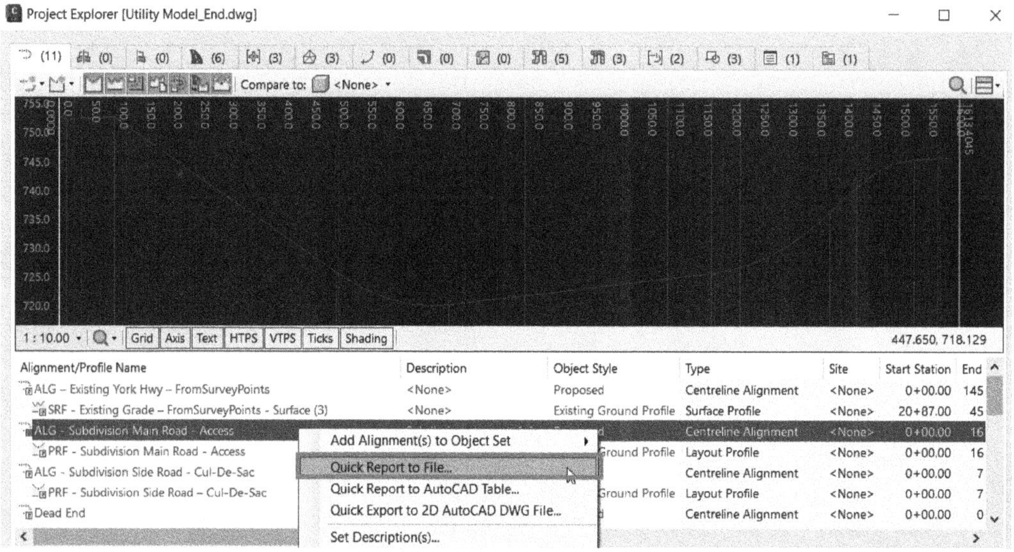

Figure 11.17 – The Quick Report to File… option

In the **Project Explorer | Select Sub-Object Type** dialog box that appears, we'll change the sub-object type from **Calculated Stations** to **Alignment Entities** this time to generate a report consisting of specific data points relevant to the stakeholder reviewing, as shown in *Figure 11.18*, and then click **Next**.

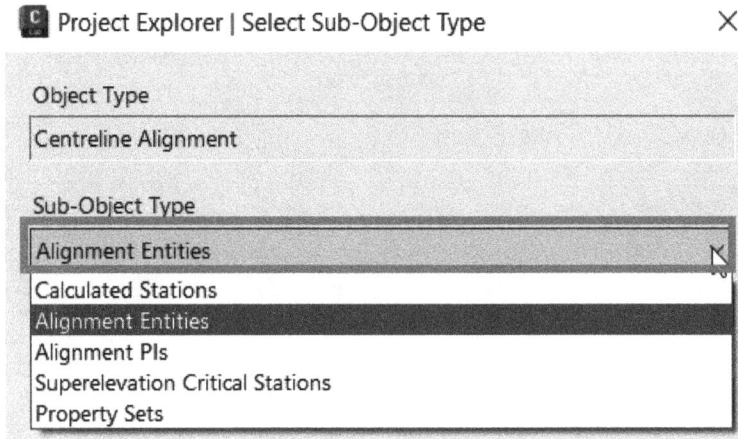

Figure 11.18 – The Project Explorer | Select Sub-Object Type dialog box

Next, in the **Project Explorer | Create Quick Report** dialog box, we'll fill out the fields, make the following selections (also displayed in *Figure 11.19*), and click **OK**:

- **File Name Template: ALG - Subdivision Main Road - Access**
- **File Type: HTML file (*.htm)**
- **Title Cell Text Template: {Object_Name}**
- **Layout Style for Quick Report: Use Layout of Project Explorer Window**
- **Use Object Table(s)**: Check the box
- **Use Sub-Object Table(s)**: Check the box

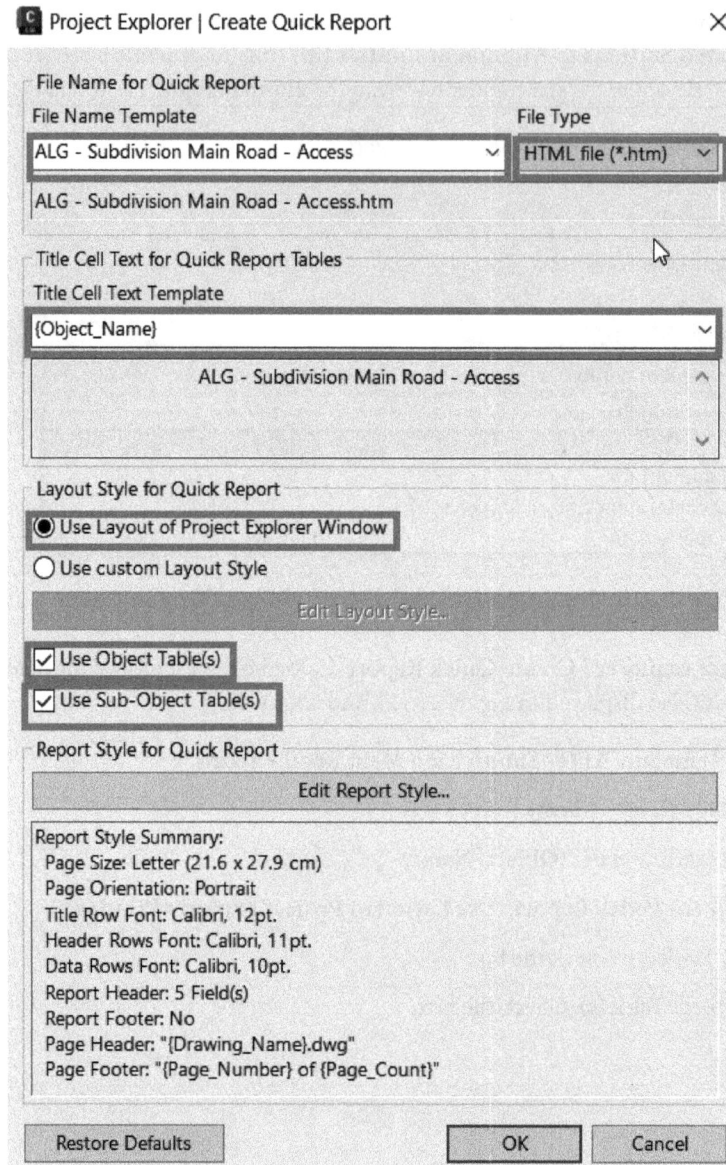

Figure 11.19 – The Project Explorer | Create Quick Report dialog box

After making all selections and clicking **OK**, we'll get a new prompt asking us where we want to save the report. For now, let's go ahead and save it with our project in the following location – `Civil 3D 2025 Unleashed\Chapter 11\Project Explorer`. After saving, another dialog box will appear, saying that our report has been generated, and it will ask us whether we want to review it, as shown in *Figure 11.20*.

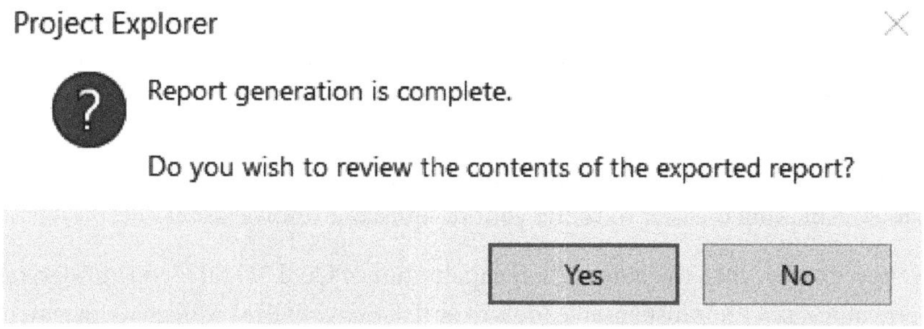

Project Explorer   ✕

? Report generation is complete.

Do you wish to review the contents of the exported report?

[Yes]   [No]

Figure 11.20 – Project Explorer's successful report creation notification

If we click **Yes**, our internet browser will open up automatically with a final result of our report, looking similar to that displayed in *Figure 11.21*.

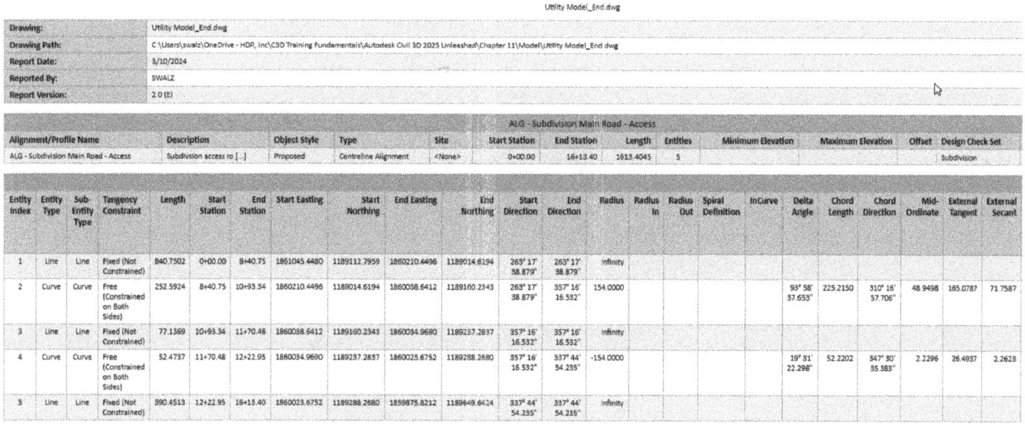

Utility Model_End.dwg

Drawing:	Utility Model_End.dwg
Drawing Path:	C:\Users\swalz\OneDrive - HDR, Inc\C3D Training Fundamentals\Autodesk Civil 3D 2025 Unleashed\Chapter 11\Model\Utility Model_End.dwg
Report Date:	3/10/2024
Reported By:	SWALZ
Report Version:	2.0 (E)

ALG - Subdivision Main Road - Access

Alignment/Profile Name	Description	Object Style	Type	Site	Start Station	End Station	Length	Entities	Minimum Elevation	Maximum Elevation	Offset	Design Check Set
ALG - Subdivision Main Road - Access	Subdivision access ro [...]	Proposed	Centreline Alignment	<None>	0+00.00	16+13.40	1613.4045	5				Subdivision

Entity Index	Entity Type	Sub-Entity Type	Tangency Constraint	Length	Start Station	End Station	Start Easting	Start Northing	End Easting	End Northing	Start Direction	End Direction	Radius	Radius In	Radius Out	Spiral Definition	InCurve	Delta Angle	Chord Length	Chord Direction	Mid-Ordinate	External Tangent	External Secant
1	Line	Line	Fixed (Not Constrained)	840.7502	0+00.00	8+40.75	1861045.4480	1189112.7959	1860210.4496	1189014.6194	263° 17' 38.879"	263° 17' 38.879"	Infinity										
2	Curve	Curve	Free (Constrained on Both Sides)	252.5924	8+40.75	10+93.34	1860210.4496	1189014.6194	1860058.6412	1189160.2343	263° 17' 38.879"	357° 16' 16.532"	154.0000					93° 58' 37.653"	225.2150	310° 16' 57.706"	48.9498	165.0787	71.7587
3	Line	Line	Fixed (Not Constrained)	77.1369	10+93.34	11+70.48	1860058.6412	1189160.2343	1860034.9690	1189237.2897	357° 16' 16.532"	357° 16' 16.532"	Infinity										
4	Curve	Curve	Free (Constrained on Both Sides)	52.4737	11+70.48	12+22.95	1860034.9690	1189237.2857	1860025.6752	1189288.2680	357° 16' 16.532"	337° 44' 54.235"	-154.0000					19° 31' 22.298"	52.2202	347° 30' 35.183"	2.2296	26.4057	2.2628
5	Line	Line	Fixed (Not Constrained)	390.4513	12+22.95	16+13.40	1860023.6752	1189288.2680	1859875.8212	1189648.6414	337° 44' 54.235"	337° 44' 54.235"	Infinity										

Figure 11.21 – The final HTML report

It's important to note that all the reports we've generated throughout this section have been created using the out-of-the-box default settings. Throughout the creation process, we can generate new templates that can be customized for your company or specific clients, which can be recalled for future projects. As you get more familiar with utilizing Project Explorer and realize the many benefits and added value it provides from a senior design and model management standpoint, we highly recommend testing out these customization capabilities to extend how your data is analyzed and what you can do with it.

Summary

Throughout this chapter, we explored how we can manage and analyze our models while utilizing Project Explorer within Autodesk Civil 3D 2025. This robust toolset significantly streamlines design processes and improves model management tasks at a high level. Project Explorer stands out as a central hub to facilitate various critical tasks, such as adjusting models, implementing design changes, reviewing work, and swiftly generating reports. Its intuitive interface ensures quick access to a multitude of functions, including reviewing, editing, and reporting on Civil 3D objects within a unified platform, all while still being able to navigate and interact with our actual modeled objects within model space, due to its modeless interface. Through detailed guidance on navigating Project Explorer's interface and exploring its functionalities, we provided some practical insights into its review and editing capabilities, thereby making it easier to refine your designs and resolve issues.

As our exploration extends into the automation capabilities of Civil 3D 2025, our understanding of Project Explorer equips us with indispensable tools to enrich our civil BIM Model Management toolkit. This understanding sets the stage for further exploration into automation with Dynamo for Civil 3D 2025 in the upcoming chapter, promising even greater efficiency and effectiveness in managing Civil 3D models, and driving consistency in design workflow processes that can be shared across an entire design team.

12

Automating Routine Workflows with Dynamo and Scripting

In the previous chapter, we began to put on our model management hat to get a taste of how Project Explorer can help us advance our careers by deploying its built-in capabilities to truly elevate our civil BIM design and analysis game. Project Explorer offers a streamlined approach that will empower us to execute critical tasks with efficiency and consistency.

In this chapter, we'll continue building our model management skills by diving into the world of automation using Dynamo for Civil 3D 2025. Utilizing Dynamo for Civil 3D 2025 has many benefits to designs teams, from streamlining design workflows to driving consistency in how we develop our civil BIM designs, to improving collaboration across all project stakeholders, to simply spending less time on routine and mundane tasks that tend to bog design teams down.

That said, in this chapter, we'll cover the following topics:

- Introduction to Dynamo and its user interface
- Understanding nodes, concepts, and packages
- Building our first Dynamo for Civil 3D 2025 script

Technical requirements

Here are the technical requirements for this chapter:

- **Operating system**: 64-bit Microsoft Windows 10
- **Processor**: 4+ GHz
- **Memory**: 32 GB RAM (we suggest going with either 64 GB or 128 GB)
- **Graphics card**: 8 GB
- **Display resolution**: 1980 x 1080 with True Color

- **Disk space**: 20 GB

- **Pointing device**: Microsoft compliant mouse

With that, let's go ahead and open up our `Survey Model_Start.dwg` file located within our `Civil 3D 2025 Unleashed\Chapter 12\Model` location. Once opened, we'll notice that the file has been zoomed into our site as displayed in *Figure 12.1*.

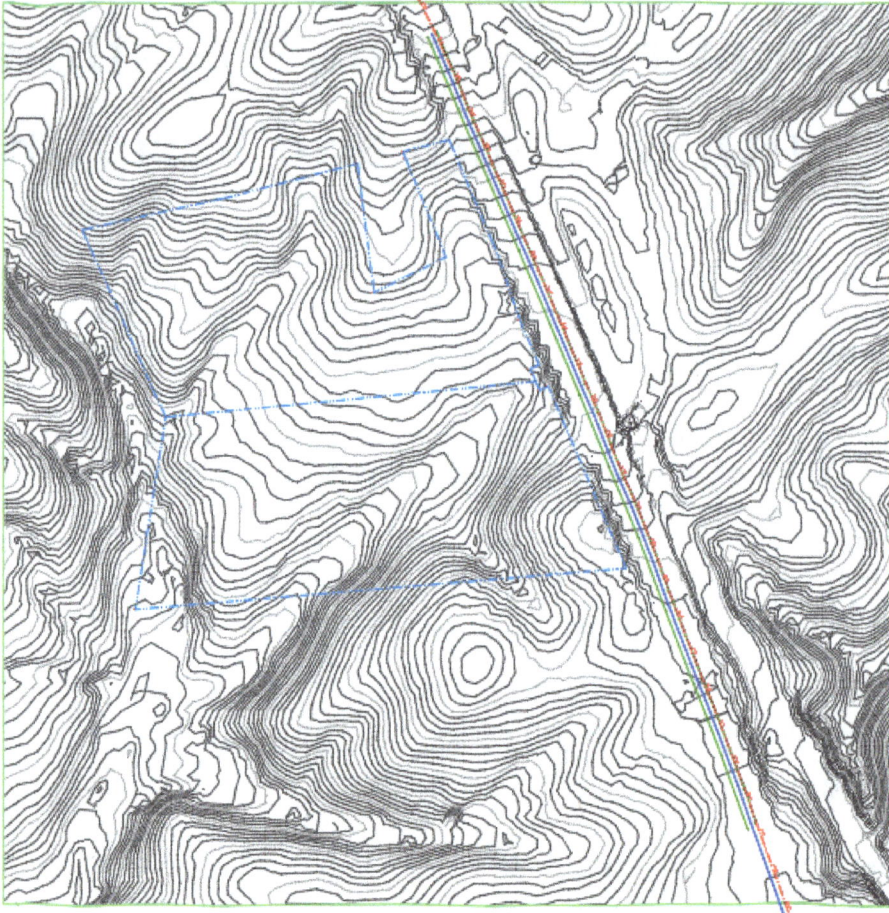

Figure 12.1 – Survey Model_Start.dwg display

With our `Survey Model_Start.dwg` file now opened, let's learn how to access Dynamo for Civil 3D 2025 and begin familiarizing ourselves with it in order to understand how we can best utilize its capabilities within our civil BIM designs.

Introduction to Dynamo and its user interface

Dynamo for Civil 3D 2025 has a ton of potential to increase efficiencies at both the individual and organization levels. Dynamo for Civil 3D 2025 offers users the means to significantly enhance their workflow efficiency by harnessing automation, customization, and scripting functionalities within the Autodesk Civil 3D environment. Dynamo for Civil 3D 2025's greatest attributes are its intuitive node-based interface and that it has a relatively low barrier to entry, unlike other common programming products.

At its core, Dynamo for Civil 3D 2025 empowers users to break free from the constraints of manual labor by automating repetitive tasks that would otherwise consume valuable time and resources. Whether it's creating intricate designs and supporting objects, generating complex geometry, updating annotations and labels, or even updating Property Set values that we created in a previous chapter, Dynamo for Civil 3D 2025 enables users to craft automated routines that will execute these tasks with precision and efficiency. By eliminating the need for manual intervention, Dynamo not only accelerates the design process but also reduces the likelihood of errors and inconsistencies arising, ultimately leading to more reliable and accurate design deliverables. With that in mind, let's jump into the user interface and see how to begin utilizing this tool for more functionality in Civil 3D.

Accessing Dynamo's interface

To access Dynamo for Civil 3D 2025, we'll hop up to our **Manage** ribbon and navigate to our **Visual Programming** panel, as shown in *Figure 12.2*.

Figure 12.2 – Accessing the Visual Programming panel within Civil 3D 2025

You'll notice that we have the following two tools available to us:

- **Dynamo**: Selecting this icon will launch Dynamo for Civil 3D as another modeless interface. With Dynamo for Civil 3D 2025 open, we will continue to have the ability to work in, and navigate around, our current file that's displayed in model space.

- **Dynamo Player**: Selecting this icon will launch our **Dynamo Player** dialog box. As we develop scripts within Dynamo for Civil 3D, we can call up completed scripts within Dynamo Player, select them, and run them on our current files quickly. I equate Dynamo Player to hitting the "*Easy Button*", which comes in handy for those individuals on your design teams that aren't familiar with Dynamo for Civil 3D 2025 just yet.

For now, since we haven't created any Dynamo for Civil 3D scripts just yet, let's go ahead and select the Dynamo icon within our Visual Programming panel. After selecting this icon, we'll be presented with the home screen of our modeless Dynamo for Civil 3D extension, looking similar to what is displayed in *Figure 12.3*.

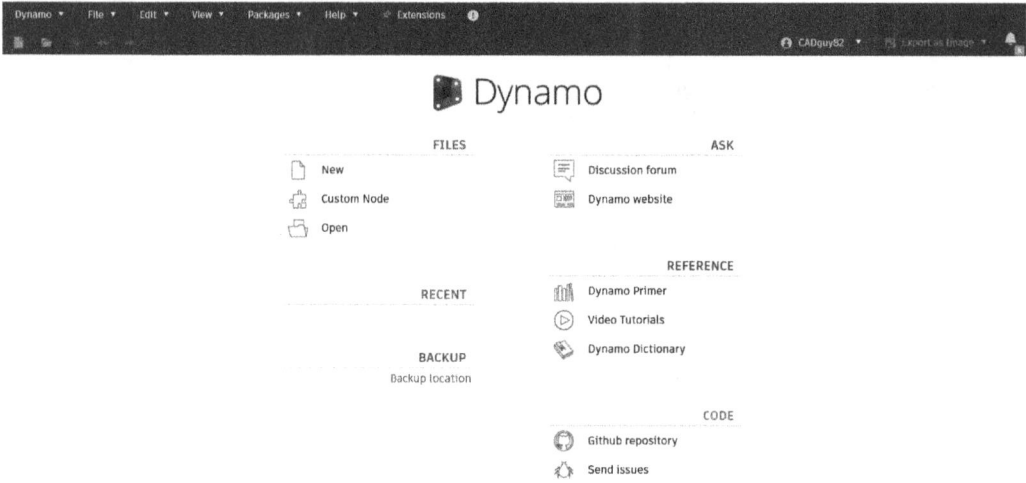

Figure 12.3 – Dynamo for Civil 3D home screen

Once Dynamo for Civil 3D is launched, we'll be presented with the home screen where we can see recently worked-on scripts and have the ability to create new scripts and custom nodes and open older scripts as needed on the left side, as shown in *Figure 12.4*.

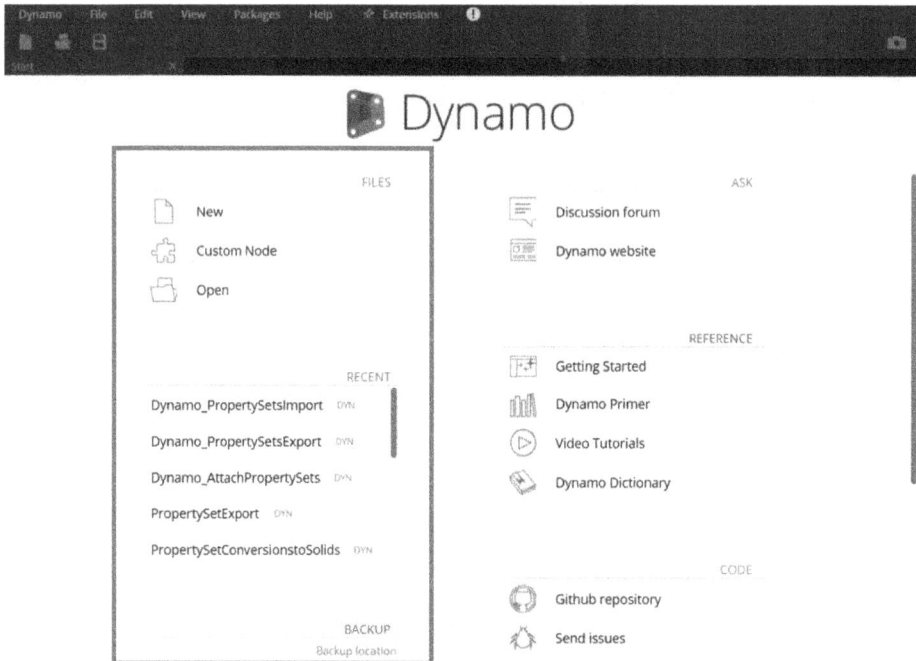

Figure 12.4 – Dynamo for Civil 3D home screen | Script creation and modification access

On the right side, we have access to a whole slew of additional resources that can get us started with understanding this type of programming and help us along in our Dynamo journey, as shown in *Figure 12.5*.

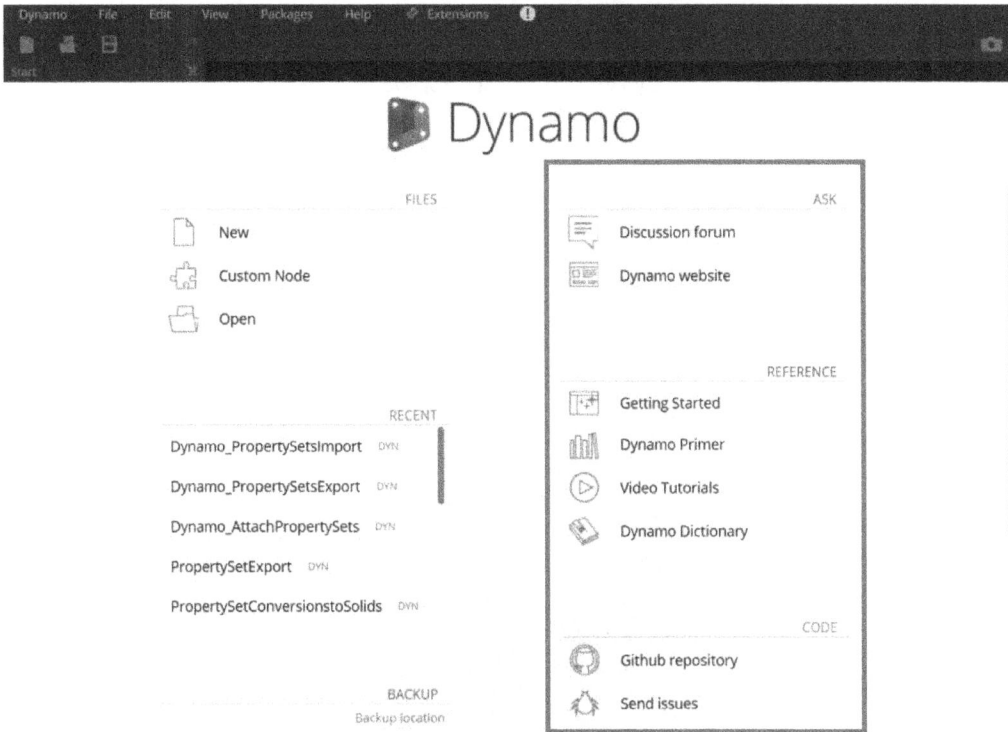

Figure 12.5 – Dynamo for Civil 3D home screen | Additional learning resources

Worth noting is that all items listed here provide excellent resources to all Dynamo for Civil 3D users, both old and new. One that we have used quite frequently in our journey has been the **Discussion forum** option. There are many Dynamo for Civil 3D users who regularly visit and interact on these boards and are willing to help in any way possible. Whenever we get stuck or a script is giving us the results we did not expect, the discussion forums are the first place we go to. If we can't find a solution via a simple search, we post a new topic detailing our situation and will get a response pretty quickly from someone in the community.

Creating a new script in Dynamo

Let's go ahead and select **New** on the left side of the Dynamo for Civil 3D home screen to begin a new session and script. As our new session is launched, we'll quickly notice that we have the following major areas displayed:

- Our standard set of drop-down menus along the top of our session is as shown in *Figure 12.6*.

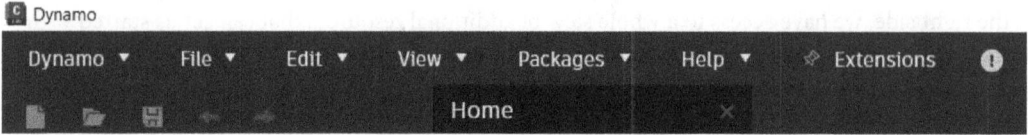

Figure 12.6 – Dynamo for Civil 3D drop-down menus

- Our **Library** section along the left side consists of out-of-the-box nodes that we'll use to define our scripts, as shown in *Figure 12.7*.

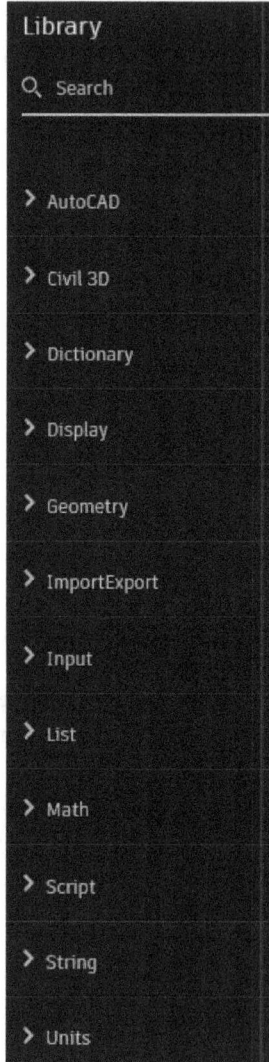

Figure 12.7 – Dynamo for Civil 3D Library

- Our **Add-ons** section below the **Library** is where we can incorporate custom nodes that can be made available to us by downloading custom Dynamo packages as shown in *Figure 12.8*.

Figure 12.8 – Dynamo for Civil 3D Add-ons

- Our graph view is where we'll build out our scripts to be applied to our design files, as shown in *Figure 12.9*.

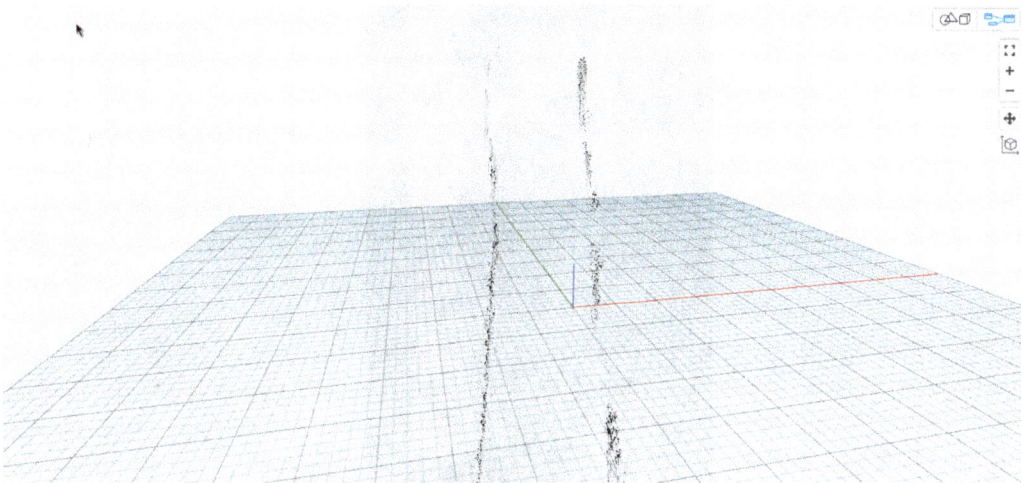

Figure 12.9 – Dynamo for Civil 3D graph view

With this basic overview of how to access Dynamo for Civil 3D from the Civil 3D environment, along with an understanding of how to create a new script and navigate the user interface, let's begin familiarizing ourselves with the library of out-of-the-box nodes we have by default along with some add-on custom nodes. We'll use these nodes to extend our capabilities and explore an almost endless number of possibilities in which we can apply scripts to our design files.

Understanding nodes, concepts, and packages

Moving over to our **Library** section along the left side of our Dynamo for Civil 3D session, let's start getting a better understanding of all the tools we have at our disposal. All of the nodes available within our Library, and even the Add-**ons se**ction, can be used to develop a script that can be deployed on our current file. To some, developing a script can sound a little intimidating and daunting. If we look at any script that has been developed by someone else, we'll see that it's really just a series of commands and inputs that specify a workflow that is deployed. So, in essence, we're actually creating a workflow, which we should all be rather familiar with at this stage.

Exploring the Library section

If we start looking closer at our **Library** section, we can see many different categories listed, as shown in *Figure 12.10*.

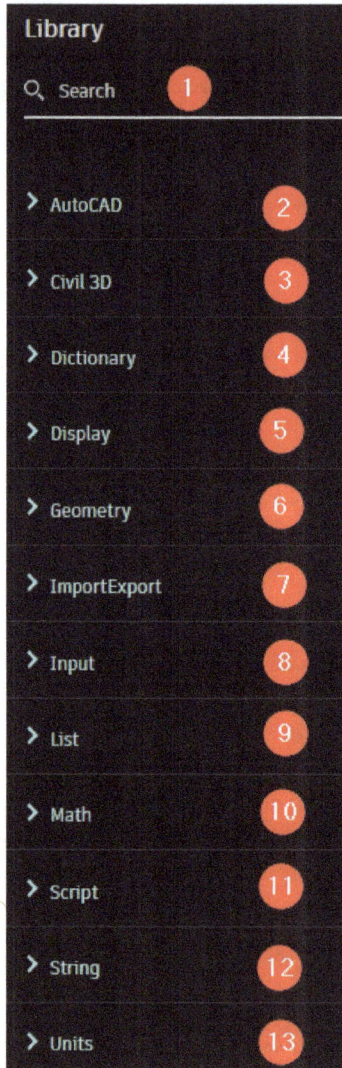

Figure 12.10 – Dynamo for Civil 3D Library

I'd recommend expanding each of these categories out to really understand what's in each one, but at a high level you can find the following capabilities and subcategories within each:

1. **Search**: Allows us the ability to quickly locate any node we need to access and add to our workflow, whether in our standard out-of-the-box Library of nodes or from our custom add-on nodes.

2. **AutoCAD**: This concerns document-level information, 2D and 3D AutoCAD-based objects, management of Property Sets, along with the ability to select specific objects based on various filters that we can apply.

3. **Civil 3D**: Basic Civil 3D objects (Alignments, Cogo Points, Corridors, and Surfaces), along with the ability to select these basic Civil 3D objects based on various filters that we can apply.

4. **Dictionary**: Allows us to analyze and look for specific keys within a list.

5. **Display**: Ability to create various charts/plots of data, identify/specify colors and ranges, and preview output values.

6. **Geometry**: Provides the ability to build Dynamo objects, as well as convert to and from objects within our design file.

7. **ImportExport**: Provides the ability to create and/or read data files (such as JSON, XLSX, CSV, etc.) and communicate with external filesystems.

8. **Input**: Provides various nodes required to define, or round out, input requirements for connecting nodes, including Integers, Booleans, Strings, Dates, Times, and Locations.

9. **List**: Allows us the ability to create, manipulate, and manage various lists that are created within our workflow.

10. **Math**: Allows us to apply mathematical equations to integers within our workflow.

11. **Script**: Provides us with the ability to extend out-of-the-box functionality with additional coding capabilities.

12. **String**: Allows us the ability to create, manipulate, and manage various strings that are created within our workflow.

13. **Units**: Provides the ability to identify, quantify, and manage units within our workflow.

In the next section, we will look into add-ons and the functionality they provide to us.

Exploring the Add-ons section

At the present time, our **Add-ons** section will likely be blank, especially if this is the first time we're accessing and using Dynamo for Civil 3D 2025. To add a custom node package to the **Add-ons** section, we use the following steps:

1. Navigate up to our top menu bar.

2. Select the **Packages** drop-down menu.

3. Select the **Package Manager...** option (as shown in *Figure 12.11*).

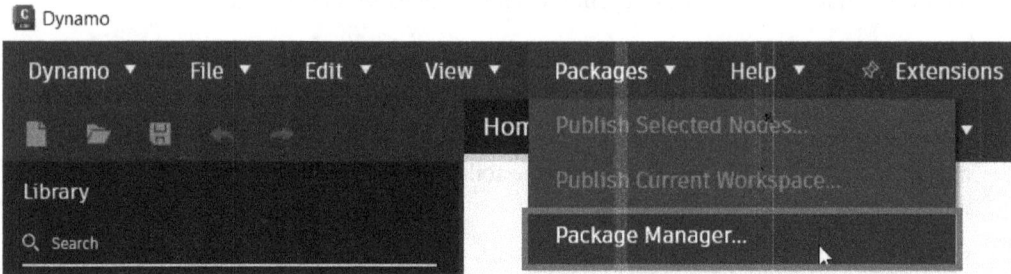

Figure 12.11 – Selecting the Package Manager option

Once selected, our **Package Manager** dialog box will appear on screen, similar to that displayed in *Figure 12.12*.

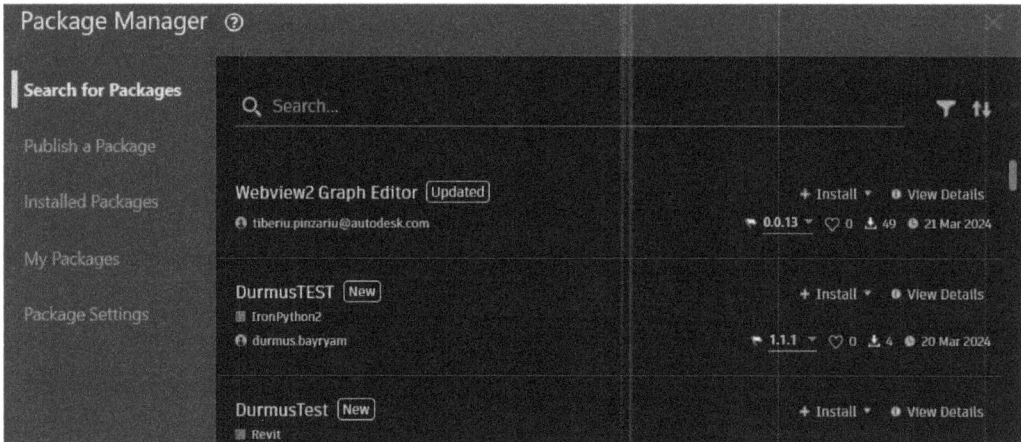

Figure 12.12 – Package Manager dialog box

Here is where we have the ability to install extensions that others within the Dynamo community have developed to extend our options and capabilities as we develop our workflows or scripts. It's important to note, though, that not all of these packages will work properly within the Dynamo for Civil 3D environment. Many that are listed are actually only applicable in the Revit environment.

4. That said, when using the **Search** bar at the top center of the **Package Manager** dialog box, it's recommended to filter using the key words `civil 3d` and apply the **Downloads** and **Descending** sorting options, as shown in *Figure 12.13*.

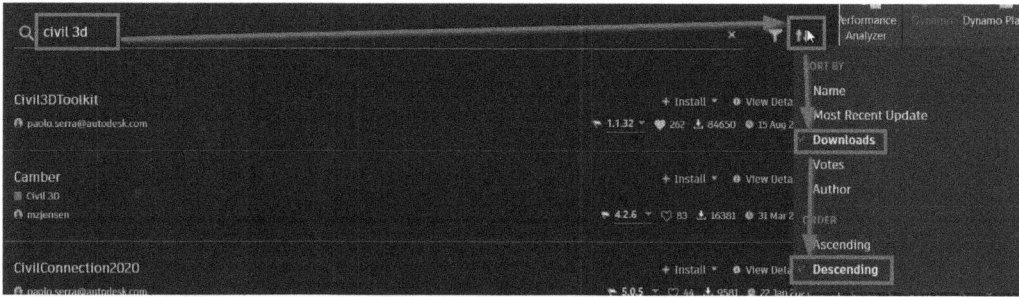

Figure 12.13 – Package Manager dialog box

Right away, we'll see the top two Dynamo for Civil 3D packages we want to make sure are installed: **Civil3DToolkit** and **Camber** (refer to *Figure 12.14*).

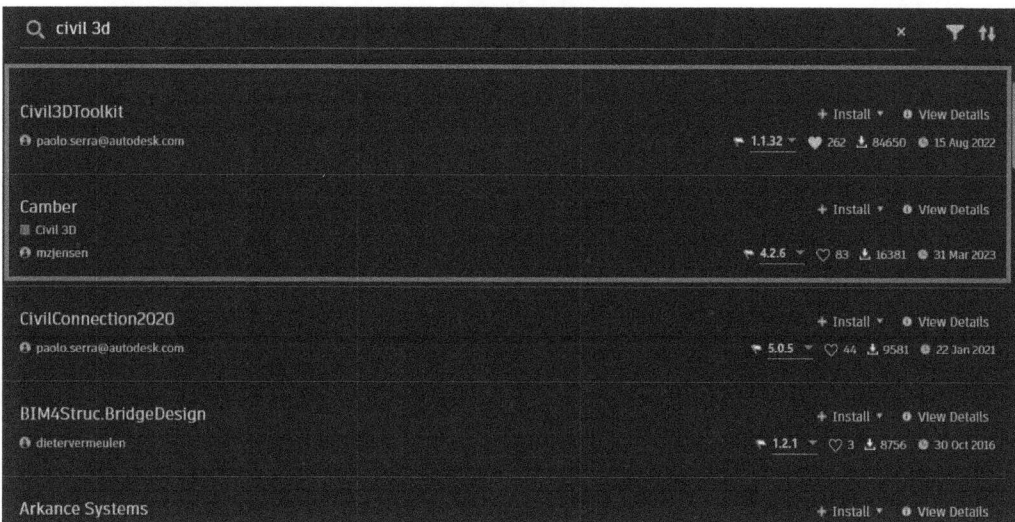

Figure 12.14 – Civil3DToolkit and Camber packages within Package Manager

Both of these add-on packages contain very extensive libraries of nodes that allow us to develop workflows around more AutoCAD- and Civil 3D-based objects and also provide us with a plethora of additional nodes that enable seamless integration, analysis, and collaboration capabilities.

5. That said, let's go ahead and click **Install** for both package and accept the proceeding dialog boxes that appear asking you to confirm the installations, including dependencies. After everything has been accepted, we will then see these two custom packages installed and available within the **Add-ons** section, listed as **Autodesk** (which is the Civil3DToolkit package) and **Camber** as shown in *Figure 12.15*.

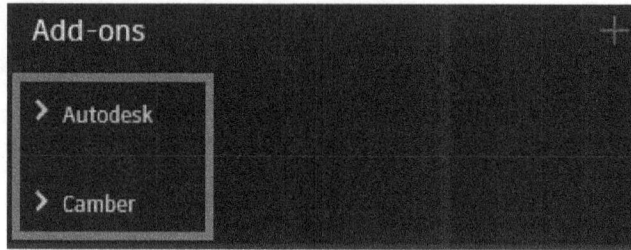

Figure 12.15 – Civil3DToolkit and Camber packages within Add-ons

At this point, we have plenty of nodes available to us to begin exploring how to put them into practice and develop new scripts that can be applied to our civil BIM design models.

Building our first Dynamo for Civil 3D 2025 script

Now that we have a general understanding of Dynamo for Civil 3D, how to access it, and an idea of what types of nodes we have available to us, let's start building our first Dynamo for Civil 3D script, or workflow. As you may have already begun to realize, Dynamo for Civil 3D has a tremendous amount of power and provides a multitude of options in terms of what we can automate. Differentiating between what we *want* to do versus what we *should* do can occasionally be a bit of a task itself. At the beginning of anyone's Dynamo for Civil 3D journey, it's recommended to keep a running list of any task or workflow that is often very labor-intensive and time-consuming.

Taking that into consideration, one of the first labor-intensive and time-consuming tasks that comes to mind is cleaning up a survey file to conform to the display standards of a client or of your own organization. Using Dynamo for Civil 3D, we can build a script to automate this task relatively quickly. Obviously, the more comprehensively we build this workflow out within Dynamo for Civil 3D, the more time we'll save ourselves at the beginning of future projects as we prepare our existing conditions file for future design integration.

Understanding the parts and areas of a node

With that, let's jump over to our **Library**, expand the **AutoCAD** category, expand the **Document** subcategory, and select our **Current** node, as shown in *Figure 12.16*.

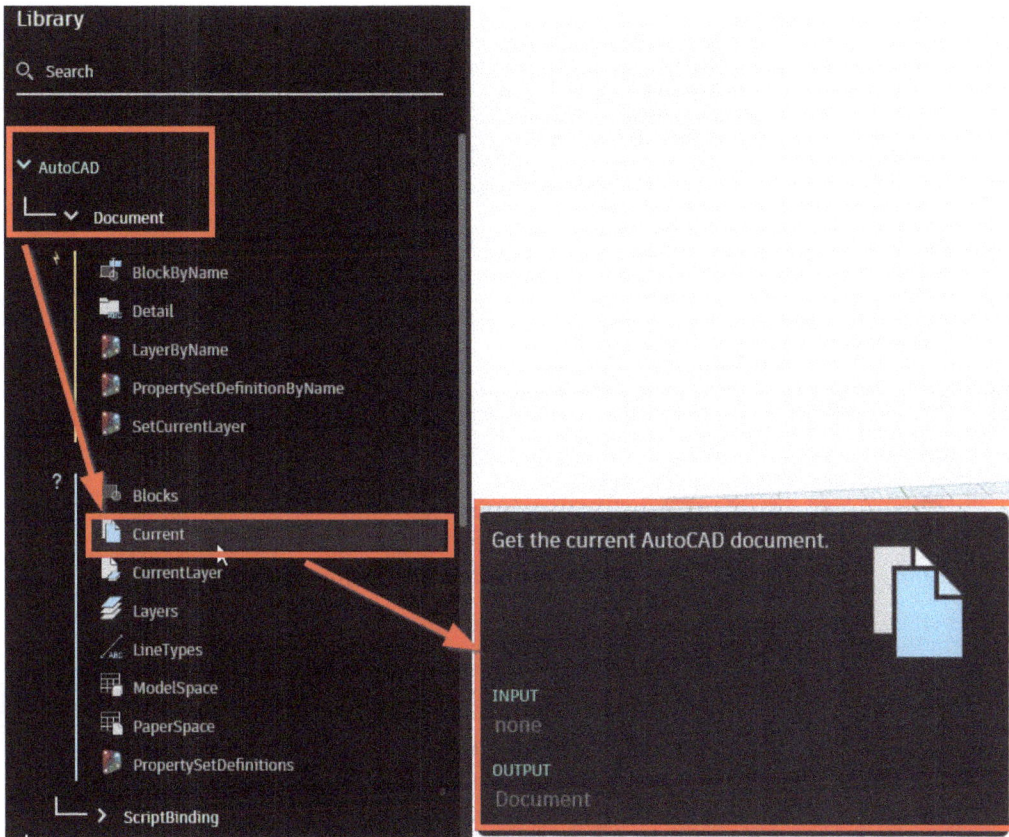

Figure 12.16 – Selecting AutoCAD | Document | Current node

As you may have noticed, before selecting any node within **Library**, we get a preview of the node that automatically appears next to it, giving a description of what that particular node represents along with its expected inputs and outputs. I highly recommend getting yourself familiar with these displays to get a better understanding of what each node represents and what you can do with them. After selecting the **Current** node, this node will appear in our graph view.

Next, with our **AutoCAD** category and **Document** subcategory still expanded, let's select the ModelSpace node as well to populate our graph view. Now that we've populated our graph view with two nodes, let's take a closer look at the different parts of each node so we know how best to construct our workflows moving forward. As shown in *Figure 12.17*, there are generally six different areas of each node with which we'll need to familiarize ourselves.

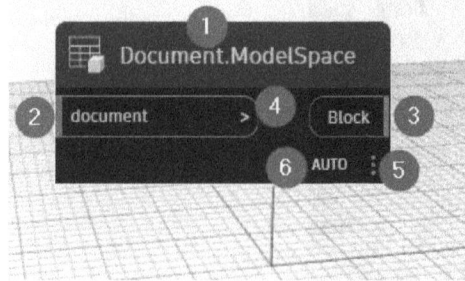

Figure 12.17 – Dynamo Node Layout

The numbers in *Figure 12.17* indicate the following:

1.   **Name**: Displays the name of the node. Double-clicking in this space allows us to quickly rename the node as well, if required.

2.   **Input**: Specifies the input connection to continue building our workflow out.

3.   **Output**: Specifies the output connection to continue building our workflow out.

4.   **Level**: Specifies the level at which we are analyzing the input.

5.   **Options**: Gives us the **Delete, Group, Preview, Freeze, Show Labels, Rename, Change Lacing, Define as an Output**, and **Help** options.

6.   **Lacing**: Defines the number of options provided as an output. Possible **Lacing** options include **Automatic, Shortest, Longest**, and **Cross Product**. **Automatic** is typically assigned as a default value to any new node placed in our graph view and will also essentially provide similar results to the **Shortest Lacing** option.

With that basic understanding of the various parts and areas of a typical node, let's go ahead and connect our **Document.Current** and **Document.ModelSpace** nodes together.

Connecting nodes together

To do this, we'll simply click with our mouse on the output of the **Document.Current** node and then click on the input of the **Document.ModelSpace** node. After performing this, we have now placed what's called a wire between the two nodes and have built the beginning of our first script, as shown in *Figure 12.18*.

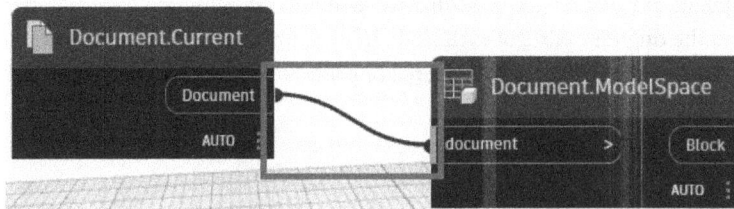

Figure 12.18 – Connecting nodes with a wire

It's important to note that these two particular nodes, **Document.Current** and **Document.ModelSpace**, will be used in a lot of our scripts and workflows that we're attempting to automate. As you have probably guessed, these two nodes alone allow the script to communicate with all components within the model space of our current file that we have opened up in Civil 3D 2025.

The **Document.ModelSpace** node will allow us to run automated workflows on objects within our model space either on an individual or grouping bases. In our case, however, we are attempting to change display styles holistically within our file. In this particular workflow that we're building out, the nodes that recognize the objects we'll be changing can actually be connected directly to the **Document.Current** node.

Continuing with the simplest approach, let's start identifying which components we want to convert to our specified display standard.

Identifying Surfaces in Dynamo

In our `Survey Model_Start.dwg` file, we have several different types of content in here that can interact with our script. For now, let's just focus on the Surfaces contained within our `Survey Model_Start.dwg` file.

That said, let's use the following steps to identify all Surfaces within our current file:

1. In our **Library**, we'll expand the **Civil 3D** category.
2. Expand the **Selection** subcategory.
3. Select the **Surfaces** node.
4. Connect the **Document.Current** node to our **Surfaces** node with a wire (as shown in *Figure 12.19*)

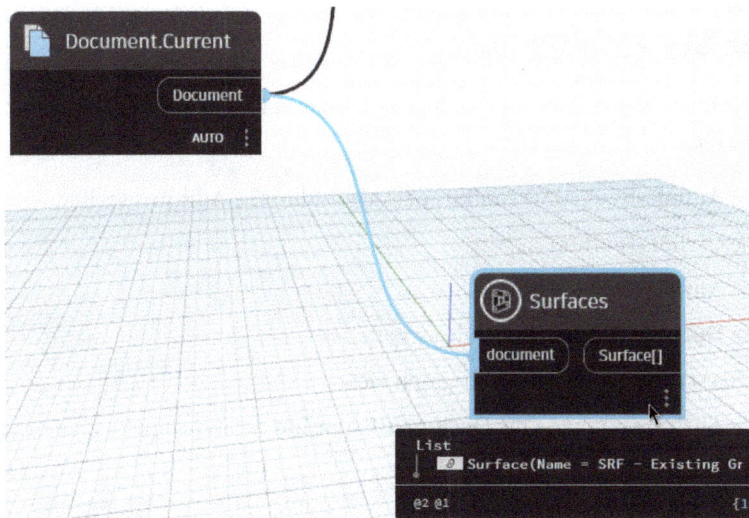

Figure 12.19 – Connecting the Document.Current node to the Surfaces node

> **Note**
>
> If we hover our mouse over the bottom of the **Surfaces** node, we can see a list preview of the expected output (as shown in *Figure 12.17*).

If we bring our attention to the lower left-hand corner of our Dynamo for Civil 3D session, we can see a selection option that, by default, is set to **Automatic** with an indication that the run is complete next to it (refer to *Figure 12.20*).

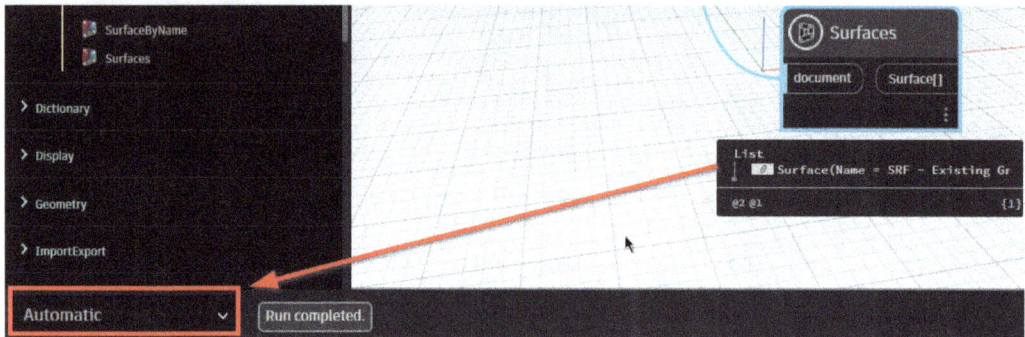

Figure 12.20 – Script running switch (Automatic or Manual)

By default, Dynamo for Civil 3D will automatically run your script on the current document as we build our scripts. The automatic running of scripts can potentially cause issues if you accidentally make the wrong connection or add the wrong node. That said, it's highly recommended that we switch this from **Automatic** to **Manual** so we can continue building our scripts without any impacts on our current file until we force it to run.

Dealing with Styles in Dynamo

Next, we'll use the following steps to get the Styles that are associated with all identified Surfaces within the current file:

1. Expand the **Autodesk (Civil3DToolkit)** package within the **Add-ons** section.
2. Expand the **Civil3DToolkit** category.
3. Expand the **AutoCAD** subcategory.
4. Expand the **DocumentExtensions** subcategory.
5. Select the **SetStyle** node to add to our graph view.
6. Connect the **Surfaces** output to the **DocumentExtension.SetStyle** input within our graph view.

These steps are shown in *Figure 12.21*:

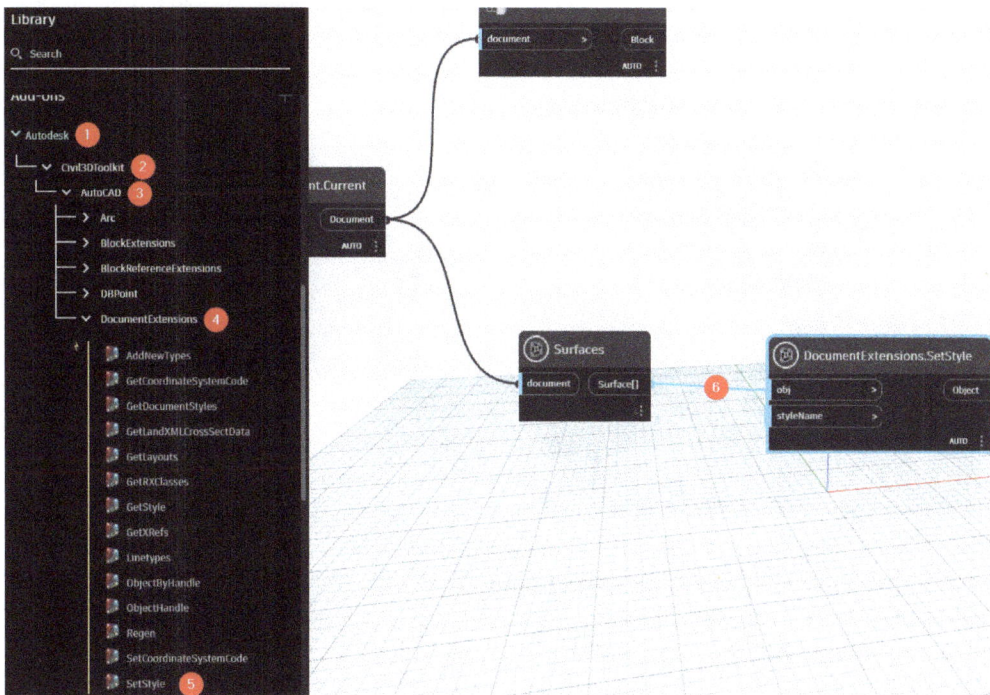

Figure 12.21 – Steps to add the DocumentExtensions.SetStyle node to our workflow

As you can see in the **DocumentExtensions.SetStyle** node, we are missing one additional input parameter. If we hover over the **styleName** input, the information about the requirements needed to connect to the input indicates that we need to add a String as shown in *Figure 12.22*.

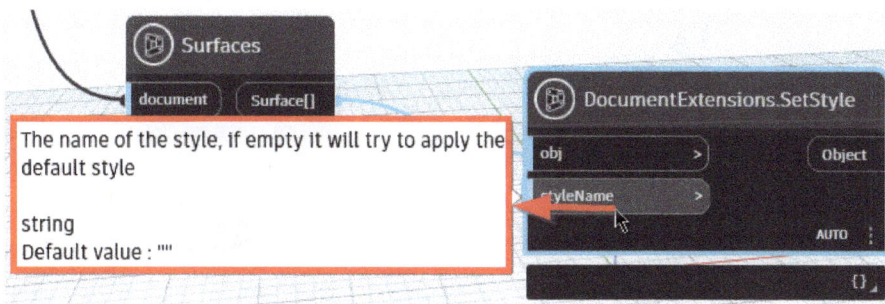

Figure 12.22 – Input parameter requirement guidance

That said, we'll need to add a String node now, which is essentially the equivalent of a manual input that we type in ourselves. Any time we require manual entry, we can simply double-click in our graph view to create a **Code Block**. Once our **Code Block** has been created, go ahead and type in `"Contours 1' and 5' (Background)";` and then connect the **Code Block** to the **styleName** input on our **DocumentExtensions.SetStyle** node as shown in *Figure 12.23*.

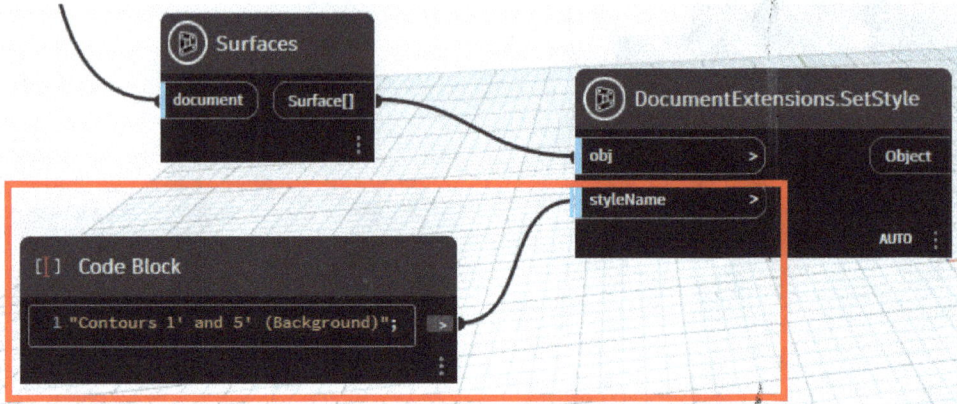

Figure 12.23 – Manual entry within our Code Block node

Finally, let's go ahead and select all nodes displayed in our graph view, navigate up to the **Edit** drop-down menu along the top of our Dynamo for Civil 3D session, and select the **Cleanup Node Layout** option at the bottom of the drop-down menu, as shown in *Figure 12.24*.

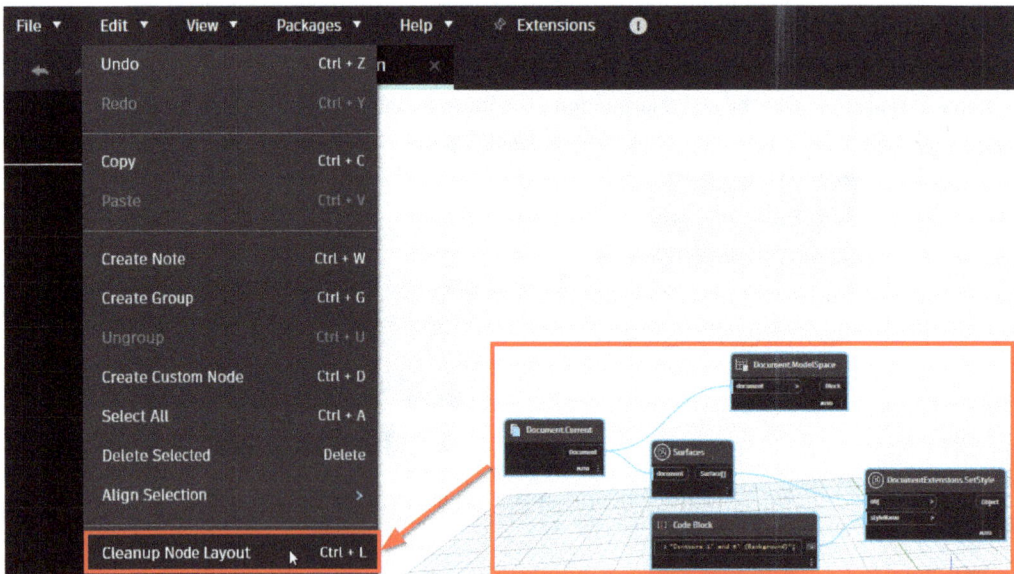

Figure 12.24 – Cleanup Node Layout

Alternatively, as a shortcut, after selecting our nodes, we can also simply hold the *Ctrl + L* keys on our keyboard to clean up the node layout.

Important note

You have probably noticed that we left the **Document.ModelSpace** node in our script at this point. Although this particular node is not required to convert our display Styles for this task, there may be a need to utilize this node as we continue to develop our script. Using the **Document.ModelSpace** node will allow us to access specific modeled objects currently located within our model space of our drawings, in the event we wanted to isolate our selection and conversion processes without affecting all objects of a particular type.

We have now created a very simple workflow within Dynamo to convert all Surface display styles to a predetermined standard using only four nodes. Go ahead and select the **Run** button (refer to *Figure 12.25*) in the lower left corner of our Dynamo for Civil 3D session and watch the magic happen in our `Survey Model_Start.dwg` file.

Figure 12.25 – Run the Dynamo for Civil 3D script

With that, we've officially created and successfully run an automated workflow on our design file. If we wanted, we could continuebuilding this particular script out to account for Alignments, Parcels, Gravity Networks, Pressure Networks, and so on, all branching out from the **Document.Current** node similar to that displayed in *Figure 12.26*.

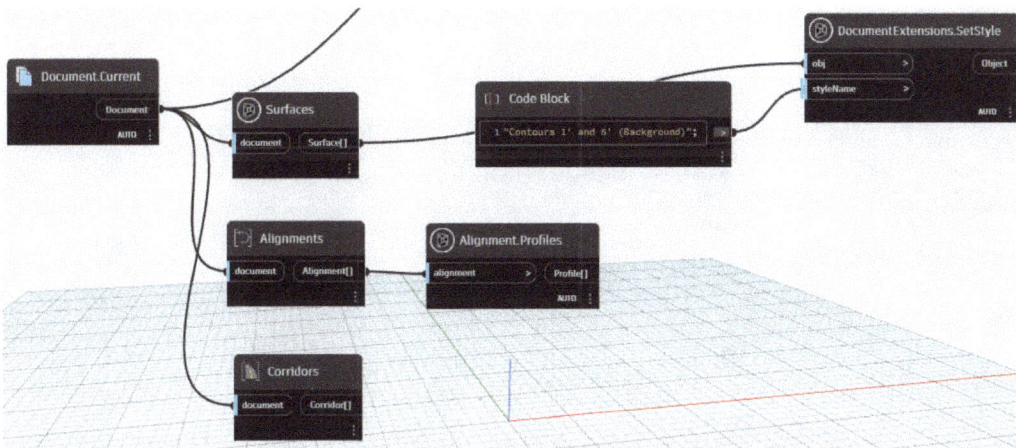

Figure 12.26 – Building out of our Dynamo for Civil 3D Script

As we add more object nodes to this script and specify the Styles we wish to apply, we are in essence gaining a tremendous amount of efficiency in how we process survey models that we may receive from another project stakeholder. These efficiencies not only save our production teams a ton of time, but also really drive consistency in workflows and ensure that nothing is accidentally overlooked or missed. It's also important to note that if we specify an object to convert that may not exist in the current file, the script will still run and apply to all applicable objects within our current file. Dynamo for Civil 3D will simply skip over any objects that are not included in the current file that we're running our script on at the time.

Accessing Dynamo script

Let's go ahead and save our Dynamo for Civil 3D script in our `Civil 3D 2025 Unleashed\`
`Chapter 12\DynamoScripts` location and close our Dynamo for Civil 3D session. Back in our
`Survey Model_Start.dwg` file, let's pull up **Dynamo Player** from the **Visual Programming**
panel in our **Manage** ribbon. Once the **Dynamo Player** dialog box appears, we'll use the following
steps to access our newly created Dynamo for Civil 3D script:

1. Select the *add folder* icon.

2. Select the *plus* sign, then navigate to and select the `Civil 3D 2025 Unleashed\Chapter`
    `12\DynamoScripts` folder.

3. Select the **OK** button.

These steps are displayed in *Figure 12.27*:

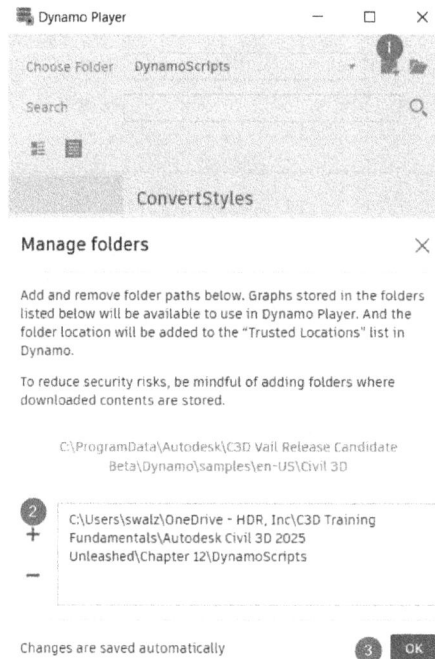

Figure 12.27 – Add the DynamoScripts folder to Dynamo Player

After selecting our `DynamoScripts` folder, we should have the newly created Dynamo for Civil 3D script listed in the view (refer to *Figure 12.28*).

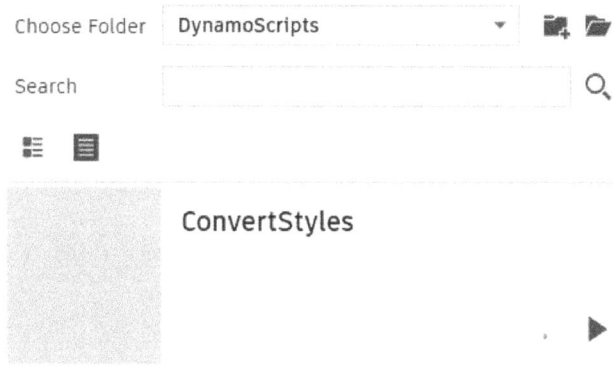

Figure 12.28 – New ConvertStyles Dynamo for Civil 3D script displayed in Dynamo Player

From here on out, we can simply pull up **Dynamo Player**, select this script, and click the Play button to run it on our files as needed.

Summary

Throughout this chapter, we explored how we can utilize Dynamo for Civil 3D 2025 to streamline design workflows that would previously take a considerable amount of time to complete. We also leveled up our game a bit and enhanced our civil BIM model management skills. Dynamo for Civil 3D 2025 offers many benefits that increase an individual's, design team's, and even an entire organization's efficiencies across the board.

We learned how to access Dynamo for Civil 3D within our design files, along with Dynamo Player to run already created scripts that we have in our library. We explored Dynamo for Civil 3D capabilities to get more comfortable with its interface and created a simple, yet very effective script that can be deployed on files in many different design situations. We're really just scratching the surface at this point though, and we highly recommend exploring ways to automate repetitive and labor-intensive tasks wherever possible.

In the next chapter, we'll continue building up our civil BIM model management skills as we explore how we can integrate our civil BIM designs into collaboration and visualization tools. Extending our civil BIM design integrations outside the design software itself will open our world up to even more possibilities to increase efficiency and become a more effective civil BIM model manager.

Part 4: Extending Infrastructure Projects beyond Civil 3D

It's a much bigger design world out there than just Civil 3D. Many projects require multidisciplined coordination and collaboration with other design teams, and we'll need to ensure that all models being developed align with each other in the real world. Stakeholder engagement is also a key component of coordinating designs to ensure that increased approval and buy-in is received as early as possible to limit rework/redesign later on down the road. This is where we will add another element to our model management career and learn what it means to have a well-coordinated design.

This part contains the following chapter:

- *Chapter 13, Preparing and Extending the Purpose of Our BIM Designs for Collaboration and Visualization*

13

Preparing and Extending the Purpose of Our BIM Designs for Collaboration and Visualization

In the previous chapter, we learned how to use Dynamo for Civil 3D 2025 to make our design work easier, faster, and more consistent. We also got better at managing civil BIM models, especially those we receive from other project stakeholders. Dynamo for Civil 3D 2025 has many benefits that help individuals, design teams, and organizations work more efficiently.

In this chapter, we'll look at extending our civil designs beyond Civil 3D and into the areas of collaboration and visualization that are often required for communicating and integrating our BIM designs in a multidisciplinary environment. By doing this, we'll find new ways to work more efficiently and become even better civil BIM model managers. The software we will be using in this chapter includes not only Civil 3D, but also Autodesk Navisworks for model collaboration, Autodesk InfraWorks for model visualization, and Autodesk Revit as an example of merging disparate model data. These software products can do much more than this chapter will show, but treat this chapter as a starting point for growing your knowledge in the world of BIM technology.

That said, in this chapter, we'll cover the following topics:

- Establishing shared coordinates in our design models
- Preparing and exporting Civil 3D BIM designs to Navisworks
- Preparing and exporting Civil 3D BIM designs to InfraWorks

Technical requirements

We will be using the same hardware and software requirements as discussed in the *Technical requirements* section of *Chapter 1*.

With that, let's go ahead and open up our Site Plan Reference.dwg file located within our Civil 3D 2025 Unleashed\Chapter 13\Reference location. Once opened, you'll notice that the file has been zoomed into our site, as displayed in *Figure 13.1*.

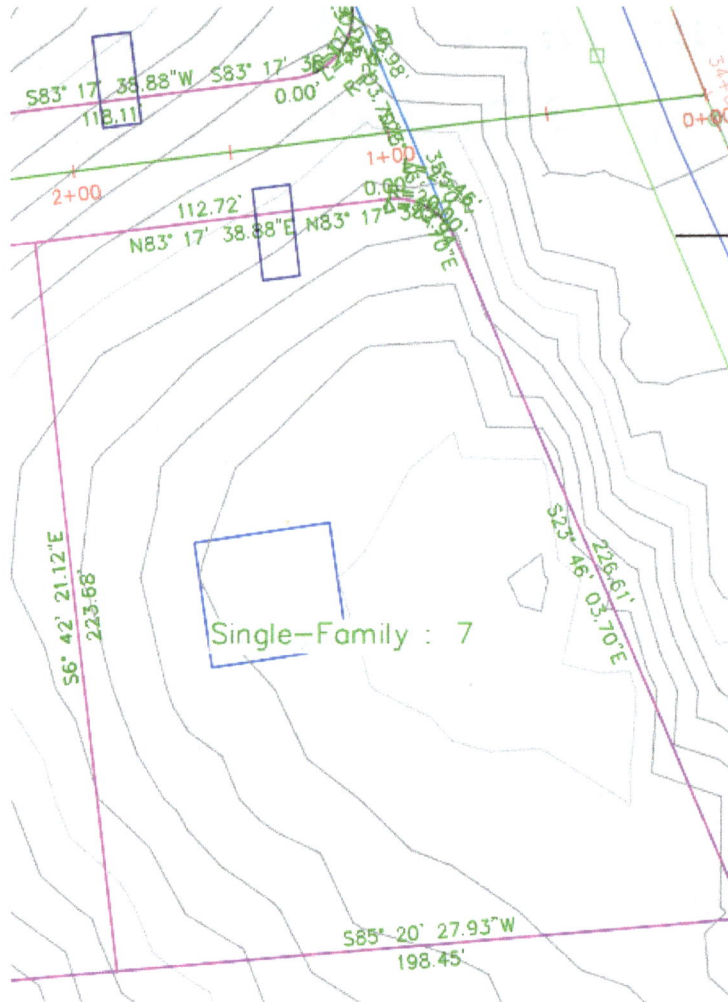

Figure 13.1 – Site Plan Reference.dwg display

With our Site Plan Reference.dwg file now opened up, let's learn how we can configure a shared coordinate system that will enable seamless model coordination across various Autodesk-based products.

Establishing shared coordinates in our design models

As you've realized by now, Autodesk Civil 3D has a plethora of tools and extensions that can really optimize and speed up the design and collaboration processes we are required to perform on our projects. We've covered how we can improve our design workflows along with our overall model management processes with a heavy focus on the civil side.

With designs often requiring multiple disciplines and multiple products to complete, it's important to recognize some of the workflows that can be used to streamline cross-product design integration. This is key to ensuring we have a well-coordinated model at the design phase and that we are limiting conflicts that could potentially arise during construction.

In that vein, it's necessary to recognize that most of this level of design model coordination is often between Autodesk Revit and Autodesk Civil 3D. Designers and modelers working on vertical designs within Autodesk Revit will quite often generate their designs without any concern for where the building is located in the real world. Autodesk Revit encourages users to design their vertical models close to 0,0,0 origin.

That's not to say that it cannot be geolocated afterward. It will just take a little extra coordination and configuration to adjust the Autodesk Revit design models to enable seamless integration and coordination between all product design models.

On the Civil 3D side, we've already had it engraved into our minds and workflows to set up our civil BIM designs to be properly projected, or georeferenced, into real-world coordinates. And if not a true real-world projection, some level of a local coordinate system projection is often required.

Within Autodesk Civil 3D 2025, there's a pretty simple workflow we can perform to ensure that all sharing and integration of design models from either Revit or Civil 3D are well coordinated. This level of coordination will allow the design models to fall right into place without any need to move or rotate the design models after importing.

Creating a shared coordinate file from Autodesk Civil 3D

Using the following steps, we'll be able to create a shared coordinate file from Autodesk Civil 3D that will be recognized within Autodesk Revit:

1. Activate the **Toolspace** button in the **Home** tab if not done so already.

2. Within our **Toolspace**, activate the **Toolbox** tab.

3. Expand the **Miscellaneous Utilities** category.

4. Expand the **Shared Reference Point** subcategory.

5. Right-click on the **Export Shared Reference Points for Autodesk Revit** option and select **Execute....**

Figure 13.2 shows these steps:

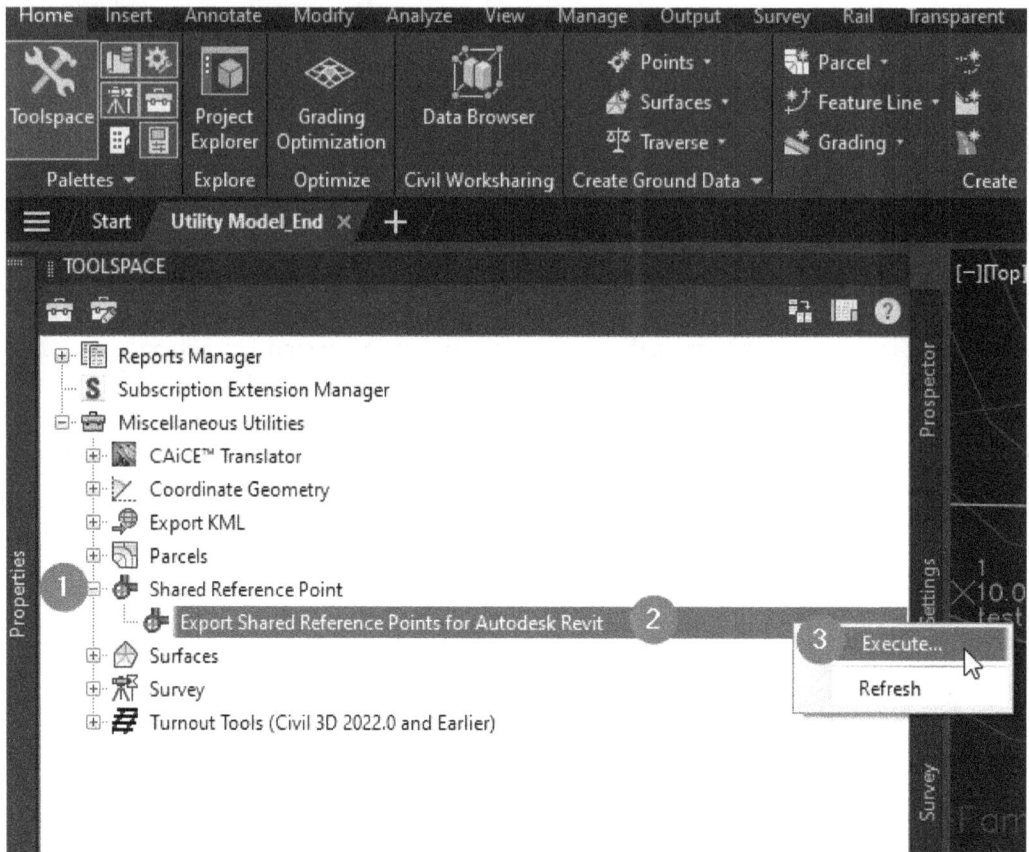

Figure 13.2 – Steps to initiate the Export Shared Reference Points command

6. Now, with the **Export Shared Reference Points for Autodesk Revit** command activated, we'll want to direct our attention to the series of command-line prompts to finish the workflow. Additionally, we'll want to navigate in our model to the parcel labeled **Single-Family : 7** in the southeastern portion of our site where we have already added a 2D footprint of a residential house, as shown in *Figure 13.3*.

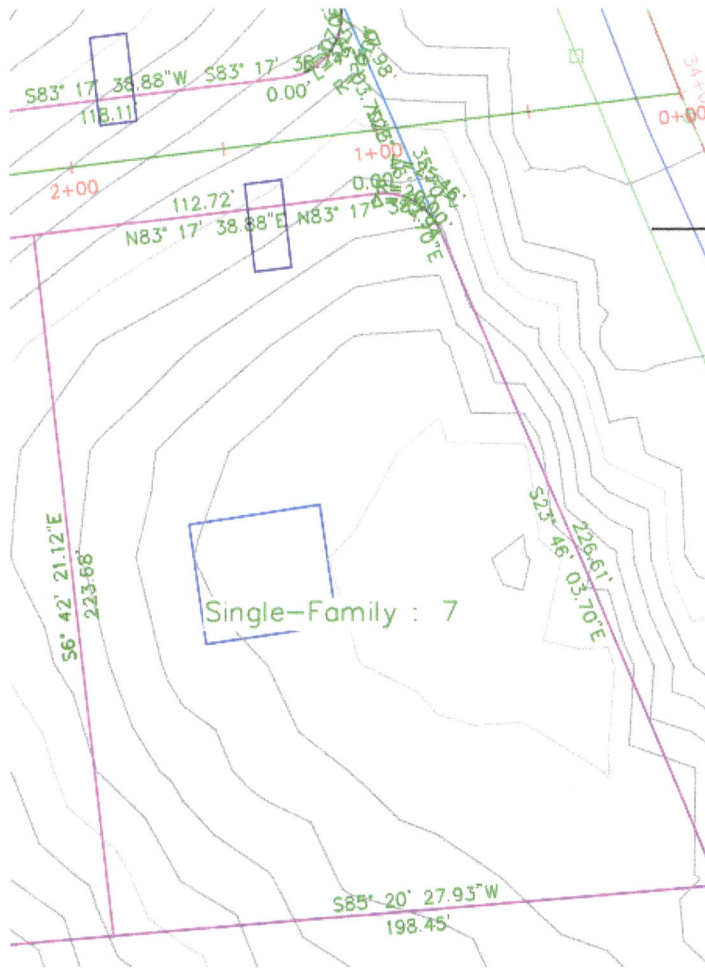

Figure 13.3 – Navigating to our Single-Family : 7 parcel

Using the next steps, we'll wrap up this workflow to define our shared coordinate file that will be used in Revit.

7. When prompted at the command line to **Select ORIGIN Point:**, we'll select the lower-left corner of our building outline. Note that it's good practice to either select a corner of a structure or a grid line to maintain consistency and have it tied to something that has meaning.

8. When prompted at the command line to **Select a point on +Y axis ('quasi-north'):**, we'll snap to the upper-left corner of our building outline, as shown in *Figure 13.4*.

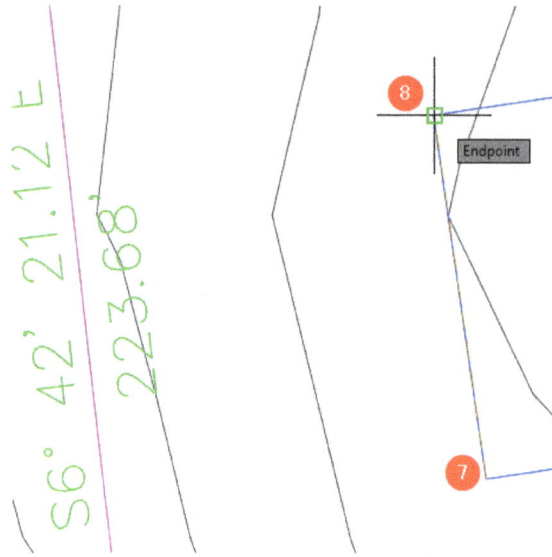

Figure 13.4 – Defining points to establish the +Y axis

9.   Once both points have been defined, we'll be prompted with a **Select Units and Confirm** dialog box, at which point we'll change our **Select DWG Units** from **Meters (M)** to **Feet (FT)** to match our files, as shown in *Figure 13.5*.

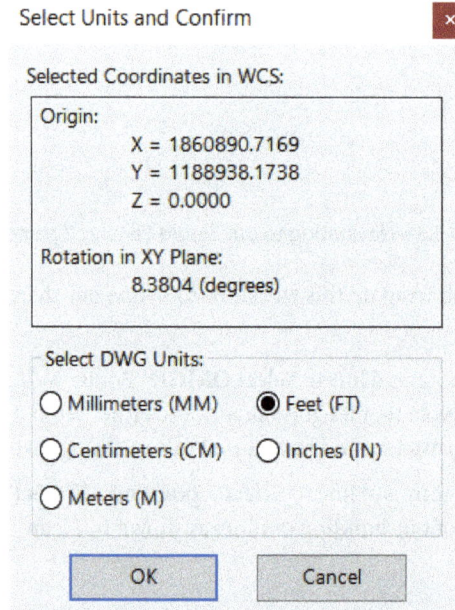

Figure 13.5 – Select Units and Confirm dialog box

10. Once changed, we can click **OK** within our **Select Units and Confirm** dialog box. It's important to note that this tool will default to **Meters** as the unit, so don't be too quick to accept the value and click **OK**.

11. Next, we'll be asked where we want to save it, at which point we'll navigate to our `Civil 3D 2025 Unleashed\Chapter 13` location and give it a name of `MySharedRefPnt.xml` and click the **Save** button, as shown in *Figure 13.6*.

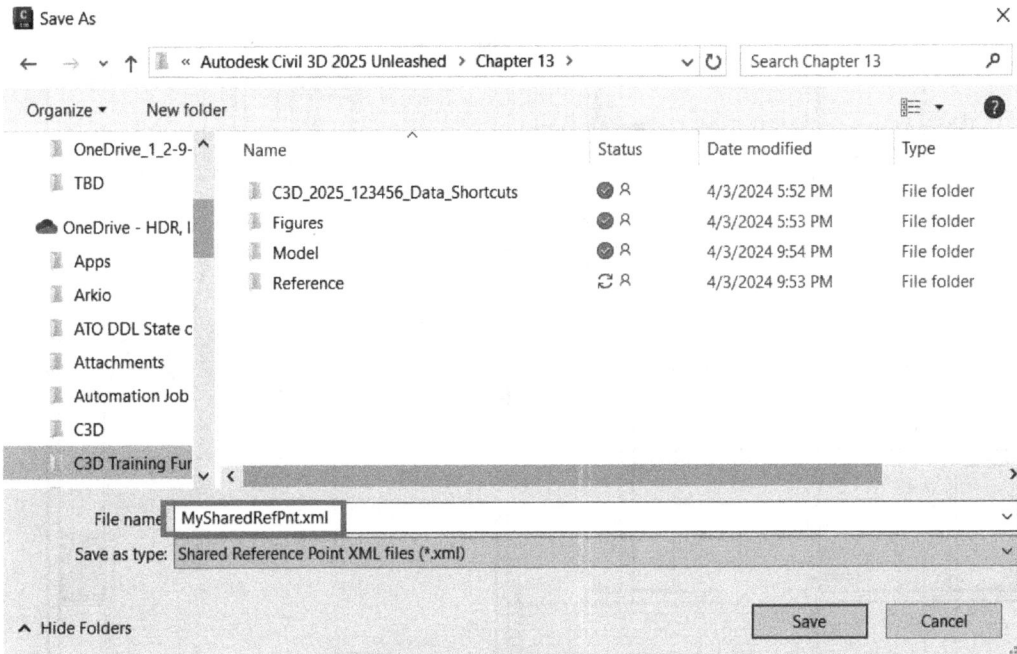

Figure 13.6 – Saving the Shared Reference Point file

> **Important note**
>
> If you have multiple Revit models that need to be integrated into our designs, we'll need to provide this Shared Reference Point file for each additional structure that requires this level of coordination.

We have now officially created a Shared Reference Point file that can be imported into Autodesk Revit to ensure all design model files are geolocated accurately. On the Autodesk Revit side, there's a series of steps the user will need to take to read the `MySharedRefPnt.xml` Shared Reference Point file we just created and verify that their models are in alignment.

Although this isn't a Revit book, understanding how this workflow comes full circle is great knowledge to have as a BIM model manager and continues to expand our skillsets and understanding of all BIM integration processes. With that, let's review the workflow required on the Autodesk Revit side that will allow for streamlined coordination of BIM designs across multiple products and platforms.

Understanding the workflow required on the Autodesk Revit side

If we were to open up the `Small Family 2 Bedroom House.rvt` file within our `Civil 3D 2025 Unleashed\Chapter 13\Revit` location inside of Revit, we would be presented with a file that looks similar to that displayed in *Figure 13.7*.

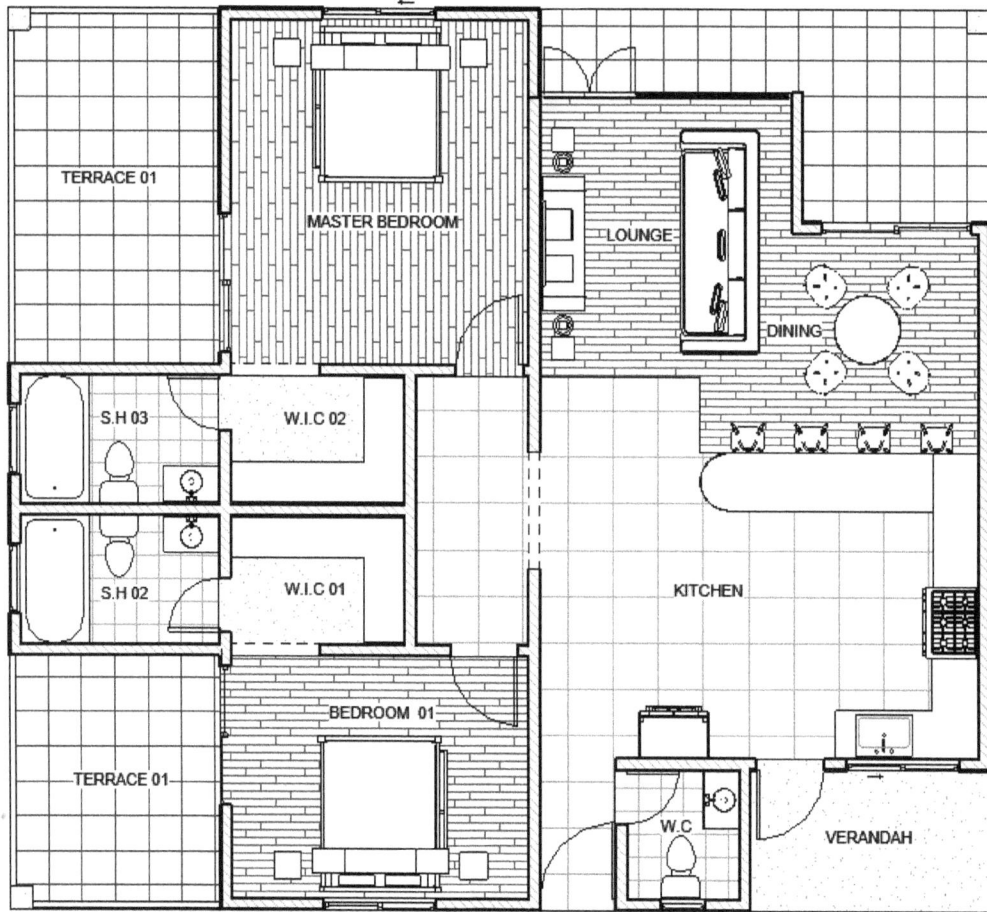

Figure 13.7 – Display of our Small Family 2 Bedroom House.rvt file

Once our `Small Family 2 Bedroom House.rvt` file is opened, we'll hop up to and activate our **Add-Ins** ribbon and select the **Import Shared Coordinates from XML file** tool within our **Shared Reference Point** panel, as shown in *Figure 13.8* (*it should be noted that the Shared Reference Point is available as an extension and will need to be installed separately from the standard Autodesk Revit installation*):

Figure 13.8 – Activating our Import Shared Coordinates from XML file tool

Once activated, we'll use the following steps to link our `MySharedRefPnt.xml` Shared Reference Point file to our `Small Family 2 Bedroom House.rvt` file:

1. When prompted to **Select ORIGIN Point to align to**, we'll select the lower-left corner of our building outline, just as we did in Civil 3D.

2. When prompted to **Select a Point on +Y (Up) Direction to align to**, we'll select the upper-left corner of our building outline, just as we did in Civil 3D.

3. When the **Open** dialog box appears, we'll select our `MySharedRefPnt.xml` file that we created earlier within our `Civil 3D 2025 Unleashed\Chapter 13` location.

4. After selecting and opening the file, we'll likely be prompted with a follow-up dialog box asking us **Are you sure to create the NEW shared Coordinates 'MySharedRefPnt'?**, at which point we'll click on the **Yes** button within this dialog box to accept.

5. One more dialog box will appear indicating that we have **Successfully set Shared Coordinates 'MySharedRefPnt'**. Go ahead and select the **OK** button to finalize the linking of our `MySharedRefPnt.xml` Shared Reference Point file to our `Small Family 2 Bedroom House.rvt` file.

To view our `Small Family 2 Bedroom House.rvt` file as if it were in its true location in the real world within the Revit environment, we'll use the following steps to update our file and view settings:

1. Activate the **Manage** ribbon.

2. Select the **Location** tool within the **Project Location** panel.

3. When the **Location and Site** dialog box appears, activate the **Sites** tab.

4. Select **MySharedRefPnt** from among our sites defined in this project list and select **Make Current**.

5. Click the **OK** button at the bottom of the **Project Location** panel dialog box.

6. In the **Properties** panel, change the **Orientation** property from **Project North** to **True North**.

The final output looks similar to that displayed in *Figure 13.9*:

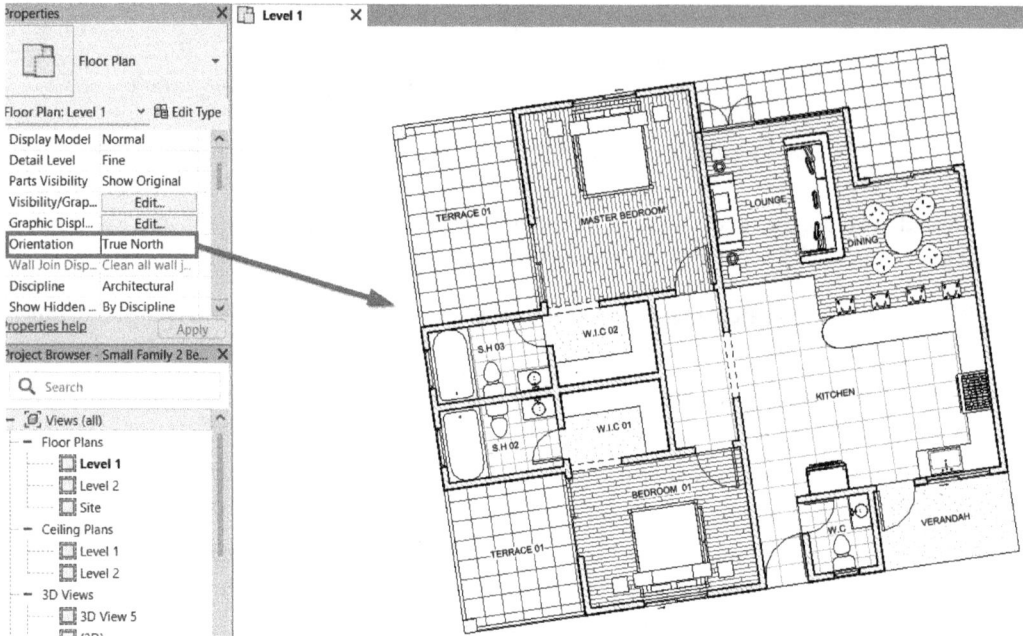

Figure 13.9 – True North orientation after linking our MySharedRefPnt Shared Reference Point file

With our `Small Family 2 Bedroom House.rvt` file now properly linked and georeferenced, the final step is understanding how to properly export Revit models to be inserted or externally referenced in our civil BIM designs in Civil 3D 2025. To do so, we'll go up to our **File** drop-down menu, select **Export**, then **CAD Formats**, and finally **DWG**, as shown in *Figure 13.10*.

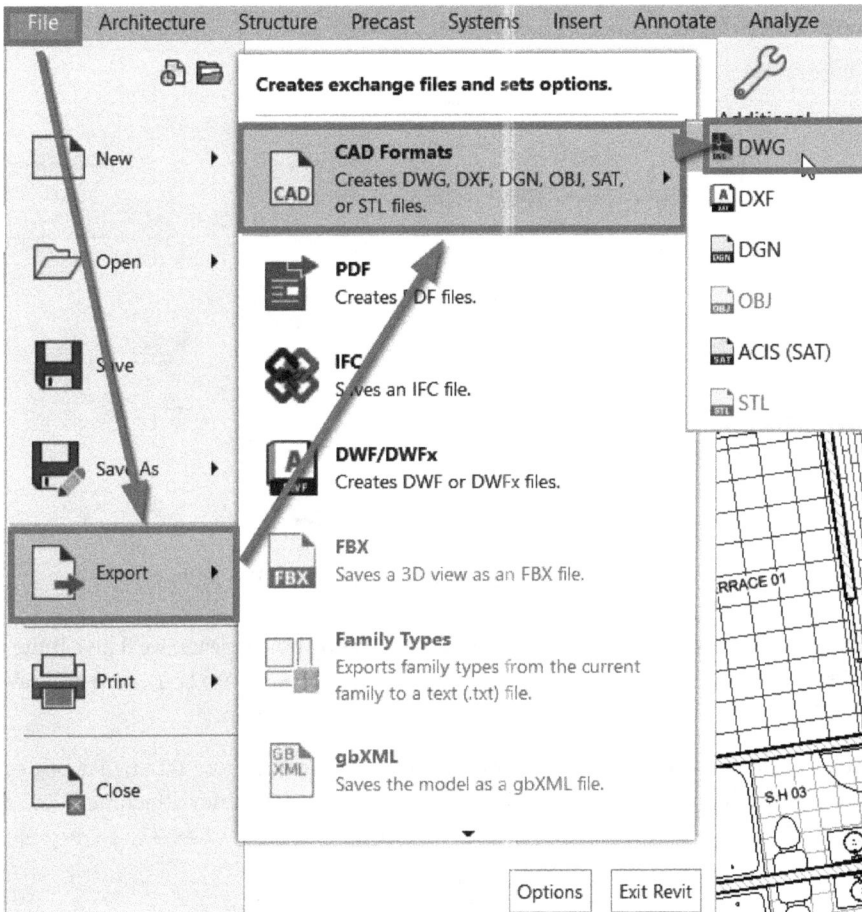

Figure 13.10 – Export RVT to DWG

When the **DWG Export** dialog box appears, we'll use the following steps to export our Small Family 2 Bedroom House.rvt file to a DWG formatted exchange:

1. Select the *ellipses* icon next to the **Select Export Setup** option.

2. Activate the **Units & Coordinates** tab in the **Modify DWG/DXF Export Setup** dialog box.

3. Set the units to **Foot**.

4. Set **Coordinate Base** to **Shared Coordinates**.

5. Select **OK** to close out of the **Modify DWG/DWF Export Setup** dialog box.

6. Select **Next** in the **DWG Export** dialog box to begin the export process.

These steps are also shown in *Figure 13.11*:

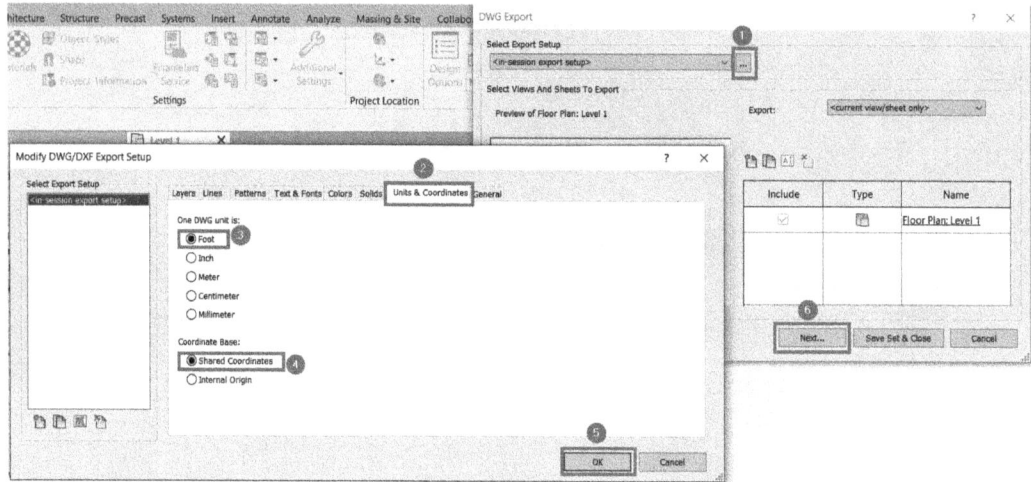

Figure 13.11 – Steps to configure the export RVT to DWG settings

When the **Export CAD Formats – Save to Target Folder** dialog box appears, we'll give it the name of Small Family 2 Bedroom House - Floor Plan - Level 1.dwg and save it to our Civil 3D 2025 Unleashed\Chapter 13\Revit location.

Hopping back over to our Civil 3D 2025 session where we have our Site Plan Reference. dwg file opened up, let's go ahead and externally reference in, as an Overlay attachment, our newly created Small Family 2 Bedroom House - Floor Plan - Level 1.dwg file, with the final result aligning on our site as shown in *Figure 13.12*.

Figure 13.12 – Displaying our Revit drawing geolocated within Civil 3D

With all design models properly georeferenced, let's begin exploring what else we can do to extend our BIM designs using a few other products.

Preparing and exporting Civil 3D BIM designs to Navisworks

Once we have all of our design models from Autodesk Revit and Autodesk Civil 3D geolocated and in alignment with each other, we can now begin to explore opportunities to extend our design models for collaboration and design review purposes. Throughout this book, we've picked up on several tips and have familiarized ourselves with many advanced tools that will allow us to customize, analyze, review, interrogate, and modify our Civil 3D design outcomes.

When it comes to multidisciplined designs where additional design authoring tools are being utilized, we can leverage additional products such as Autodesk Navisworks to aid in many different areas and extend the possibilities with what we can do with our BIM designs holistically. Autodesk Navisworks brings many benefits and added value into the mix, including capabilities such as the following:

- **Advanced clash detections**: With Navisworks' advanced clash detection capabilities, designers can pinpoint conflicts between Civil 3D elements and other disciplines early in the design phase. Whether it's identifying clashes between Civil 3D-modeled underground utilities and Revit-modeled structural elements or other instances, Navisworks helps prevent expensive fixes and delays by catching these problems well before construction starts.

- **Visualizations**: Navisworks empowers designers with unparalleled visualization tools, enabling the creation of immersive 3D renderings, animations, and walkthroughs of combined project models. These visualizations serve as potent communication tools, elucidating design intent to clients, stakeholders, and regulatory bodies with clarity and conviction.

- **Quantifications and estimations**: Navisworks' robust quantification and estimation tools further augment project efficiency, allowing users to extract crucial data such as quantities, areas, and volumes from merged project models. By seamlessly integrating Civil 3D models into Navisworks, designers can generate precise quantity reports and conduct cost analyses swiftly and accurately.

- **Construction planning and simulation**: Navisworks extends its utility into construction planning and simulation through its 4D scheduling capabilities. By synchronizing project models with construction schedules, designers can visualize the project's evolution over time, incorporating site logistics, phasing plans, and construction sequences seamlessly into the simulation. Important to note is that as we begin populating our Civil 3D design models with construction-phase custom attributes through the application of property sets, we can leverage these custom attributes to specify when these elements will actually be constructed in our simulations.

- **Design review and coordination meetings**: Navisworks centralizes collaboration efforts, providing a hub for stakeholders to review, comment, and resolve design issues efficiently. Utilizing Navisworks' clash detection tools, teams can orchestrate virtual coordination meetings, swiftly identifying clashes, assigning responsibilities, and tracking resolutions in real time.

On small and large projects alike, Autodesk Navisworks will certainly aid in delivering a well-coordinated and comprehensive design to your clients within budget and schedule constraints.

Creating a Navisworks export

With that, let's open up our `Utility Model_Start.dwg` file located within our `Civil 3D 2025 Unleashed\Chapter 13\Model` location to familiarize ourselves with the workflow required to share our civil BIM designs in Autodesk Navisworks.

Once opened, we'll want to go down into our command line, type `NWCOUT`, and hit the *Enter* key on our keyboard. Typing this command will bring up our **Export to Autodesk Navisworks Exporter 2025** dialog box, where we'll want to navigate to our `Civil 3D 2025 Unleashed\Chapter 13\Navisworks` location, give our exported file a name of `Utility Model_Start.nwc`, and click the **Save** button. It should be noted that to access the `NWCOUT` command, Autodesk Navisworks 2025 or the Navisworks NWC Export Utility will need to be installed prior to initiating this command.

It is worth mentioning, Navisworks has several different file types for different use cases. In our exercises, we will be working with Navisworks files as a `.nwc`. Navisworks also has two other file formats, `.nwd` and `.nwf`, which serve different functions depending on our use of the data.

We've now officially created a Navisworks export that can be opened directly within Navisworks itself. Again, although this isn't a Navisworks book, understanding how this workflow comes full circle is great knowledge to have as a BIM model manager and continues to expand our skillsets and understanding of all BIM integration processes. With that, let's review the workflow required on the Autodesk Navisworks side that allows for streamlined coordination of BIM designs across multiple products and platforms.

Understanding the workflow required on the Autodesk Navisworks side

Launching Autodesk Navisworks, the following steps are how we open up our newly created `Utility Model_Start.nwc` file:

1. Select the *Open* icon at the top of our Navisworks session.
2. Navigate to our `Civil 3D 2025 Unleashed\Chapter 13\Navisworks` location.
3. Change the file type to `Navisworks Cache (*nwc)`.
4. Select the `Utility Model_Start.nwc` file.
5. Click on the **Open** button.

These steps are also displayed in *Figure 13.13*:

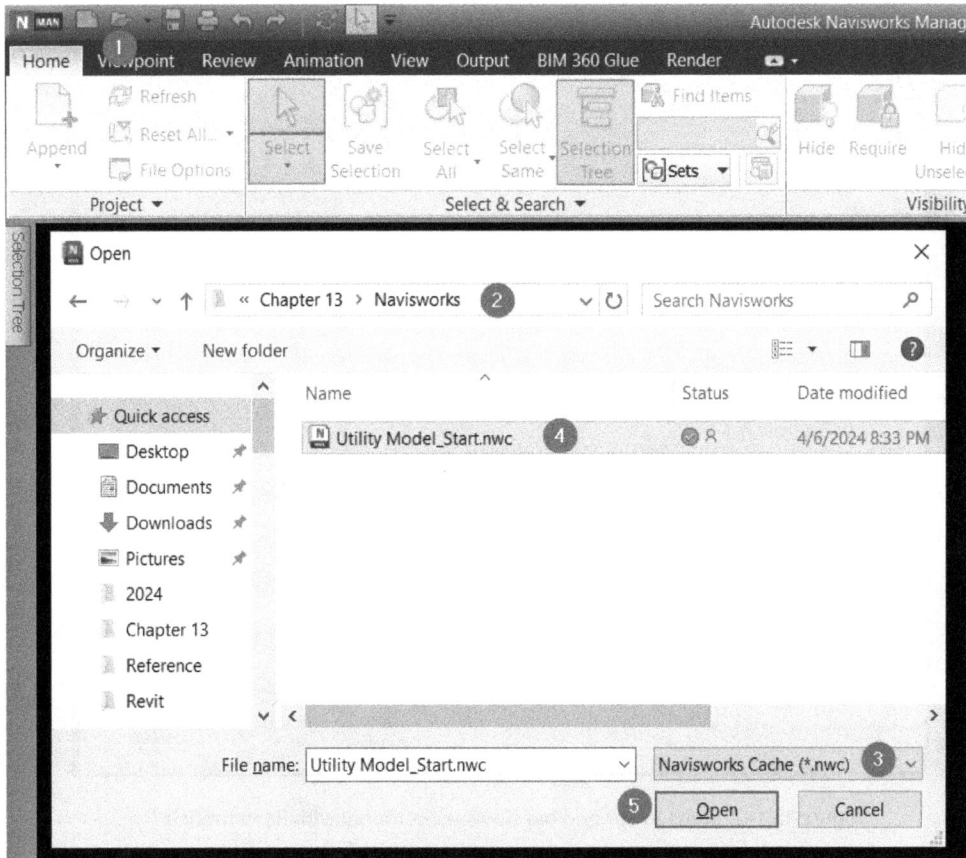

Figure 13.13 – Steps to open up our Utility Model_Start.nwc File in Navisworks

Once opened, we should be able to see our full design from Civil 3D 2025. To pull in our updated Revit model now that we have it reading the correct coordinates, we can use the following steps to add our Revit model to Navisworks as well to begin creating a coordinated model where we bring in all design models:

1. Activate the **Home** ribbon if not done so already.
2. Select the **Append** tool within the **Project** panel.
3. Navigate to our `Civil 3D 2025 Unleashed\Chapter 13\Revit` location.
4. Change the file type to `Revit (*.rvt, *.rfa, *.rte)`.
5. Select the `Small Family 2 Bedroom House.rvt` file.
6. Click on the **Open** button.

These steps are also displayed in *Figure 13.14*:

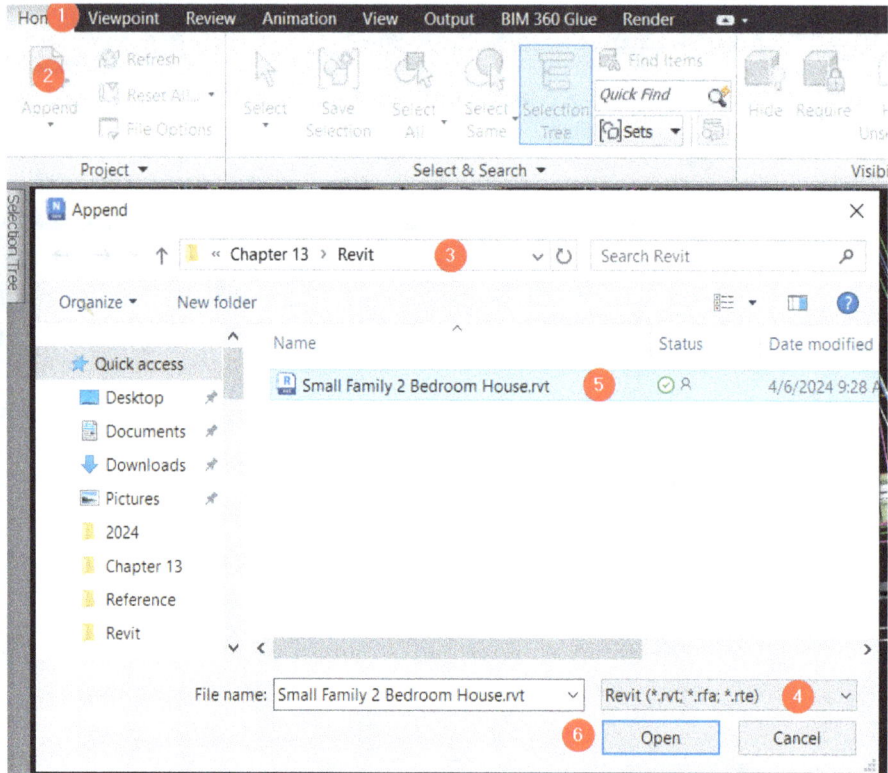

Figure 13.14 – Steps to append our Navisworks model with Revit models

After all of the files have been appended to our current model, we should see a federated model within Autodesk Navisworks similar to that displayed in *Figure 13.15*.

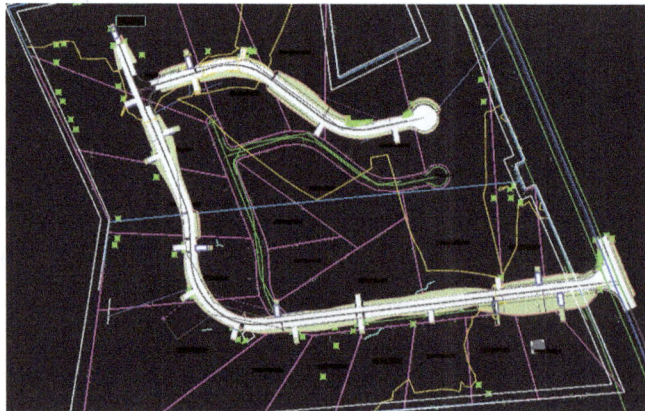

Figure 13.15 – Display of a federated model within Autodesk Navisworks

With all of our design models now brought into Navisworks as one consolidated and federated model, we can begin taking advantage of all of the tools and features that will open our teams and stakeholders up to all of the added value and benefits mentioned earlier.

Continuing through the additional areas into which we can extend the use of our civil BIM designs, let's explore what Autodesk InfraWorks can do to help us out next.

Preparing and exporting Civil 3D BIM designs to InfraWorks

As with all of our coordination requirements, we'll want to ensure that all of our design models coming from all design authoring tools are geolocated and in alignment with each other. Setting this up earlier in the design process makes the sharing of models and integration into other products so much easier for us later on as we introduce new products and processes.

When it comes to multidisciplined designs where additional design authoring tools are being utilized, we can leverage products such as Autodesk InfraWorks to aid in many different areas and extend the possibilities of what we can do with our BIM designs. Autodesk InfraWorks has many benefits and added value of its own, including the following capabilities:

- **Visualizations**: InfraWorks offers several visualization features and capabilities, enabling all project stakeholders to view the proposed designs within a 3D setting. These visualizations provide clear communication of design intent and objectives to clients, stakeholders, and the general public, ultimately removing any guesswork from the equation.

- **Collaboration and communication**: InfraWorks aids in enhancing seamless project team and stakeholder engagement by allowing instant sharing and coordination of design models and data, which will reduce errors and rework, and improve the design process overall. Bi-directional integration with design authoring tools such as Autodesk Revit and Civil 3D makes it easy to move from a conceptual design to detailed design processes and the generation of plan sets. Additionally, InfraWorks can be accessed on mobile devices through cloud-based services, allowing project stakeholders to review designs from anywhere, which enhances flexibility and collaboration, especially during field inspections and construction.

- **Conceptual and alternative design and analysis**: InfraWorks enables engineers and designers to quickly create conceptual designs for infrastructure-focused projects. Its intuitive interface and pre-built templates offer a low barrier of entry to the product itself and facilitate the rapid creation and iteration of design alternatives, helping clients and project stakeholders explore various alternative design options quickly and efficiently.

- **Contextual designs**: InfraWorks supports a multitude of file formats derived from a multitude of products used across the AEC industry, ranging from Autodesk to Bentley solutions, to common model formats such as FBX and OBJ files, to GIS. The ability to integrate any and all of these types of files and data allow us to truly generate a cohesive federated model for visual inspection and coordination. We have the ability to incorporate existing infrastructure, terrain data, environmental constraints, land use information, and so on into our designs, ensuring that the proposed solutions are feasible and compatible with the built environment.

- **Quantification and cost estimation**: InfraWorks facilitates the quantification of project and design components and provides tools for cost estimation. This helps in budgeting and cost forecasting, allowing design teams to evaluate the financial feasibility of proposed designs.

- **Simulation and analysis**: InfraWorks offers simulation and analysis capabilities for various aspects of infrastructure projects, including traffic flow, drainage, mobility, and sun/shadow analysis. Engineers can assess the performance and behavior of their designs under different conditions, ensuring compliance with regulatory requirements and project specifications. We can also perform scenario analysis to evaluate how these simulations might potentially impact design alternatives in terms of project performance and cost. This helps in making informed decisions and optimizing designs early on to meet project objectives.

Overall, Autodesk InfraWorks enhances the design process holistically, especially in the earlier phases of a project when conceptual and alternative design scenarios are being considered. InfraWorks will greatly aid in providing some clarity and a real-world visual perspective that our design authoring tools lack. InfraWorks is also great at providing design staff with a simple platform that assists with quick visual inspection checks to ensure our models are in alignment and well coordinated throughout the entire design process.

> **Important note**
>
> One of InfraWorks' greatest assets is its ability to generate this model federation environment that merges designs from various products. Once consolidated, we can export our InfraWorks federated model to an FBX file that can be brought into higher-end rendering and visualization tools such as TwinMotion, Lumion, Unreal, or Unity. This major benefit should not go unnoticed as it will allow you and your teams to really separate yourselves from the competition when it comes to project pursuits, public engagements, and overall marketing campaigns.

With that, let's open up our `Utility Model_Start.dwg` file located within our `Civil 3D 2025 Unleashed\Chapter 13\Model` location to familiarize ourselves with the workflow required to share our civil BIM designs in Autodesk InfraWorks.

Exploring bidirectionality between Civil 3D and Autodesk InfraWorks

Starting in Civil 3D 2025, let's get ourselves familiar with the **InfraWorks** ribbon within Autodesk Civil 3D to better understand how we can seamlessly integrate our civil BIM designs into InfraWorks. Going back up to the top of our Civil 3D 2025 session, let's activate the **InfraWorks** ribbon. Once activated, we have the following tools and options available to us:

1. **Launch Autodesk InfraWorks**: Provides quick access to and launches Autodesk InfraWorks directly from your Civil 3D 2025 session, rather than launching it from a shortcut or searching for it in your Windows programs.

2. **Exchange Settings**: Allows us to specify how InfraWorks models are imported into Civil 3D at an object level, where we can specify display styles, object layers, and so on that InfraWorks modeled objects will be placed on during the import process.

3. **Open Model**: Allows us to open an Autodesk InfraWorks file as a standalone DWG file.

4. **Import IMX**: Allows us to import an Autodesk InfraWorks file to amend our current file.

5. **Export IMX**: Allows us to export all Civil 3D modeled objects in our current file to an InfraWorks file.

6. **Product Page**: Launches the Autodesk InfraWorks main page that provides detailed feature and functionality information.

These tools and options are also displayed in *Figure 13.16*:

Figure 13.16 – InfraWorks ribbon in Civil 3D 2025

Now that we have a basic understanding of what tools we have available to us inside of Civil 3D to integrate our design models into Autodesk InfraWorks, let's start putting them to use.

Connecting Civil 3D and Autodesk InfraWorks

We want to begin by using the **Export IMX** tool within the **Export** panel of our **InfraWorks** ribbon (listed as tool *5* in *Figure 13.16*).

Once the **Export IMX** tool has been activated, a **Save to IMX** dialog box will appear where we'll want to navigate to our `Civil 3D 2025 Unleashed\Chapter 13\InfraWorks` location, give it the name of `Utility Model_Start.imx`, and click the **Save** button.

Next, we'll click the **Launch Autodesk InfraWorks** button within the **Run** panel of our **InfraWorks** ribbon (listed as tool *1* in *Figure 13.16*). When Autodesk InfraWorks has been launched, we'll select **New** along the left hand side and then make selections and fill out fields as follows and click the **OK** button:

- **Name**: `Residential Subdivision`

- **Work Local**: Select this option and navigate to our `Civil 3D 2025 Unleashed\Chapter 13\InfraWorks` location

- **Coordinate System: SC83F**

These fields are also displayed in *Figure 13.17*:

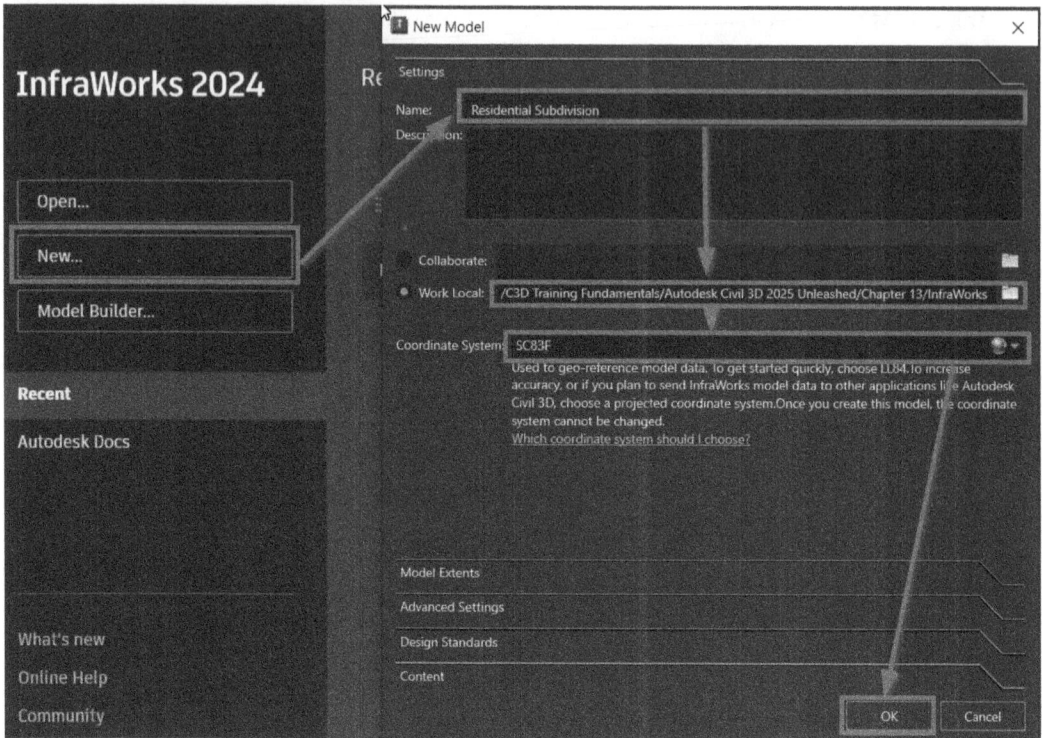

Figure 13.17 – Creating a new model in Autodesk InfraWorks

With our new residential subdivision InfraWorks model created and now opened up, we'll hop over to the right side of our session where we have our **DATA SOURCES** dialog box pinned, click the **Add Data Source** dropdown, and select **Autodesk IMX**, as shown in *Figure 13.18*.

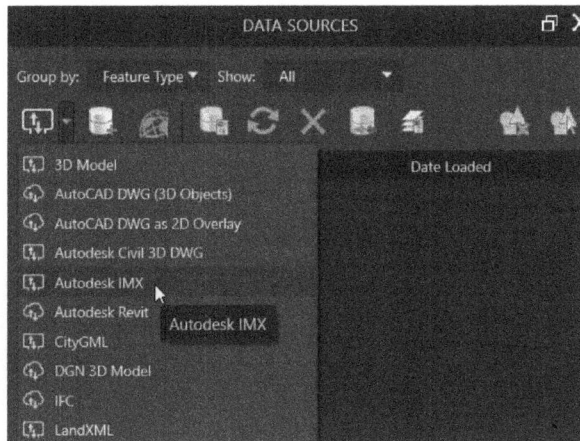

Figure 13.18 – Import Autodesk IMX file

When the **Select Files** dialog box appears, we'll navigate to our `Civil 3D 2025 Unleashed\`
`Chapter 13\InfraWorks` location and open our `Utility Model_Start.imx` file. After
selecting our IMX file to import, a **Choose Data Sources** dialog box will appear. Let's keep everything
selected for now and click the **OK** button, as shown in *Figure 13.19*.

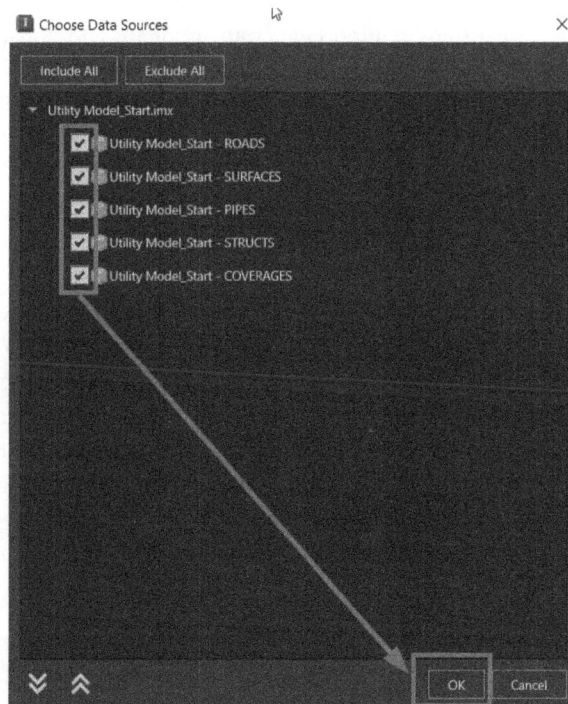

Figure 13.19 – Select the data sources to import

InfraWorks then processes the files, listing them in the **Configured in the Data Sources** dialog box that's still pinned on the right side of our Autodesk InfraWorks session. If we right-click on all of these and select **Refresh**, these items will change their status from **Configured** to **Imported** and will appear in our model as displayed in *Figure 13.20*.

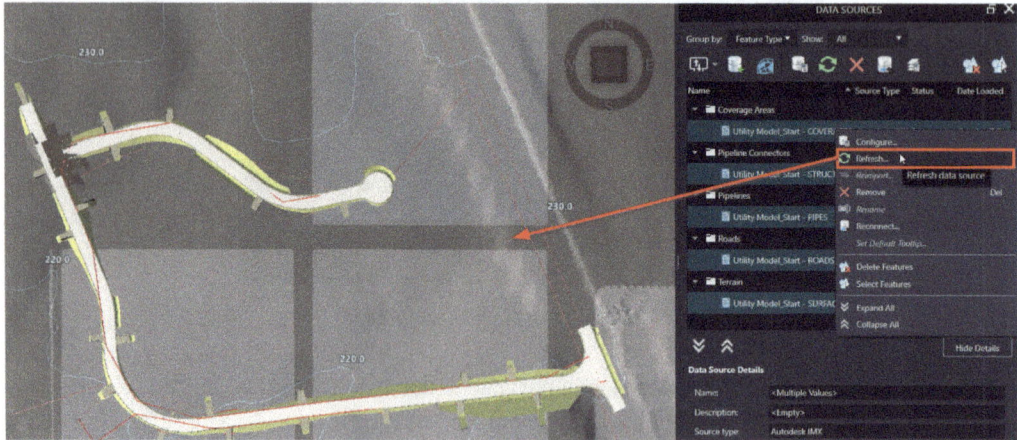

Figure 13.20 – Data sources connected in InfraWorks

We would essentially follow similar steps to import data sources for any additional design models that need to be integrated into our designs, including our Revit models. With all of our design models now brought into InfraWorks as one consolidated and federated model, we can begin taking advantage of all of the tools and features that will open our teams and stakeholders up to all of the added value and benefits mentioned earlier.

Summary

In this final chapter of our journey, we delved into the complexities of elevating collaboration and visualization products and capabilities within civil BIM designs, a crucial step in advancing your technology career. We continuously built on our foundational knowledge of Autodesk Civil 3D 2025 and even embraced a multidisciplinary approach where it makes sense. We began by understanding the ease of establishing shared coordinates in our design models to ensure seamless coordination across disciplines, products, and platforms. With Autodesk Civil 3D 2025, we mastered the creation of Shared Reference Point files, enabling precise geolocation of design models in Autodesk Revit, underlining the need for meticulous coordination to mitigate conflicts during construction.

Transitioning to both Navisworks and InfraWorks, we reviewed at a very basic level how we can carry our designs into these incredibly powerful tools. Both Navisworks and InfraWorks allow us to truly extend our civil BIM designs and ensure that we have a well-coordinated model leading into construction.

Mastering all workflows covered throughout this book equips you with the necessary skillsets to become an effective BIM model manager, leading design teams while increasing project efficiency, generating optimized design outcomes, and delivering well-coordinated, comprehensive designs within budget and schedule constraints. Ultimately, our exploration showcased the transformative potential of integrating BIM designs across multidisciplinary environments, fostering collaboration, visualization, and innovation throughout the design process, enriching your technology career journey.

Index

‹packt›

packtpub.com

Subscribe to our online digital library for full access to over 7,000 books and videos, as well as industry leading tools to help you plan your personal development and advance your career. For more information, please visit our website.

Why subscribe?

- Spend less time learning and more time coding with practical eBooks and Videos from over 4,000 industry professionals

- Improve your learning with Skill Plans built especially for you

- Get a free eBook or video every month

- Fully searchable for easy access to vital information

- Copy and paste, print, and bookmark content

Did you know that Packt offers eBook versions of every book published, with PDF and ePub files available? You can upgrade to the eBook version at packtpub.com and as a print book customer, you are entitled to a discount on the eBook copy. Get in touch with us at customercare@packtpub.com for more details.

At www.packtpub.com, you can also read a collection of free technical articles, sign up for a range of free newsletters, and receive exclusive discounts and offers on Packt books and eBooks.

Other Books You May Enjoy

If you enjoyed this book, you may be interested in these other books by Packt:

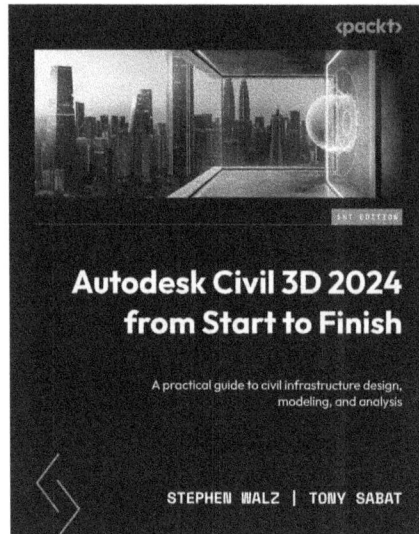

Autodesk Civil 3D 2024 from Start to Finish

Stephen Walz, Tony Sabat

ISBN: 978-1-80323-906-4

- Understand civil project basics and how Autodesk Civil 3D helps achieve them
- Connect detailed components of your design for faster and more efficient designs
- Eliminate redundant workflows by creating intelligent objects to handle design changes smoothly
- Collaborate with distributed teams efficiently and produce designs swiftly and effectively
- Optimize 3D usage and decision-making, using a model-based approach on the impact of your designs and accelerate your career

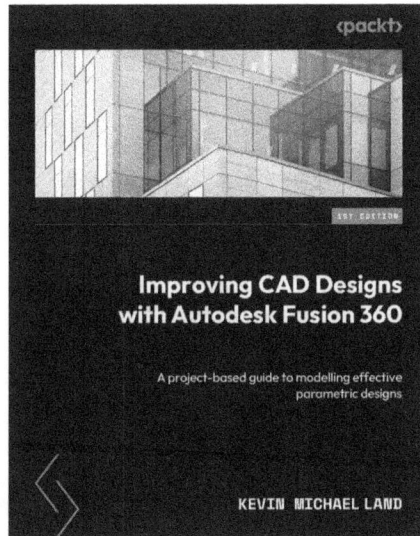

Improving CAD Designs with Autodesk Fusion 360

Kevin Michael Land

ISBN: 978-1-80056-449-7

- Gain proficiency in Fusion 360 user interface, navigation, and functionality
- Create and transform simple 2D sketches into 3D models
- Manipulate and control parametric 2D sketches using dimensions
- Become familiar with drafting on paper and taking measurements with calipers
- Create a bicycle assembly part with Fusion 360
- Use the form environment to create organic shapes
- Render a 3D model and understand how to apply materials and lighting
- Generate 2D assembly model drawings for documentation purposes

Packt is searching for authors like you

If you're interested in becoming an author for Packt, please visit `authors.packtpub.com` and apply today. We have worked with thousands of developers and tech professionals, just like you, to help them share their insight with the global tech community. You can make a general application, apply for a specific hot topic that we are recruiting an author for, or submit your own idea.

Share Your Thoughts

Now you've finished *Autodesk Civil 3D 2025 Unleashed*, we'd love to hear your thoughts! Scan the QR code below to go straight to the Amazon review page for this book and share your feedback or leave a review on the site that you purchased it from.

`https://packt.link/r/1835467741`

Your review is important to us and the tech community and will help us make sure we're delivering excellent quality content.

Download a free PDF copy of this book

Thanks for purchasing this book!

Do you like to read on the go but are unable to carry your print books everywhere?

Is your eBook purchase not compatible with the device of your choice?

Don't worry, now with every Packt book you get a DRM-free PDF version of that book at no cost.

Read anywhere, any place, on any device. Search, copy, and paste code from your favorite technical books directly into your application.

The perks don't stop there, you can get exclusive access to discounts, newsletters, and great free content in your inbox daily

Follow these simple steps to get the benefits:

1. Scan the QR code or visit the link below

https://packt.link/free-ebook/9781835467749

2. Submit your proof of purchase
3. That's it! We'll send your free PDF and other benefits to your email directly

www.ingramcontent.com/pod-product-compliance
Lightning Source LLC
Chambersburg PA
CBHW081053220326
41598CB00038B/7075